Compendium of Plant Genomes

Series editor

Chittaranjan Kole
Nadia, West Bengal
India

More information about this series at http://www.springer.com/series/11805

Henry T. Nguyen
Madan Kumar Bhattacharyya
Editors

The Soybean Genome

Springer

Editors
Henry T. Nguyen
Division of Plant Sciences and National Center for Soybean Biotechnology
University of Missouri
Columbia, MO
USA

Madan Kumar Bhattacharyya
Department of Agronomy
Iowa State University
Ames, IA
USA

ISSN 2199-4781 ISSN 2199-479X (electronic)
Compendium of Plant Genomes
ISBN 978-3-319-64196-6 ISBN 978-3-319-64198-0 (eBook)
DOI 10.1007/978-3-319-64198-0

Library of Congress Control Number: 2017947456

Printed on acid-free paper

This Springer imprint is published by Springer Nature
The registered company is Springer International Publishing AG
The registered company address is: Gewerbestrasse 11, 6330 Cham, Switzerland

This book series is dedicated to
my wife Phullara, and our children Sourav, Carena,
and Devleena

Chittaranjan Kole

Preface to the Volume

Soybean [*Glycine max* (L.) Merr.] is the most important legume crop, which is a great source of both protein and oil. Soybean seeds contain approximately 40% protein and 20% oil. Soybean is an important source of protein in animal and fish feed in addition to human nutrition. In recent years, soybean is also becoming a source of biodiesel. Soybean root fixes nitrogen through symbiosis with a rhizobacterium, *Bradyrhizobia japonicum,* improving soil health. More than 30% of the world's soybean crop is produced in the USA and is valued at around $40 billion annually. Brazil and Argentina are other two major soybean-growing countries, followed by China and India.

Soybean was originated in East Asia. It was first domesticated over 8000 years ago in China, 5000 years ago in Japan, and 3000 years ago in Korea. It was introduced to Asian countries, such as India, Indonesia, the Philippines, Vietnam, Thailand, Cambodia, Malaysia, Burma, and Nepal between 1 AD and 1600 AD. To the major soybean-growing countries, soybean was introduced only in the recent years viz to the USA in 1765, Argentina in 1882, and Brazil in early 1950s. Over the last few years, there is a growing interest in expanding soybean cultivation in Africa. The soybean has a very narrow genetic base; most cultivars can be traced to the same handful progenitor lines. In recent years, wild species are utilized through hybridization to broaden the genetic base of modern soybean cultivars.

Considering economic importance and narrow genetic base of soybean, molecular genetics and genomics approaches are becoming vital to ensure steady increases in yield potential to meet the food and nutritional demands of over 9 billion people by 2050. Fortunately, with the advent of second- and now third-generation sequencing platforms, we are able to identify and use the genetic potentials of available germplasm in designing new soybean cultivars that are expected to meet the everincreasing nutritional demands of billions of people under the changing growing conditions, anticipated from climate change. The objective of this book is to bring attention of the readers including students to the recent advances in soybean genetics, breeding, and genomics along with resources essential for highly needed genetic improvement in soybean.

The book comprises 13 chapters with Chaps. 1 and 2 describing the economic importance and botanical aspects of soybean, respectively; with the last chapter (Chap. 13) providing the description and navigation of the

SoyBase; the toolbox for molecular markers, genetic and physical maps, mutants, metabolic pathway, gene expression, a Gbrowse for the soybean reference genome, literature searches to facilitate soybean genomic, genetic, and breeding research. The next six chapters (Chaps. 3–8) provide in-depth reviews on molecular markers, molecular maps, structural and comparative genomics, genome-wide association, and application of molecular resources in breeding soybean. Molecular markers and molecular maps essential for developing predictive selection programs for complex traits such as seed, protein, and oil yield are described in Chap. 3. The chapter reviews classical markers, then RFLP, RAPD, AFLP, SSR, and then SNP molecular markers and maps. The chapter also reviews the use of second-generation sequencing in generating tens of thousands of SNP markers for developing SNP maps from crosses involving wild species and cultivars.

Chapter 4 describes detection, description, and development of structural variation in the soybean genome and how to incorporate these polymorphisms in ongoing soybean research and genetic improvement. Genome assemblies and structural variations among a large collection of soybean genotypes obtained from resequencing are presented in Chap. 5. The Chap. 6 reviews comparative genomics in soybean and that of cultivars with accessions of a wild relative for identifying genes involved in defense response, cell growth, and photosynthesis.

In the USA, soybean germplasm collection is composed of nearly 22,000 accessions including 19,648 modern and landrace cultivars (*G. max*), 1168 wild relatives of soybean (*G. soja*), and 1184 perennial wild species. Most lines of the collection have been genotyped for 52,041 SNP loci making it feasible to conduct genome-wide association studies (GWAS). Chapter 7 reviews GWAS conducted for identification of candidate genes for various agronomical traits, seed composition, seed weight, nitrogen traits, photochemical reflectance index, resistance to soybean cyst nematode, and brown stem rot. The Chap. 8 describes progresses made in implementing tightly linked molecular markers in breeding and introgressing quantitative trait loci/genes that control important agronomic traits.

The next four chapters (Chaps. 9–12) review the recent advances in functional genomics of soybean through analyses of mutants created by the chemical mutagen ethyl methanesulfonate, gene silencing, or transposon-induced mutation. The Chap. 9 reviews application of targeted induced local lesions in genomes (TILLING) to identify mutations within soybean genes of interest. TILLING facilitates conducting both forward and reverse genetics in plant species, and this chapter reviews what has been accomplished in soybean with an example of functional characterization of a soybean cyst nematode resistance gene *Rhg4* encoding serine hydroxymethyltransferase (SHMT) involved in one carbon metabolism. The chapter also documents—the identification of suitable mutants through TILLING for improving quality of oil in soybean and the approach can be applicable also to any traits of interest. Chapter 10 describes the recent advances in virus-induced gene silencing, gene silencing through RNAi in stable transgenic plants, and gene editing systems in soybean.

In Chap. 11, landscape of the transposable elements (TEs) including retrotransposons and type II or DNA transposons in soybean is described. As in other crop species, retrotransposons with long terminal repeat retrotransposons are a major component of the soybean genome that are preferentially accumulated in the pericentromeric regions of all chromosomes and are inactive; but they can be activated under certain stressful conditions. In Chap. 12, application of heterologous transposon systems and an endogenous type II transposon element, *Tgm9*, in tagging and functional characterization of soybean genes is described.

The 13 chapters included in this book have been prepared by experts. We greatly appreciate their contributions. We expect that this book will be a useful reference—for graduate students as well soybean researchers and also researchers of other crop species.

We are grateful to all our colleagues for their contribution. We wish to record our thanks and appreciations for Prof. Chittaranjan Kole, the Series Editor, for his assistance and guidance right from the inception till publication of this book.

Columbia, USA — Henry T. Nguyen
Ames, USA — Madan Kumar Bhattacharyya

Contents

The Economic Evolution of the Soybean Industry

1

Chad Hart

Abstract

Since the crop's humble beginnings in China, the global soybean industry has grown to be a multi-billion dollar enterprise. Since World War II, the USA has been the dominant market for both production and consumption. For the 2015 growing season, US farmers planted over 80 million acres and produced nearly 4 billion bushels of soybeans. With market prices averaging $10 per bushel, that translates to $40 billion in raw production value each harvest. However, the soybean plant was not always one of the dominant crops in the US agricultural landscape. Large-scale soybean production in the USA is a relatively recent phenomenon. Roughly 55% of the world's soybean crop is directed to feed use currently. Over the course of the past 20 years, another use for soybeans has emerged in the energy sector, biofuel, specifically biodiesel production. Soybean production costs have changed dramatically over the past 20 years. In general, crop production costs can be broken down into four major categories: machinery, land, labor, and crop inputs (seed, chemicals, and fertilizer). Land costs make up nearly half of the total cost of soybean production. Soybean, like many row crops, is a bulk commodity. This means that soybeans produced by one farmer/practice/variety are not differentially marketed or priced from soybeans produced by others (with some rare exceptions for food-grade non-genetically modified varieties or organic practices). Or to put it another way, a soybean is a soybean no matter who produces it or how it is produced. So soybean producers are transacting in a competitive market, where producers can enter and exit the market fairly easily and there is very little room for differentiation. But there is a similar situation for the entities that purchase soybeans as well. So the soybean market is made up of many sellers (producers) and buyers (country elevators, crushing facilities, river terminals, and exporters). In competitive

C. Hart (✉)
Iowa State University, 478F Heady Hall, 518 Farm House Lane, Ames, IA 50011, USA
e-mail: chart@iastate.edu

H.T. Nguyen and M.K. Bhattacharyya (eds.), *The Soybean Genome*,
Compendium of Plant Genomes, DOI 10.1007/978-3-319-64198-0_1

markets like this, prices are generally equal to production costs. If prices exceed costs, the resulting profits will inspire additional production from existing and new producers, leading to larger supplies and lower prices. If costs exceed prices, the resulting losses will drive some producers out of the business, leading to supply reduction and price improvement.

1.1 Introduction

Since the crop's humble beginnings in China, the global soybean industry has grown to be a multi-billion dollar enterprise. Since World War II, the USA has been the dominant market for both production and consumption. For the 2015 growing season, US farmers planted over 80 million acres and produced nearly 4 billion bushels of soybeans. With market prices averaging $10 per bushel, that translates to $40 billion in raw production value each harvest. However, the soybean plant was not always one of the dominant crops in the US agricultural landscape. Large-scale soybean production in the USA is a relatively recent phenomenon. Like many things in US history, soybeans are not native, but imported and adopted by US producers over time.

The soybean plant was domesticated roughly 3000 years ago in China. And for the first 2900 years, it remained mainly a Chinese crop. Soybeans were first brought to the USA just prior to the Revolutionary War. However, significant production of soybeans in the USA did not begin until the mid-1900s. The change in production coincides with a change in use for the crop. Prior to World War II, soybeans were mainly seen as a forage crop, providing feed for grazing animals. One of the side effects of the war was the significant disruption of agricultural trade throughout the world. At the start of World War II, the USA imported nearly half of the edible fats and oils it used. The war severely curtailed that trade, creating pressure to develop domestic sources for edible fats and oils. Soybean production grew to fill the void. Similar pressure led to the development of canola in Canada (Gibson and Benson 2005).

But the need for edible fats and oils is not the only feature of soybean that helped the crop become the 2nd largest crop in the USA, and the soybean plant has several other aspects that made it attractive to US farmers over the past 75 years. Crushing soybeans for oil also provides meal, which has become a staple of livestock rations (continuing the connection between soybean and livestock, but moving the relationship from forage to feed). The timing and production practices for soybean are similar to that of corn, the largest crop in the US, and farmers developed a rotation with the two crops. That rotation benefits corn, as soybean's legume properties maintain nitrogen levels in the soil.

The United States Department of Agriculture (USDA) first began tracking soybean acreage and production in 1924. At that time, roughly 1.5 million acres were planted to soybeans. And since much of the crop was used as forage, less than 30% of the crop was harvested, resulting in production of approximately 5 million bushels. Early yield figures were in the 11–13 bushel per acre range. Acreage and production slowly grew through the 1930s. As noted earlier, the first major shift for soybeans occurred during World War II. Soybean plantings topped 10 million acres for the first time, and crop usage shifted from forage to harvested production. In 1940, just under 46% of the US soybean crop was harvested. By 1945, over 82% of the crop was harvested. The need for edible fats and oils changed the dynamics for soybeans. Since then, forage usage of soybean has become a minimal activity. As Fig. 1.1 shows, US farmers have continued to devote more acreage to soybeans, exceeding 50 million acres in the early 1970s and expanding to over 80 million acres now.

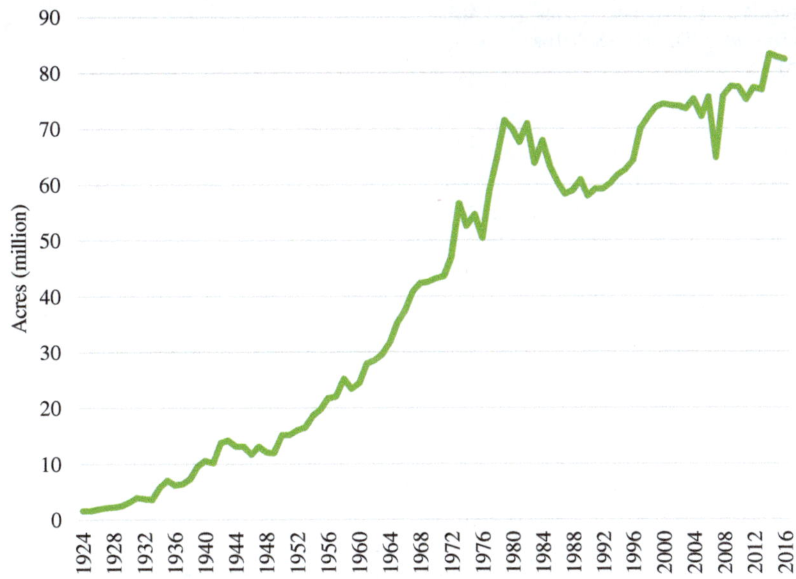

Fig. 1.1 US soybean planted area (*Source* USDA-NASS 2016a, b)

As soybeans were first introduced in the southeast USA, most of the early production originated there. However, over time, soybean production has shifted to the upper Midwest. By the time USDA started tracking soybeans, Illinois was already firmly established as a leading production state. But a majority of the top 10 soybean-producing states in the 1920s were in the southeast (Tennessee, North Carolina, Alabama, Virginia, Mississippi, and Georgia). Iowa did not enter the top 10 until 1930. Gradually, soybean acreage moved northwest and that shift continues today. While Illinois and Iowa compete for the top spot, Minnesota and North Dakota have emerged as strong soybean production states and the further state southeast in the soybean top 10 states is Missouri.

As Fig. 1.2 shows, soybean yields have steady increased over the past 90 years. USDA first estimated national soybean yields in 1924. At that time, the national average yield was 11 bushels per acre. Since then, the national average soybean yield has had a rough annual growth rate of 0.35 bushels per year. The growth in yield, combined with the increase in planted area, resulted in tremendous growth in soybean production. In 2015, the national soybean crop set a record yield of 48 bushels per acre, with annual production approaching 4 billion bushels.

In terms of global production, the USA has been the dominant producer of soybeans for quite some time. Consistent official global agricultural statistics were first captured in the 1960s. By that time, the USA was already the top soybean-producing country in the world, harvesting over half of the global crop. Commercial soybean production did not really begin in South America until the late 1960s. But since then, South American production has ramped up significantly. While the USA is currently still the top producing country, Brazil will likely surge past in the next 10 years. Argentina is the 3rd largest source of soybeans. And Bolivia, Paraguay, and Uruguay are also strong soybean-producing countries. And China is the 4th largest producer, but their production is well behind their current usage.

As Fig. 1.3 shows, the global production of soybean is very concentrated among the USA, Brazil, and Argentina. However, consumption (shown in Fig. 1.4) is more distributed throughout the globe. The USA was the largest consumer of soybeans until 2007. Since then, China has been the dominant buyer in the global soybean market. China consumes roughly 30% of global production and is the largest soybean buyer for the USA, Brazil, and Argentina. So the global soybean market is driven by three major

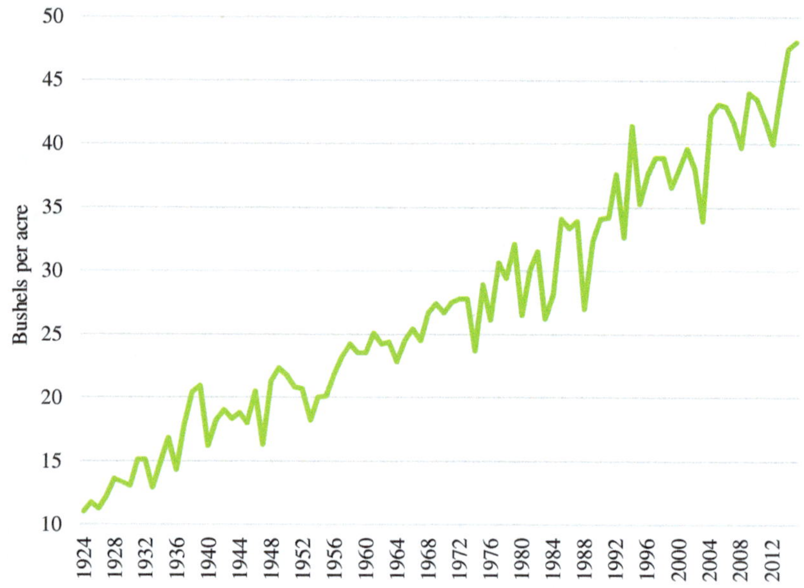

Fig. 1.2 US soybean yields (*Source* USDA-NASS 2016a, b)

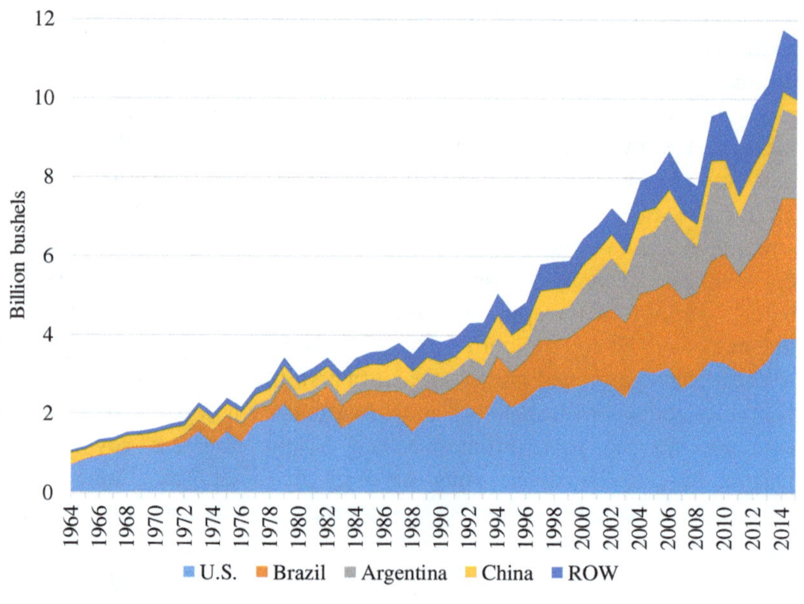

Fig. 1.3 World soybean production (*Source* USDA-FAS 2016)

producers (USA, Brazil, and Argentina) and one major consumer (China).

From the 1960s to the late 1980s, food demand for soybeans led the market. However, as global demand for meat has risen, feed demand for soybeans has now become the leading use. Roughly 55% of the world's soybean crop is currently directed to feed use. So soybean usage has circled back to its original use as livestock feed. Over the course of the past 20

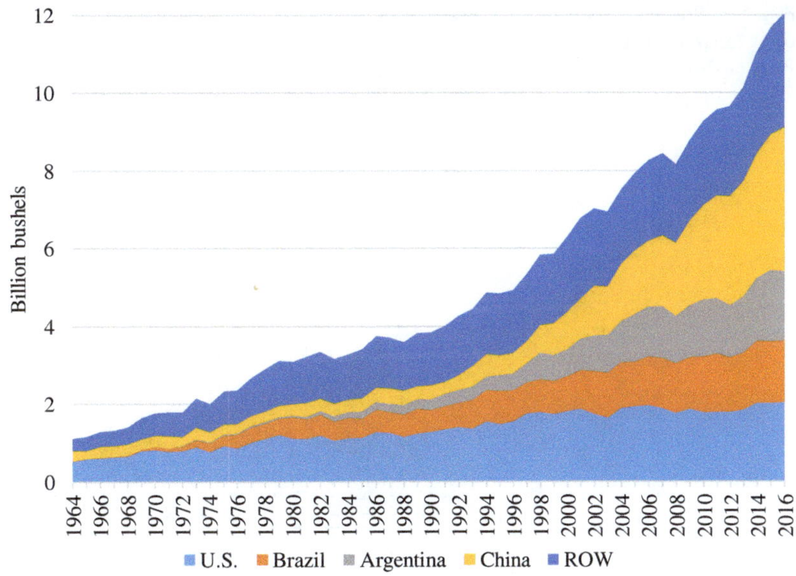

Fig. 1.4 Global soybean consumption (*Source* USDA-FAS 2016)

years, another use for soybeans has emerged in the energy sector, biofuel, specifically biodiesel, production. Biodiesel can be produced from a variety of feedstocks: animal fats, palm oil, rapeseed or canola oil, etc. In the USA, biodiesel production is largely dependent on soybean oil as the feedstock. Currently, roughly 25% of the soybean oil used in the USA is converted to biodiesel.

Over the course of 100 years, soybean has evolved from a minor forage crop to the 2nd largest row crop in the USA. And as demand for soybeans continues to grow, it may, 1 day, challenge corn as the most planted crop.

1.2 Production Economics

Soybean production costs have changed dramatically over the past 20 years. The growing popularity of the crop and the competition for farmland among alternative crops have led to significant cost increases. In the early 1970s, Iowa soybean producers could raise an acre of soybeans for less than $100. Given current conditions, soybean production costs exceed $500 per acre. In general, crop production costs can be broken down into four major categories: machinery, land, labor, and crop inputs (viz. seed, chemicals, and fertilizer). Land costs make up nearly half of the total cost of soybean production. Many farming operations rent land for row crop production, representing a direct land cost for the crop. On owned land, the rental rate represents the opportunity cost foregone by not renting the land to another farmer. As a number of crops could be grown on the land, land costs per crop are consistent across crops (i.e., corn land costs are equal to soybean land costs). Crop input costs are the next largest category, making up roughly 30% of production costs. Seed costs have risen over time due to increased plant populations on soybean fields (farmers are planting more soybean seeds per acre) and new innovations in seed technology (via plant breeding and genetic modification). Fertilizer and chemical costs have varied significantly as prices and usage shift. Machinery costs have grown more slowly, but still account for roughly 15–20% of costs. Labor costs have been the most consistent, but represent less than 10% of costs.

Production costs have nearly doubled over the past 10 years. The largest increase occurred in land costs as farm incomes rose to record levels,

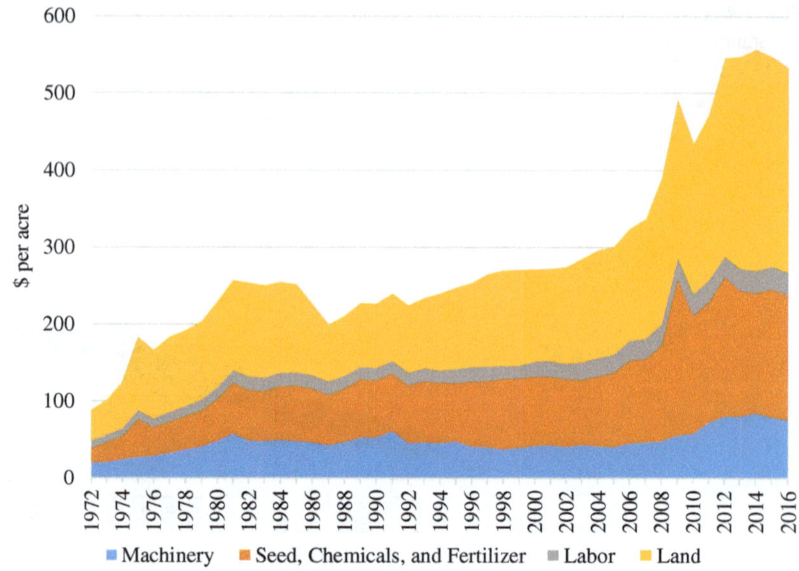

Fig. 1.5 Soybean production costs (*Source* Johanns and Plastina 2016)

driving strong competition among farmers for rented land. A typical Iowa land rent in 2006 was $145 per acre per year. By 2014, rents had risen to $287 per acre per year. Over the same time period, crop input costs rose from $107 per acre to $156 per acre. As Fig. 1.5 shows, there was a spike in crop input costs for the 2009 growing season. Fertilizer prices skyrocketed that year as global supplies were short and demand was strong.

Seed and chemical costs have been impacted by the large-scale adoption of genetically modified soybean varieties. The first genetically modified varieties entered the marketplace in the mid-1990s. The National Agricultural Statistics Service of the USDA began tracking adoption in 2000 by capturing the percentage of the crop planted to the varieties. By that time, over half of the US soybean crop came from genetically modified varieties. In 2007, that percentage rose to over 90%. And since 2014, the percentage of the soybean crop from genetically modified varieties has held steady at 94%. For soybeans, the major modification was the inserted tolerance to herbicides, specifically glyphosate (or as it is commonly called by its product name, Roundup). This change simplified weed control during soybean production and reduced chemical costs. However, as weeds develop resistance, those cost savings are eroding.

Figure 1.6 outlines soybean production costs for the 2016 crop year in Iowa. Given a yield target of 50 bushels per acre, the total cost per acre to produce soybeans is $533.30. That equates to $10.67 per bushel. Pre-harvest machinery costs (covering the pre-planting preparation and tillage practices) are $38.50 per acre. Seed costs are $53.60 per acre. Fertilizer and lime expenses add up to nearly as much as seed, $53.05 per acre. Other pre-harvest expenses contribute, such as crop insurance and interest payments, roughly $56 per acre. Once harvest begins, additional machinery costs of $37 per acre are incurred. Approximately 2.25 h of labor per acre are required over the course of the production cycle adding $29.25 to the costs. But the largest cost component is the land charge. For 2016, that is $266 per acre. As the figure shows, production costs change with the yield target, mainly for two reasons. First, the quality of the soil impacts potential yields. More productive

Fig. 1.6 Soybean production budget for Iowa during the 2016 crop year (*Source* Plastina 2016)

Herbicide Tolerant Soybeans Following Corn

	45 bu. per acre		50 bu. per acre		55 bu. per acre	
	Fixed	Variable	Fixed	Variable	Fixed	Variable
Preharvest Machinery [1/]	$21.10	$17.40	$21.10	$17.40	$21.10	$17.40
Seed, Chemical, etc.	Units		Units		Units	
Seed @ $53.60 per 140 k.	140	$53.60	140	$53.60	140	$53.60
Phosphate @ $0.45 per lb.	36	16.20	40	18.00	44	19.80
Potash @ $0.35 per lb.	68	23.80	75	26.25	83	29.05
Lime (yearly cost)		8.80		8.80		8.80
Herbicide [2/]		32.20		32.20		32.20
Crop insurance		6.90		7.80		8.70
Miscellaneous		9.00		10.00		11.00
Interest on preharvest variable costs (8 months @ 5.15%)		5.76		5.98		6.20
Total		$156.26		$162.63		$169.35
Harvest Machinery						
Combine	$15.90	$6.80	$15.90	$6.80	$15.90	$6.80
Grain cart	6.20	2.70	6.20	2.70	6.20	2.70
Haul	1.93	1.34	2.15	1.49	2.36	1.64
Handle (auger)	0.78	0.75	0.87	0.83	0.95	0.91
Total	$24.81	$11.58	$25.11	$11.82	$25.41	$12.05
Labor						
2.25 hours @ $13.00	$29.25		$29.25		$29.25	
Land						
Cash rent equivalent	$225.00		$266.00		$296.00	
Total fixed, variable						
Per acre	$300.16	$185.25	$341.46	$191.84	$371.76	$198.80
Per bushel	$6.67	$4.12	$6.83	$3.84	$6.76	$3.61
Total cost per acre	$485.41		$533.30		$570.56	
Total cost per bushel	$10.79		$10.67		$10.37	

[1/] *Chisel plow, tandem disk, field cultivate, plant, and two sprays. See the Estimated Machinery Costs table.*
[2/] *Estimates do not include any insecticide or fungicide costs.*

soils generally translate into higher yields, and those higher yields often result in higher land rental costs. Second, fertilization schemes shift higher with the yield target.

1.3 Marketing

Soybean, like many row crops, is a bulk commodity. This means that soybeans produced by one farmer/practice/variety are not differentially marketed or priced from soybeans produced by others (with some rare exceptions for food-grade non-genetically modified varieties or organic practices). Or to put it another way, a soybean is a soybean no matter who produces it or how it is produced. So soybean producers are transacting in a competitive market, where producers can enter and exit the market fairly easily and there is very little room for differentiation. But there is a similar situation for the entities that purchase soybeans as well. So the soybean market is made up of many sellers (producers) and buyers (country elevators, crushing facilities, river terminals, and exporters). In competitive markets like this, prices are generally equal to production costs. If prices exceed costs, the resulting profits will inspire additional production from existing

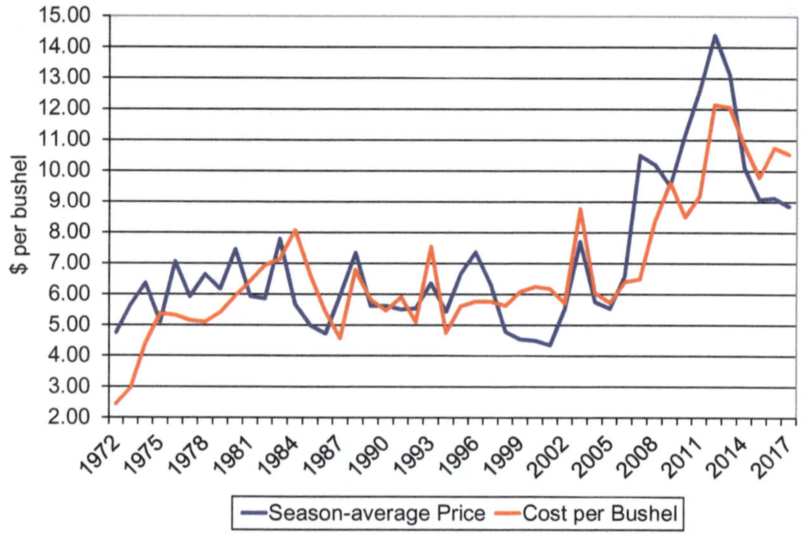

Fig. 1.7 Iowa soybean prices and costs per bushel (*Source* USDA-NASS 2016a, b; Johanns and Plastina 2016)

and new producers, leading to larger supplies and lower prices. If costs exceed prices, the resulting losses will drive some producers out of the business, leading to supply reduction and price improvement.

Over the past 45 years, that balance between prices and costs has held on average. The cost line in Fig. 1.7 below shifts due to changes in the costs per acre, but also changes in soybean yield. In a weather-stressed year, such as 1993 (with floods affecting the central USA) or 2012 (with drought doing the same), costs jump to the reduced yields. But prices rise in those years as well. In general, the soybean market, like most agricultural markets, goes through multi-year profitability swings. Prices will exceed costs for a few years, followed up by a reversal where costs exceed prices for a few years. In the early to mid-1970s, soybean exports surged. Prices followed and the profitability of soybeans led to increased soybean plantings through the late 1970s and 1980s. Soybean supplies eventually caught up to demand, costs rose, and prices fell, leading to a soybean contraction in the late 1980s. Similar swings have continued since then. During the last 10 years, the soybean market has experienced record high prices, again spurred on by strong exports. Starting in 2007, cash soybean prices rose above $10 per bushel. And prices stayed above $10 per bushel over several years after that. The profitability generated by those high prices led farmers to increase soybean planting above 80 million acres. The recent combination of record acreage and yield has lowered prices significantly, putting pressure on the soybean industry to contract.

Given the numerous participants in the soybean market and the competitive nature of the market, soybean pricing has developed a pathway to link local market prices to global ones. Global soybean prices are mainly determined at large commodity exchanges (with the largest for soybeans being the Chicago Mercantile Exchange). The futures' prices established at these exchanges reflect the global supply and demand situation and provide all market participants current and future price signals for soybean sales and production decisions. Local soybean prices are derived from the futures prices, with the difference between the futures prices and local cash prices referred to as the basis. The basis reflects local supply and demand conditions and incorporates the transportation costs between the local market and the commodity exchange. The basis is often negative (i.e., the local cash price is below the futures price) in local markets, where soybean supplies are large, for example, in Iowa and Illinois. But the basis can become positive when soybeans are in short supply, for example, when a drought limits soybean production. This pricing formula means that local market prices are often quoted as two pieces of information, the futures price, capturing the global picture, and the basis, highlighting local changes.

References

Gibson L, Benson G (2005) Origin, history, and uses of soybean (Glycine max). http://agron-www.agron.iastate.edu/Courses/agron212/Readings/Soy_history.htm. Accessed 25 June 2016

Johanns A, Plastina A (2016) Historical costs of crop production. http://www.extension.iastate.edu/agdm/crops/pdf/a1-21.pdf. Accessed 9 Sept 2016

Plastina A (2016) Estimated costs of crop production in Iowa—2016. http://www.extension.iastate.edu/agdm/crops/pdf/a1-20.pdf. Accessed 10 Sept 2016

United States Department of Agriculture-Foreign Agricultural Service (USDA-FAS) (2016) Production, supply, and distribution database. https://apps.fas.usda.gov/psdonline/. Accessed 27 June 2016

United States Department of Agriculture-National Agricultural Statistics Service (USDA-NASS) (2016a) Acreage. Various issues. http://usda.mannlib.cornell.edu/MannUsda/viewDocumentInfo.do?documentID=1000. Accessed 9 Sept 2016

United States Department of Agriculture-National Agricultural Statistics Service (USDA-NASS) (2016b) Quick stats database. https://www.nass.usda.gov/Quick_Stats/. Accessed 25 June 2016

Botany and Cytogenetics of Soybean

2

R.J. Singh

Abstract

Soybean [*Glycine max* (L.) Merr.], an economically important dicot legume, is a member of the family Fabaceae and belongs to the genus *Glycine* Willd. Based on classical and molecular taxonomy, the genus *Glycine* has been divided into two subgenera; the subgenus *Soja* (Moench) F.J. Hermann includes soybean and its wild annual progenitor *G. soja* Sieb. & Zucc. Both species contain $2n = 40$ chromosomes, are cross-compatible, produce fertile F_1 plants, and belong to the primary gene pool. The subgenus *Glycine* consists of 26 wild perennial species. Vegetative and reproductive morphology of soybean has been examined extensively. The cytogenetic knowledge of soybean lags far behind that of other model economically important crops (viz. rice, maize, wheat, tomato), because its somatic chromosomes are symmetrical and only one pair of satellite chromosomes can be identified. Molecular linkage maps have been associated with specific chromosomes, and soybean genome has been sequenced. The soybean breeders, worldwide, are confined to crossing within the primary gene pool; thus, genetic base of soybean is very narrow. Wild perennial *Glycine* species of the tertiary gene pool have been recently exploited to broaden the genetic base of modern soybean cultivars.

2.1 Introduction

The soybean [*Glycine max* (L.) Merr.; $2n = 40$] is an economically very important leguminous seed crop for feed and food products that is rich in seed protein (about 40%) and oil (about 20%). Taxonomy of the genus *Glycine* Willd. is well defined based on morphological features, cytogenetics,

R.J. Singh (✉)
USDA-Agricultural Research Service, Soybean/Maize Germplasm, Pathology, and Genetics Research Unit, Department of Crop Sciences, University of Illinois, 1101 West Peabody Drive, Urbana, IL 61801, USA
e-mail: ramsingh@illinois.edu

H.T. Nguyen and M.K. Bhattacharyya (eds.), *The Soybean Genome*, Compendium of Plant Genomes, DOI 10.1007/978-3-319-64198-0_2

and molecular methods (Chung and Singh 2008). Vegetative (Lersten and Carlson 2004) and reproductive (Carlson and Lersten 2004) morphological features of soybean have been extensively described. However, soybean is not considered a model plant for cytogenetic studies because of the large number of chromosomes ($2n = 40$) (Karpechenko 1925; *Soja hispida*; syn. *G. max*), their small and similar chromosome size (1.42–2.84 μm) (Sen and Vidyabhusan 1960), and the lack of morphological distinguishing landmarks (Singh 2003). Using primarily restriction fragment length polymorphism (RFLP) and simple sequence repeat (SSR) loci, 20 molecular linkage groups (MLGs) have been developed (Song et al. 2004; Xia et al. 2007) but not all linkage groups have been associated with the respective chromosomes (Zou et al. 2003b). The cytogenetic knowledge of the soybean lags far behind many crops such as maize, barley, rice, wheat, tomato, brassicas, pea, and faba bean (Singh 2003; Singh et al. 2007a, b).

The genetic base (number of ancestral varieties that contributed to the development of modern commercial varieties) and diversity are narrow for public soybean cultivars being grown worldwide. Soybean breeders have been confined to improving soybeans using land races and occasionally used *G. soja* Sieb. & Zucc. Soybean breeders have not exploited wild perennial *Glycine* species for broadening the genetic base of soybean (Chung and Singh 2008). Despite the apparent limitation of having a narrow genetic base for world soybean production, soybean breeding has continued to make significant progress.

The objective of this chapter is to document brief information on vegetative and reproductive features (botany) of soybean and describes cytogenetics of the genus *Glycine*. Cytogenetics covers handling of soybean chromosomes, genomes of the *Glycine* species, origin of polyploid complexes, chromosomal aberrations and wide hybridization.

2.2 Botany

2.2.1 Taxonomy

The taxonomy of wild annual and cultivated soybean is as follows:

Order	Fabales
Family	Fabaceae (Leguminosae)
Subfamily	Papilionoideae
Tribe	Phaseoleae
Subtribe	Glycininae
Genus	*Glycine* Willd.
Subgenus	*Soja* (Moench) F.J. Herm.
Species	*Glycine soja* Sieb. & Zucc.
Species	*Glycine max* (L.) Merr.

The taxonomy of the genus *Glycine* to which soybean belongs has been revised many times. Hermann (1962) divided the genus *Glycine* into three subgenera (Tables 2.1 and 2.2): The subgenus *Leptocyamus* included six wild perennial species indigenous to Australia, South China, South Pacific Islands, Philippines, and Formosa (Taiwan). The subgenus *Glycine* contained two species (*G. petitiana* from Ethiopia and *G. javanica* from India and Malaya (Malaysia)). *Glycine javanica* included two subspecies: the subspecies *G. micrantha* with four varieties and subspecies *G. pseudojavanica* with one variety, and all were indigenous to Africa (Tables 2.1). He included cultigen soybean [*G. max* (L.) Merr.] and *G. ussuriensis* Regel and Maack. in the subgenus *Soja*.

Verdcourt (1966) questioned the validity of *G. javanica* L. since it has $2n = 22$ or 44 chromosomes and the chromosomes (morphology) are larger than those of other species of the genus *Glycine*. He kept the generic name and proposed *G. wightii* (R. Grah. Ex Wight and Arn.) Verdcourt as the species name. He changed the names of the genus *Glycine* L. assigned by Hermann (1962) to *Glycine* Willd., and the names of two of the subgenera of *Glycine*: subgenus *Leptocyamus* (Benth.) Hermann became

Table 2.1 Systematic classification of the *Glycine* L. (Hermann 1962)

Species	Distribution
Subgenus *Leptocyamus*	
1. *Glycine clandestina* Wendl	Australia; Formosa (Taiwan), Micronesia
1a. var. *sericea* Benth	Australia
2. *G. falcata* Benth	Australia
3. *G. latrobeana* (Meissn.) Benth	Australia
4. *G. canescens* F.J. Herm	Australia
5. *G. tabacina* (Labill.) Benth	Australia; S. China; S. Pacific Islands
6. *G. tomentella* Hayata	Australia; S. China; Philippines; Formosa (Taiwan)
Subgenus *Glycine*	
1. *G. petitiana* (A. Rich.) Schweinf	Ethiopia
2. *G. javanica* L	India; Malaya (Malaysia)
2a. ssp. micrantha (Hochst.) F.J. Herm	Trop. Africa
2b. var. *claessensii* (De Wild.) Hauman	Uganda to Nyasaland (Republic of Malawi)
2c. var. *paniculata* Hauman	Belgian Congo (Democratic Republic of the Congo)
2d. var. *longicauda* (Schweinf.) Bak	Ethiopia to Angola
2e. var. *moniliformis* (hochst.) F.J. Herm	Ethiopia to Eritrea
2f. subsp. *pseudojavanica* (Taub.0 Hauman	Belgian Congo (Democratic Republic of the Congo) to Angola
2g. var. *laurentii* (De Wild.) hauman	Belgian Congo (Democratic Republic of the Congo)
Subgenus *Soja*	
1. *G. ussuriensis* Regal and Maack	Asia
2. G. max (L.) Merr	Cultigen

Table 2.2 Revision of the genus *Glycine* by Verdcourt (1966)

Genus *Glycine* Willd.
Subgenus *Glycine*
1. *Glycine clandestina* Wendl.
1a. var. *sericea* Benth
2. *G. falcata* Benth
3. *G. latrobeana* (Meissn.) Benth
4. *G. canescens* F.J. Herm
5. *G. tabacina* (Labill.) Benth
6. *G. tomentella* Hayata
Subgenus *Bracteata* Verdcourt
1. *G. wightii* (R. Grah. ex Wight and Arn.) Verdcourt
Subgenus *Soja*
1. *G. soja* Sieb. & Zucc
2. *G. max* (L.) Merr

a synonym of *Glycine* subgenus *Glycine*; subgenus *Soja* (Moench) Hermann was unchanged. Lackey (1977) (Table 2.2) later removed *G. wightti* (Arnott) Verdcourt from the genus *Glycine* and designated it *Neonotonia wightii* (Wight and Arn.) J.A. Lackey.

Currently, the genus *Glycine* consists of two subgenera. The subgenus *Glycine* consists of 26 wild perennial species indigenous to Australia and various surrounding Pacific Islands. Of the 26 perennial species, *G. tomentella* Hayata constitutes four cytotypes ($2n$ = 38, 40, 78, 80) and *G. hirticaulis* and *G. tabacina* have accessions with $2n$ = 40 and 80 chromosomes (Table 2.3). *Glycine tomentella* accessions with $2n$ = 80 chromosomes is distributed in Australia, Papua New Guinea, the Philippines, Indonesia, and Taiwan, while 80-chromosome *G. tabacina* (Labill.) Benth. has been collected from Australia, Tonga,

Table 2.3 Taxonomy of the Genus *Glycine* Willd

Species	Mol Group	2*n*	Genome symbol N	Genome symbol C	PI-number	G number	Distribution	Species described since Table 2.2
Subgenus *Glycine*								
1. *G. albicans* Tindale and Craven		40	I	A		2049	Aust.: WA	Tindale and Craven (1988)
2. *G. aphyonota* B. Pfeil		40	I_3	A		2589	Aust.: WA	Pfeil and Craven (2002)
3. *G. arenaria* Tindale		40	H	A	505204	1305	Aust.: WA	Tindale (1986b)
4. *G. argyrea* Tindale		40	A_2	A	505151	1420	Aust.: Q	Tindale (1984)
5. *G. canescens* F.J. Hermann		40	A	A	440932	1853	Aust.: Q, NSW, V, SA, NT, WA	
6. *G. clandestina* Wendl		40	A_1	A	440958	1126	Aust.: Q, NSW, V, SA, T	
7. *G. curvata* Tindale		40	C_1	C	505166	1849	Aust.: Q	Tindale (1986a)
8. *G. cyrotoloba* Tindale		40	C	C	440962	1184	Aust.: Q, NSW	Tindale (1984)
9. *G. falcata* Benth		40	F	A	505179	1155	Aust.: Q, NT, WA	
10. *G. gracei* B.E. Pfeil and Craven		40	?	?		3124	Aust.: NT	Pfeil et al. (2006)
11. *G. hirticaulis* Tindale and Craven		40	H_1, (??)	A, (A)	IL1246	2876	Aust.: NT	Tindale and Craven (1988)
		80			IL943	1956	Aust.: NT	
12. *G. lactovirens* Tindale and Craven		40	I_1	A	IL1247	2720	Aust.: WA	Tindale and Craven (1988)
13. *G. latifolia* (Benth.) Newell and Hymowitz		40	B_1	B	378709	1697	Aust.: Q, NSW	Newell and Hymowitz (1980)
14. *G. latrobeana* (Meissn.) Benth		40	A_3	A	483196	1385	Aust.: V, SA, T	
15. *G. microphylla* (Benth.) Tindale		40	B	B	440956	1867	Aust.: Q, NSW, V, SA, T	Tindale (1986a, b)
16. *G. montis-douglas* B.E. Pfeil and Craven		40	?	?			Aust.: NT	Pfeil et al. (2006)
17. *G. peratosa* B.E. Pfeil and Tindale		40	A_5	A		2916	Aust.: WA	Pfeil et al. (2001)
18. *G. pescadrensis* Hayata		80	AB_1	A	440996	1433	Aust.: Q, NSW; Taiwan, Japan	Pfeil et al. (2006)
19. *G. pindanica* Tindale and Craven		40	H_2	A	595818	2951	Aust.: WA	Tindale and Craven (1993)
20. *G. pullenii* B. Pfeil, Tindale and Craven		40	H_3	A		2599	Aust.: WA	Pfeil and Craven, 2002
		40	A_4	A	440954	1874	Aust.: NSW, SA, WA	Pfeil et al. (2002)

(continued)

Table 2.3 (continued)

Species	Mol Group	2*n*	Genome symbol N	Genome symbol C	PI-number	G number	Distribution	Species described since Table 2.2
21. *G. rubiginosa* Tindale and B.E. Pfeil								
22. *G. stenophita* B. Pfeil and Tindale		40	B_3	B	378705	2600	Aust.: Q, NSW; (Japan ??)	Doyle et al. (2000)
23. *G. syndetika* B.E. Pfeil	D4	40	A6		441000	1300	Aust.: Q	Pfeil et al. (2006)
24. *G. dolichocarpa* Tateishi and Ohashi		80	D_1A	?			Taiwan	Tateishi and Ohashi (1992)
25. *G. tabacina* (Labill.) Benth		40	B_2	B	373990	1317	Aust.:Q, NSW	
		80	BB_1;BB_2; B_1B_2	B	373992	1314	Aust.:Q, NSW, V SA; West Central and South Pacific Islands	
26. *G. tomentella* Hayata	D1, D2	38	E	A	440998	1858	Aust.: Q	
	D3	40	D	A	505222	1749	Aust.: Q, WA, PNG	
	D5B	40	H_2	A	505294	1943	Aust.: WA	
	D5A	40	D_2	A	505203	1303	Aust.: WA, NT	
	T1	78	D_3E	A	441001	1133	Aust.: Q, NSW; PNG	
	T5	78	AE	A	509501	1487	Aust.: NSW	
	T6	78	E H_2	A	505286	1945	Aust.: WA	
	T2	80	D A_6	A	441005	1188	Aust.: Q; Taiwan	
	T3	80	D D_2	A	483219	1927	Aust.: Q, NT, WA; PNG; Timor	
	T4	80	D H_2	A	330961	1348	Aust.: Q, NT, WA; Philippines, Taiwan	
Subgenus *Soja* (Moench) F.J. Hermann								
G. soja Sieb.& Zucc.		40	G	G	51762		China, Japan, Russia, Korea, Taiwan	
G. max (L.) Merr		40	G_1	G			Cultigen; worldwide	

N nuclear; *C* chloroplast; *PI* Plant introduction; *G* CSIRO (Commonwealth Scientific and Industrial Research Organization) number; *Aust* Australia; *IL* Illinois; *Q* Queensland; *NSW* New South Wales; *NT* Northern Territory; *SA* South Australia; *V* Victoria; *WA* Western Australia; *T* Tasmania; *PNG* Papua New Guinea

Vanuatu, Ryukyu Islands, and Taiwan. Tateishi and Ohashi (1992) examined *Glycine* of Taiwan and recognized four species: *G. dolichocarpa* Tateishi and Ohashi, *G. tomentella*, *G. tabacina*, and *G. max* subsp. *formosana* (Hosokawa Tateishi and Ohashi). *Glycine dolichocarpa* contains $2n = 80$ chromosomes, and *G. tomentella* and *G. tabacina* of Taiwan contain $2n = 80$ chromosomes (Chung and Singh 2008). It is likely that *G. dolichocarpa* is a variant of *G. tomentella*, possibly of Australian origin, and migratory birds may have played a major role in its dispersion (Hymowitz et al. 1990).

The newly described species since 1986 (Table 2.3) are distributed in a very restricted region of Australia. For example, *G. latrobeana* is listed nationally as rare, and in Tasmania, seven populations have been recorded (Lynch 1994). Only 19 wild perennial *Glycine* species are being maintained in the soybean collection at Urbana, Illinois (http://www.ars-grin.gov/cgi-bin/npgs/html/site_holding.pl?SOY).

The subgenus *Soja* still contains *G. soja* (a wild progenitor) and *G. max* (cultigen). Both species contain $2n = 40$ chromosomes, are cross-compatible, and produce vigorous fertile F_1 hybrids, and gene exchange between them is possible. Broich and Palmer (1980) used cluster analysis techniques to examine phenotypic variation among *G. max*, *G. soja*, and *G. gracilis*. *Glycine max* and *G. soja* were found to be morphologically distinct, and *G. gracilis* was found to be conspecific with *G. max*. Thseng et al. (1999) identified a new species *G. formosana* Hosokawa from Taiwan based on pod morphology, allozyme, and DNA polymorphisms and concluded that the newly defined species is different from *G. soja* though they did not hybridize both species. It is likely that *G. formosana* is a variant of *G. soja*. Thus, we have not included *G. formosana* in Table 2.3.

Fig. 2.1 Germination of Dwight soybean: **a** epigeal germination with two cotyledons leaves, **b** cotyledon leaves with emerging primary leaves, **c** simple primary leaves, **d** trifoliolate leaf, **e** three trifoliolate leaves, **f** six trifoliolate leaves

Table 2.4 Soybean vegetative and reproductive stages

Vegetative growth stages	Description
VE (emergence)	Cotyledons have been pulled through the soil surface
VC (unifoliolate leaves)	Unifoliolate leaves expand, one node
V1 (first trifoliolate)	One set of trifoliolate leaves completely unfolded, two nodes
V2 (second trifoliolate)	Two sets of trifoliolate leaves unfolded, three nodes
V4 (four trifoliolate)	Four trifoliolate leaves have unfolded
V(n) (nth trifoliolate)	V stages continue eith the unfolding of trifoliolate leaves; the final number of trifoliolate depends on the soybean variety and the environmental conditions

2.2.2 Morphology

Soybean is an annual plant. It exhibits taproot growth initially, followed later by the development of a large number of secondary roots. The roots establish a symbiotic relationship with the nitrogen fixing bacterium, *Bradyrhizobium japonicum*, through the formation of root nodules. Soybean has four different types of leaves: the seed leaves (first pair of simple cotyledons; VE stage) (Fig. 2.1a, b; epigeal germination; Table 2.4), simple primary leaves (Fig. 2.1c; VC stage; Table 2.4), pinnately trifoliolate leaves (Fig. 2.1d; V1 stage; Table 2.4), and the prophylls (a pair of 1 mm long simple leaves at the base of each lateral branch) [Lersten and Carlson (2004)]. Two sets of trifoliolate (V2) and four sets of trifoliolate (V4; Fig. 2.1e) leaves and continues to produce trifoliolate leaves (Vn; Fig. 2.1f) depending upon the environmental conditions. The stem type may be determinate, semi-determinate, or indeterminate.

Fig. 2.2 Reproductive organs (identified) of Dwight soybean: **a** complete mature axillary flowers, **b** terminal flowers, **c** a determinate Dwight soybean plant with developing pods (*arrow*)

Fig. 2.3 Reproductive organs of Dwight soybean: **a** complete mature flower, **b** mature androecium and gynoecium, **c** a mature gynoecium with stigma, style, and ovary, **d** mature anthers with five anthers on longer filament (outer whorl), four anthers on shorter filament (inner whorl), and one free anther (*arrow*) always below the stigma

Soybean plants enter into reproductive stages following vegetative growth (Table 2.5). Axillary buds develop (Fig. 2.2a) into clusters of flowers (Fig. 2.2b). From 20 to 80% of the flowers, developed early, abscise (Carlson and Lersten 2004) and produce initially few pods (Fig. 2.2c). Generally, the earliest and latest flowers produced abort most often. Soybean has a typical papilionaceous flower with a tubular calyx of five unequal sepals and a five-part corolla. The corolla consists of a standard (posterior banner petal), two lateral wings and two anterior keel petals contacting with each other, but not fused (Fig. 2.3a). The stamens are clustered around the stigma, ensuring self-pollination (Fig. 2.3b). The gynoecium consists of an ovary, style, and stigma (Fig. 2.3c). As many as four ovules develop in the ovary. Nine stamens are arranged in two whorls; the outer whorl contains five stamens, and inner whorl contains four stamens (Fig. 2.3d). The two whorls of nine anthers align themselves into a single whorl on a staminal tube. The larger and older anthers alternate with the smaller and younger anthers in sequence around the developing gynoecium. The single free stamen (the10th) the last to appear (Fig. 2.3d). Soybean is highly self-pollinated with natural crossing usually below 1% because the stamens are elevated so that the anthers form a ring around the stigma. Thus, pollen is shed

Table 2.5 Soybean reproductive growth stages

R1	Beginning to bloom, at least one flower is present on the main stem
R2	Full bloom, flowers are found on any of the top two nodes
R3	Beginning of pod set, pods are 4.8 mm long on one of the top four nodes
R4	Full pod, pods are 19 mm long on one of the top four nodes
R5	Beginning seeds, seeds are 3.2 mm long on one of the top four nodes
R6	Full seeds, pods are completely filled by seeds on one of the top four nodes
R7	Beginning of maturity, one mature pod found on plant
R8	Full maturity, 95% pods have reached mature pod color

Fig. 2.4 A mature Dwight soybean plant: **a** a physiological mature Dwight plant showing tan and partial green pods and leaves are near to senescence, **b** four mature pods of Dwight soybean with one, two, three, and four seeds (*top* to *bottom*), **c** a Dwight soybean three-seeded open pod, **d** a mature Dwight soybean seed observed through dissecting microscope; organs are identified

directly onto the stigma surface ensuring self-pollination (Carlson and Lersten 2004).

When the pollen grains are shed onto the stigma, they germinate and the pollen tubes travel through style and enter into the filiform apparatus; perhaps as many as 90% of the tubes atrophy and die before reaching the distal end of the ovary (Carlson and Lersten 2004). Before pollen tube reaches toward ovule, the generative cell divides and forms two male gametes (sperm nuclei). Finally, the pollen tube grows through the micropyle of the ovule and enters the filiform apparatus of the degenerated synergid. The pollen tube tip bursts and releases two sperm nuclei. One sperm nucleus fuses with the egg nucleus and forms a diploid zygote, while the second sperm unites with the secondary nucleus forming the primary endosperm nucleus. The time from pollination to fertilization takes about 8–10 h (Carlson and Lersten 2004).

The inflorescence of each node of the soybean plant may develop into one to more than 20 pods (Fig. 2.4a). The soybean pod is similar to that of other legumes. A pod usually contains 1–3 seeds and rarely 4 seeds (Fig. 2.4b), except for plants that have the *na* allele that produces narrow leaflets and a much high proportion of 4-seeded pods. Separation of the two halves of the pod is preceded by the appearance of clefts through the parenchyma of the dorsal and ventral sutures. After separation, the halves twist spirally around the axis (Fig. 2.4c; Carlson and Lersten 2004).

The seeds mature about 50–80 days after fertilization depending upon the variety and environmental factors. The soybean seed is devoid of

Fig. 2.5 Harvesting of roots for mitosis from seedlings grown in a sand bench of greenhouse: **a** a correct seedling stage; about 6 days old, **b** a seedling carefully uprooted covered with loose sand showing whitish-cream color roots, **c** seedlings in a container with cold tap water, **d** *labeled* Eppendorf tube (1.5 ml) ready to harvest the roots

endosperm and contains two large fleshy cotyledons, hypocotyl, micropyle, hilum with a central fissure and a raphe (Fig. 2.4d; Carlson and Lersten 2004).

2.3 Cytology

2.3.1 Handling of Soybean Chromosomes

Mitotic and meiotic chromosomes of soybean are difficult to count because of the small size and the presence of cytoplasm stains in the simple aceto-carmine staining method. The ice-cold water pretreatment routinely used for the cereals fails to arrest a large number of cells at metaphase stage (Singh 2003). Palmer and Heer (1973) developed a root tip squash technique for soybean chromosomes: pretreating the roots with para-dichloro-benzene at 15 °C for 1½ to 2 h. The protocol is time-consuming with many steps which are not feasible when a large number of plants must be identified cytologically. The following method is simple, repeatable, and reproducible and produces excellent chromosome spreads at prometaphase and metaphase stages of mitosis (Singh 2003).

2.3.1.1 Mitotic Chromosomes

1. Germinate soybean seed in sand bench.
2. Select actively growing secondary roots from 1 week old seedlings (Fig. 2.5a).
3. Make sure not to break the actively dividing roots (Fig. 2.5b).
4. Wash off sand or soil in clean water (Fig. 2.5c) and cut 1–2 mm long root tips using forceps, transfer root tips into 1.5-ml Eppendorf tubes containing clean water (Fig. 2.5d).
5. Bring Eppendorf tubes to the laboratory, remove water, and add 0.05% (w/v) 8-hydroxyquinoline.
6. Place tubes at 16 °C heating block in a refrigerator and pretreat roots for 2–3 h.
7. Fix roots in a freshly prepared fixative consisting of 3 parts absolute ethyl alcohol (200 proof): 1 part glacial acetic acid or propionic acid for 24 h at room temperature. Store roots at −20 °C.
8. Remove fixative, wash roots once with distilled water, hydrolyze in 1 N HCl at 60 °C for 12 min, remove 1 N HCl, and wash roots once with distilled water.
9. Stain roots in Feulgen stain for 45–60 min at room temperature.

10. Remove Feulgen stain, add cold distilled water, and store in a refrigerator (4 °C).
11. Cut purple colored tip, place on a clean slide, add a drop of Carbol Fuchsin, and prepare slide using a root tip squash method. Soybean chromosomes stain very well with Carbol Fuchsin, leaving the cytoplasm clear; better than with aceto-carmine or propiono-carmine stains.
12. An alternative is to stain the roots in Carbol Fuchsin overnight in a refrigerator (4 °C), remove the stain, wash root tips with cold distilled water 3–4 times, and store the roots in distilled water in a refrigerator (4 °C). Cut tip by a sharp razor blade, apply a drop of 45% (v/v) acetic acid, and prepare chromosome spread slide by the squash method.

Some useful tips are: (1) Try to collect roots with creamy yellow tips from young seedlings; (2) make sure roots do not dry; (3) remove water completely from the Eppendorf tubes, as water will dilute the concentration of 8-hydoxyquinoline; (4) place only dark purple tips on the slide, add 1–2 drops of Carbol Fuchsin, apply cover glass slowly, heat slide but do not boil, tap slowly three to four times, heat again, and squeeze out excess stain by applying light pressure with the thumb; make sure not to move the cover glass otherwise the cells will be rolled.

The fluorescence in situ hybridization (FISH) technique has been used to identify a primary trisomic containing a nucleolus organizer chromosome (Skorupska et al. 1989; Griffor et al. 1991) and to determine the distribution of rDNA loci in the genus *Glycine* (Singh et al. 2001). Findley et al. (2010) used a fluorescence in situ hybridization system for karyotyping soybean chromosomes. The protocol was published by Singh (2003) and can be modified by researchers depending upon materials, preferences, and priorities.

2.3.1.2 Meiotic Chromosomes

1. Fix soybean buds undergoing meiosis in a proportion of 1 part propionic acid (propiono) + 3 parts 100% (v/v) ethanol + 1 g/100 ml fixative ferric chloride. Propionic acid is preferred over glacial acetic acid because propionic acid produces a clear cytoplasm, whereas the cytoplasm becomes stained when glacial acidic acid is used.
2. Fix buds for 24 h at room temperature; wash in 70% (v/v) ethanol 2 times; and store in 70% (v/v) ethanol in a refrigerator (4 °C).
3. Stain anthers with desired meiotic stages in 1% (w/v) propiono-carmine for 1 week in a refrigerator (4 °C) and prepare chromosome slides using the squash method (Singh 2003).

Some useful tips are:

1. Identify the meiotic stage using one anther and transfer the remaining 9 anthers into 1% propiono-carmine. Meiosis in soybean anthers is asynchronous, but prechecking helps in finding suitable stages.
2. Collect very young (how young? depends upon experience) soybean buds between 9 and 10 AM when the greenhouse is still cool; if greenhouse is hot or buds are collected after 10 AM, the anthers contain cells at pachynema to tetrad.
3. Place one-to-two one-week stained anthers on a clean slide, apply a drop of 45% acetic acid, and cover with a cover glass; heat but *do not boil*, tap 1 or 2 times, and squeeze out excess 45% (v/v) acetic acid by thumb pressure.

2.3.1.3 Karyotype Analysis

Karpechenko (1925) determined the $2n = 40$ as the chromosome number of *S. hispida* Mönch (now known as *G. max*), and it was verified by Fukuda (1933), Veatch (1934), and Sakai (1951). Fukuda (1933) measured chromosome diameter of *G. max*, *G. gracilis*, and *G. soja*, and arranged chromosomes 1–20 in order of diameter. The range was 1.038 μm (chromosomel) to 0.650 μm (chromosome 19 and 20), and there was no great difference in the size of chromosomes among three species. Veatch (1934) showed a photomicrograph of cross section of metaphase-I cell with 20 bivalents.

Mitotic metaphase chromosomes of the *Glycine* species are symmetrical and lack

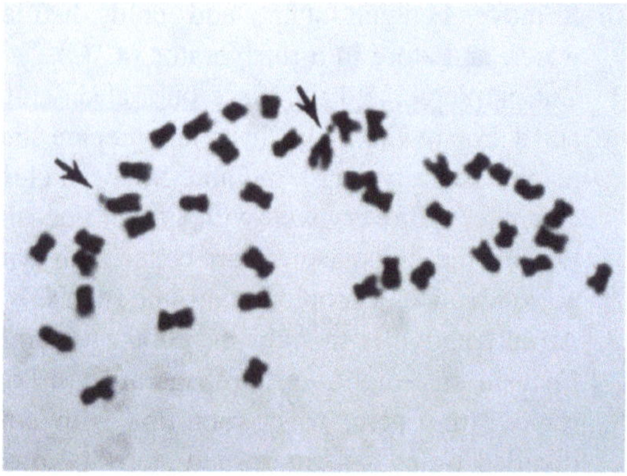

Fig. 2.6 A mitotic metaphase cell of soybean with $2n = 40$ chromosomes. One pair of chromosomes containing a nucleolus organizer region (NOR) can be distinguished (*arrows*), while 38 chromosomes are almost similar

morphological landmarks. Only a pair of nucleolus organizer chromosomes is occasionally visible (Fig. 2.6), and this has been verified by FISH (Skorupska et al. 1989; Griffor et al. 1991; Singh et al. 2001; Krishnan et al. 2001). *Glycine cyrtoloba* (C-genome) and *G. curvata* (C_1-genome) produce only curved pods, a distinct morphological trait which distinguishes these species from other *Glycine* species (Tindale 1984). These species also express 2 pairs of nucleolus organizer chromosomes at mitotic metaphase which are visible by Feulgen staining and 4 signals by FISH, while other species of the genus *Glycine* express only one pair. It is feasible that the second pair either is silent or has lost its NOR activity.

Few attempts using mitotic metaphase chromosomes of soybean (*G. max*) to construct chromosome map were not successful. The length of the mitotic chromosomes of soybean ranged from 1.42 to 2.84 μm (Sen and Vidyabhusan 1960); they grouped the chromosomes into 2 long pairs (2.61–2.84 μm), 4 short pairs (1.42–1.85 μm), and 14 medium-sized pairs (1.97–2.37 μm). They suggested that all the chromosomes contain a median or nearly median kinetochore, and they could not identify satellite chromosomes. Ladizinsky et al. (1979a) karyotyped Giemsa-stained soybean chromosomes and recorded a single band in the majority of the soybean chromosomes. They concluded that this technique has limited value for identifying individual chromosomes. Yanagisawa et al. (1991) separated 40 soybean mitotic metaphase chromosomes into 5 groups (A, B, C, D, and E) containing 7 categories, by using the chromosome image analyzing system (CHIAS). They also used the N-banding technique to identify cytological chromosome markers such as satellite chromosomes. Ohmido et al. (2007) supported idiograms of soybean developed by Yanagisawa et al. (1991) and reported a pair of nucleolus organizer chromosomes by conventional staining and FISH technique. They did not observe condensation patterns for chromosome14.

Recently, Clarindo et al. (2007) developed a method based on using DNA synthesis inhibiting, anti-drying techniques, and digital image analysis of prometaphase and metaphase chromosomes. Chromosome lengths ranged from 1.99 to 1.26 μm for metaphase chromosomes, and the range was 3.35–1.84 μm for prometaphase chromosomes. They identified six metacentric (1, 2, 9, 10, 17, and 19) and fourteen submetacentric (3, 4, 5, 6, 7, 8, 11, 12, 13, 14, 15, 16, 18, and 20) chromosomes. Furthermore, they grouped 40 chromosomes into 14 groups (1–2, 3–4, 5–6, 7, 8, 9–10, 11–12, 13, 14, 15–16, 17, 18, 19, 20). Based on these karyotypes, they proposed that soybean is of tetraploid origin. Findley et al. (2010) karyotyped *G. max* and *G. soja* metaphase chromosomes by using fluorescence in situ hybridization. They used genetically anchored bacterial artificial chromosome (BAC) clones to identify all 20 chromosomes.

Ahmad et al. (1983) utilized a quantitative method (scatter diagram) and grouped soybean chromosomes as 12 metacentric, 7 submetacentric, and 1 subtelocentric and chromosome length ranged from 2.8 μm to <1.6 μm. In *G. soja*,

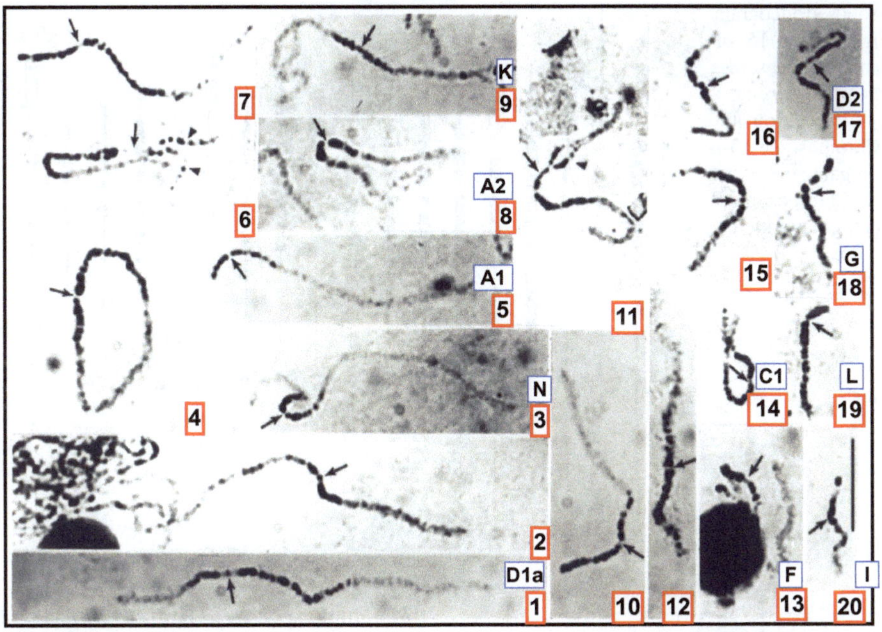

Fig. 2.7 Photomicrographs of the pachynema chromosome complement of *G. max* × *G. soja* F_1 hybrid. Each figure shows a different chromosome, 1–20. *Arrows* indicate centromere location. The *letter* above the *number* represents the molecular linkage group (MLG)

Ahmad et al. (1984) reported 12 metacentric and 2 submetacentric chromosomes, but no subtelocentric chromosomes. The *G. soja* chromosome complement was found to be about 6–7% smaller than that of *G. max*. In contrast, pachynema chromosomes of an F_1 hybrid of *G. max* and *G. soja*, described below, do not support this result. Singh and Hymowitz (1988) identified individual soybean pachynema chromosomes on the basis of length, euchromatin and heterochromatic distribution, and association of a chromosome with the nucleolus. The heterochromatin was distributed proximal to and on either side of the kinetochore on the long and short arms. It is interesting to note that 35.84% of the soybean chromosomes (genome) are composed of heterochromatin (gene-poor regions) and 64.16% chromosomes (genome) are euchromatic (gene-rich regions). Chromosomes of *G. max* and *G. soja* paired perfectly, with minor structural differences (Fig. 2.7). One reciprocal translocation differentiated materials used in this study, and one of the chromosomes involved in the interchange was a satellite chromosome.

Pachynema chromosomes were arranged from 1 to 20 based on total lengths. The range was 39.79 μm (chromosome 1) to 10.63 μm (chromosome 20). Each pair was unique with characteristic landmark features: (1) Seven short arms (chromosome 5, 7, 10, 13, 18, 19, 20) are heterochromatic; (2) chromosome 2 contains the least (14.22%) heterochromatin, chromosome 5 contains the shortest heterochromatic short arm and was easy to identify from other chromosomes, chromosome 7 consists of 59.44% heterochromatin, and chromosomes 18 and 19 were difficult to distinguish, but these chromosomes were very easy to separate from the clump of pachynema chromosomes; (3) chromosome 13 was difficult to isolate although it was expected to separate from the other groups. The satellite was completely imbedded in the nucleolus, and the short arm was completely heterochromatic; (4) it should be noted that a standard chromosome (cytological) idiogram of soybean was constructed based on pachynema chromosomes (Figs. 2.7 and 2.8) (Singh and Hymowitz 1988).

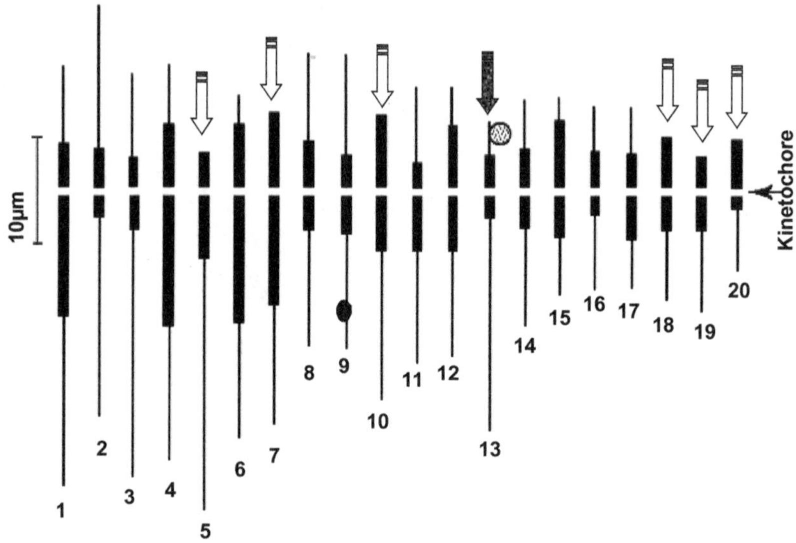

Fig. 2.8 Proposed ideogram, based on Figs. 14 and 15, of the pachynema chromosomes of soybean. *Dotted arrows* show totally heterochromatic short arm. Chromosome 13 is a nucleolus organizer chromosome. Chromosome 9 has a heterochromatic knob on the long arm

2.4 Cytogenetics

2.4.1 Genomes of the *Glycine* Species

2.4.1.1 Classical Taxonomy

Classical taxonomy and extensive plant exploration in Australia and surrounding Islands have played a major role in the identification and nomenclature of new species in the subgenus *Glycine* (Tables 2.1, 2.2, and 2.3). *Glycine clandestina* ($2n = 40$) has been observed to be a morphologically highly variable species (Hermann 1962). Newell and Hymowitz (1980) revised the subgenus *Glycine* by proposing a new species, *Glycine latifolia* (Benth.) Newell and Hymowitz. Stems of wild perennial species are twining, climbing, or procumbent and exhibit distinct morphologically traits. Short pod *G. clandestina* (Singh and Hymowitz 1985b) was removed and named *G. microphylla* (Benth.) Tind. (Tindale 1986b), and curved pod *G. clandestina* were classified as *G. cyrtoloba* (Tindale 1984) and *G. curvata* (Tindale 1986a). Costanza and Hymowitz (1987) observed the presence of adventitious roots in B-genome species (Table 2.3) that includes *G. microphylla*, *G. latifolia*, and *G. tabacina*. This morphologically distinguishing trait is absent in other *Glycine* species. Tindale and Craven (1988) described three new species (*Glycine albicans* Tind. and Craven, *Glycine lactovirens* Tind. and Craven, and *Glycine hirticaulis* Tind. and Craven). *Glycine hirticaulis* contains accessions with $2n = 40$ and 80 chromosomes. Newly described species have restricted geographical habitats in Australia, and they have proven difficult to maintain under greenhouse conditions in Urbana, Illinois.

Diploid ($2n = 40$) *G. tomentella* accessions have been separated into five groups [D3 (A, B, C,), D4, D5, D6, and D7] based on isozyme similarities (Doyle and Brown 1985; Doyle et al. 1986). Later, accessions of the D6 group from Western Australia were classified as *G. arenaria* Tind. (Tindale 1986b). Singh and Hymowitz (1985b) studied meiotic chromosome pairing between D4 group *G. tomentella* (PI441000) and *G. clandestina* ($2n = 40$; PI440948; A_1-genome) and observed 9–18II at metaphase-I. The PI441001 accession contains long pods and narrow leaves, and its meiotic pairing suggests that it is closer to A-genome species (Singh et al. 1992a; Kollipara et al. 1995). Pfeil et al. (2006) removed PI441000 from *G. tomentella* and named it *G. syndetica*. Singh et al. (2007b) assigned to it genome symbol A_6. Currently, we have 26 classified wild perennial *Glycine* species because of extensive plant exploration and taxonomic and molecular studies (Table 2.3).

Glycine tomentella with $2n = 38$ chromosomes is morphologically indistinguishable from 40-, 78-, and 80- chromosome *G. tomentella*. However, it differs genomically from 40-chromosome *G. tomentella* (Kollipara et al. 1993). Taxonomists should closely observe 38-chromosome *G. tomentella* and determine whether this warrants specific species recognition.

Crossing Affinity

Intra- and interspecific crossability is an excellent method to establish the species relationships. Interspecific crosses involving parental species with similar genomes usually set normal pods and F_1 seeds, and the hybrids are fertile, while in crosses between genomically dissimilar species, seed abortion is common, and hybrids if obtained are sterile (Newell and Hymowitz 1983; Grant et al. 1984b; Singh and Hymowitz 1985b, c; 1987, 1988, 1989, 1992a, b; Kollipara et al. 1993). Normal pod set and fertile F_1 hybrids are expected from crosses between morphologically and genomically similar species. However, this is not always true. For example, *G. cyrtoloba* and *G. curvata* carry curved pods and their morphological features are nearly identical (Tindale 1984, 1986a). In one study involving a *G. cyrtoloba* and *G. curvata* cross, a total of 748 flowers were pollinated, but all of the gynoecia died 2–3 days after pollination (DAP) and no pod set was recorded (Singh et al. 1992a). Pod abortion was recorded in a cross between 38- and 40-chromosome *G. tomentella* (Singh et al. 1988) even though both cytotypes were morphologically similar.

2.4.1.2 Meiotic Chromosome Pairing

The degree of chromosome pairing in interspecific hybrids provides an important cytogenetic context for interpreting phylogenetic relationships among diploid species, enhances our understanding of the evolution of the genus, and provides information about the ancestral species. Generally, species with similar genomes exhibit complete or almost complete chromosome pairing (intragenomic chromosome pairing) in their hybrids (Fig. 2.9a). Chromosome migration to anaphase-I poles is normal, and sometimes, species differ only by chromosomal interchanges or by paracentric inversions (Fig. 2.9b).

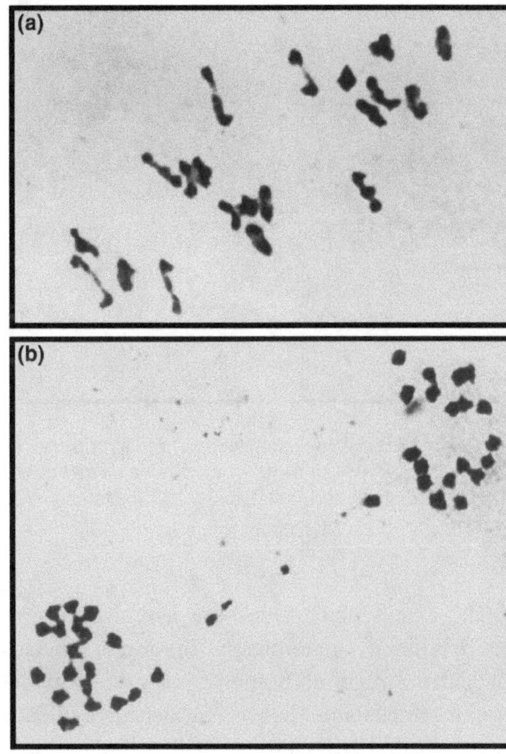

Fig. 2.9 Meiosis in intragenomic F_1 hybrid of *G. latifolia* (B_1B_1; $2n = 40$) × *G. microphylla* (BB; $2n = 40$). **a** Metaphase-I, showing 20 bivalents, (From Singh et al. 2007b); **b** Anaphase-I, showing a chromatin bridge and an acentric fragment (paracentric inversion) in an interspecific hybrid of *G. clandestina* (A_1A_1; $2n = 40$) *G. canescens* (AA; $2n = 40$)

Putievsky and Broué (1979) laid the foundation of genomic relationships in the genus *Glycine*. They produced 19 intraspecific and 30 interspecific F_1 hybrids among *G. canescens*, *G. clandestina*, *G. tomentella* ($2n = 78$, 80), *G. falcata*, and *G. tabacina* ($2n = 40$, 80). In the genus *Glycine*, all F_1 hybrids from crosses among A-genome (*G. canescens*, *G. argyrea*, *G. clandestina*, *and G. syndetika*) and B-genome (*G. microphylla*, *G. latifolia*, *and G. tabacina*) species displayed 20 bivalents at metaphase-I in the majority of the sporocytes (Putievsky and Broué 1979; Newell and Hymowitz 1983; Grant et al. 1984a, b; Singh and Hymowitz 1985b, c; Singh et al. 1988, 1992a, b; Fig. 2.10).

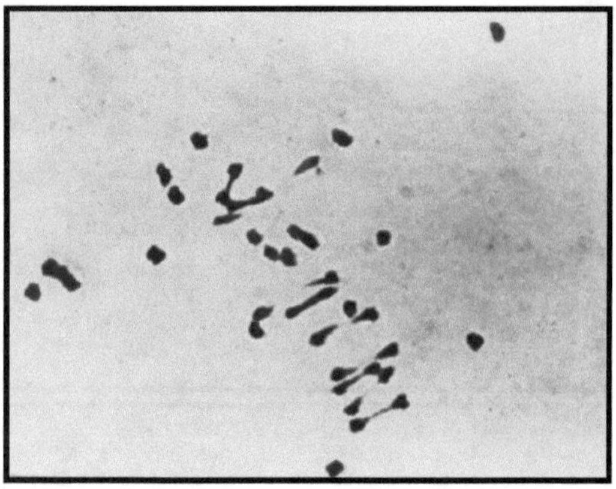

Fig. 2.10 Meiosis in intergenomic F_1 hybrid of *G. latifolia* (B_1B_1; $2n = 40$) × *G. canescens* (AA; $2n = 40$); metaphase-I, showing 20 univalents + 10 bivalents

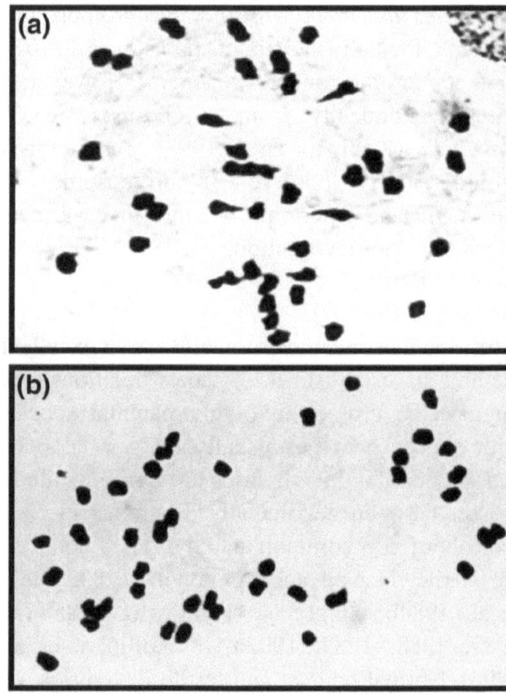

Fig. 2.11 Meiosis in interspecific *Glycine* hybrids. **a** Metaphase-I, showing 31univalents + 4 loosely associated rod-shape bivalents in *G. tomentella* (EE; $2n = 38$) *G. canescens* (AA; $2n = 40$). **b** Metaphase-I, showing 40 univalents in [*G. clandestina* (A_1A_1; $2n = 40$) × *G. canescens* (AA; $2n = 40$)] × *G. falcata* (FF; $2n = 40$)

The extent of chromosome association with the hybrids of genomically dissimilar species elucidates structural homology in the parental chromosomes and hence furnishes evidence regarding the progenitor species (Singh 2003; Singh 2017; Singh and Hymowitz 1985b). Usually, the F_1 hybrids generated from genomically unlike parents (different biological species) are germinated through in vitro techniques. Hybrid seed inviability, seedling lethality, and vegetative lethality are common occurrence in intergenomic crosses (Newell and Hymowitz 1983; Singh et al. 1988, 1992a). In general, such hybrids are weak, slow in vegetative and reproductive growth, and sterile. In the subgenus *Glycine*, A- and B-genome species hybrids show an average chromosome association of 19.7I + 10.2II (A_3 B_1) and 20.9I + 9.5II (A × B_1) (Singh et al. 1988). This suggests strongly that one genome is common in the A- and B-genome species and that it may be the common progenitor species with $2n = 20$ chromosomes. However, since *Glycine* species with $2n = 20$ chromosomes have not been identified, we cannot identify the species.

Variable (semi-homologous–homoeologous) and minimal chromosome pairing are common in intergenomic F_1 hybrids. An erroneous conclusion may be drawn if genome designations of species are based on classical taxonomy only. For example, aneudiploid ($2n = 38$) *G. tomentella* is morphologically similar to 40, 78, and 80 chromosome tomentellas. In contrast, limited chromosome pairing was observed between 38- and 40-chromsome *G. tomentella* (D_3) and aneudiploid *G. tomentella* (E-genome) (Singh et al. 1998) and *G. canescens* (A-genome) (Fig. 2.11a). *Glycine falcata* is morphologically distinct among 26 wild perennial species of the subgenus *Glycine* and 2 species of subgenus *Soja*. Chromosome pairing results (B_1 × F, 37.8I + 1.1II; A × F, 38.7I + 0.6II) support the uniqueness of genome (F) of *G. falcata* because it showed minimal chromosome synapsis with A- and B-genomes (Fig. 2.11b). Putievsky and Broué (1979) reported distant relationship between *G. falcata* (F) and *G. clandestina* (genome A_1) with an average chromosome association of 36.1I + 1.85II + 0.05III. Newell and Hymowitz (1983) obtained non-viable hybrids in *G. falcata* × *G. canescens* and *G. falcata* × *G. tomentella* ($2n = 40$). Cytogenetic studies demonstrate that *G. falcata* does not have a common progenitor present in A, B, C, D, and E-genome species of the subgenus *Glycine* and the origin of this species may be independent or the genomes are completely differentiated.

Based on classical taxonomy, *G. soja* and *G. max* are different species (Hermann 1962). Both species carry $2n = 40$ chromosomes, hybridize readily, produce viable, vigorous, and fertile hybrids, and some lines differ by a reciprocal translocations (Karasawa 1936; Palmer et al. 1987; Singh and Hymowitz 1988) or by paracentric inversions (Ahmad et al. 1977, 1979). Pachynema chromosome pairing was completely normal all along the lengths of the long and short arms with the exception of chromosomes 6 and 11 (Singh and Hymowitz 1988). Therefore, *G. soja* and *G. max* have now been assigned genome symbols G and G_1, respectively (Singh et al. 2007b).

2.4.1.3 Genomic Affinity by Molecular Techniques

During the past two decades, the literature on genomic relationships (plant phylogenetic relationships) has been dominated by molecular studies, including nuclear [seed protein electrophoresis, isozyme variation, restriction fragment length polymorphism (RFLP), amplified fragment length polymorphism (AFLP), random amplification of polymorphic DNA (RAPD), simple sequence repeat (SSR), sequences variation in the gene such as ITS region of rDNA], extra-nuclear (chloroplast and mitochondrial) DNA variation, and genomic in situ hybridization (GISH) by multicolor FISH. Molecular techniques are extremely powerful tool especially for determining species relationships where production of interspecific or intergeneric hybrids is not feasible by conventional methods (Singh 2003). Molecular tools verified the cytogenetically based conclusion that *G. max* and *G. soja* are genomically similar (Doyle and Beachy 1985; Doyle 1988; Kollipara et al. 1995, 1997; Zhu et al. 1995). *Glycine max* and *G. soja* appeared to be identical by Doyle (1988), and the sequence divergence for internal transcribed spacer (ITS) region of rDNA was 0.2% (Kollipara et al. 1997).

Broué et al. (1977) were the first to use isozyme technique to establish genomic relationships among *G. canescens*, *G. clandestina*, and *G. tomentella*. Kollipara et al. (1997) determined phylogenetic relationships among 16 species of the subgenus *Glycine* and two species of the subgenus *Soja* from nucleotide sequence variation in the ITS region of nuclear ribosomal DNA. This study helped to assign genome symbols to five species: H to *G. arenaria*, H_1 to *G. hirticaulis*, H_2 to *G. pindanica*, I to *G. albicans*, and (I_1) to *G. lactovirens*. Cytogenetic relationships among these five species have not been determined because only a few accessions are available and they are difficult to grow in the greenhouse. These genome designations have been verified by histone H3-D gene sequences, and genomes were also assigned to *G. aphyonota* (I_3), *G. peratosa* (A_5), *G. pullenii* (H_3), and *G. stenophita* (B_3) (Brown et al. 2002; Doyle et al. 2002). The ITS region (nrDNA) is a multigene family. However, in the soybean, the nrDNA has been mapped to a single locus on the short arm of chromosome 13 based on the location of the nucleolus organizer region by pachynema chromosome analysis (Singh and Hymowitz 1988) and also by FISH (Singh et al. 2001).

Of the 26 wild perennial *Glycine* species, *G. tomentella* is a unique species because it consists of four cytotypes ($2n = 38, 40, 78, 80$). Aneudiploid ($2n = 38$) *G. tomentella* is distributed in a restricted region of Queensland, Australia. The diploid ($2n = 40$) cytotype is distributed widely in Australia (Queensland, Northern Territory, Western Australia) and Papua New Guinea. Isozyme banding patterns grouped the aneudiploids into two isozyme groups (D1 and D2), and the diploids form six isozyme (D3A, D3B, D3C, D4, D5, D6) groups (Doyle and Brown 1985). Cytogenetics revealed that D1 and D2 isozyme groups carry a similar genome and are distinct from other isozyme groups (Fig. 2.12). Singh

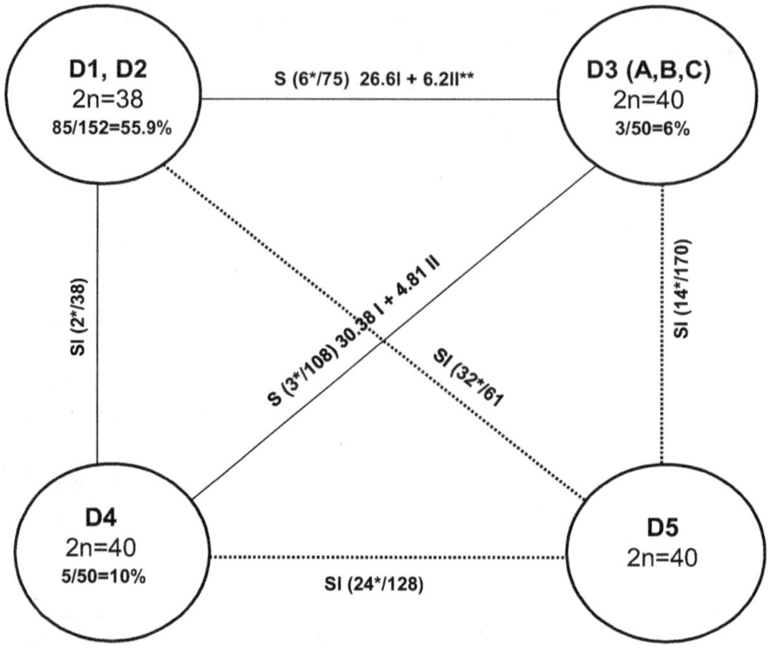

Fig. 2.12 Summary of genomic relationships among five groups of aneudiploid ($2n = 38$; D1 and D2) and diploid ($2n = 40$; D2 (A, B, C), D4, and D5) *G. tomentella* based on crossability rate and meiotic chromosome pairing in F_1 hybrids. The within-group crossability rate (%) is shown inside the *circle*. The between-group crossability rates (number of pod set/total number of flowers pollinated) are shown in the parentheses. *, number of aborted pods; **, Singh et al. (1988); *S* sterile; *SI* seed inviability

et al. (1988) assigned it the E-genome symbol. The D4 isozyme group *G. tomentella* contains PI 441000 (D_3-genome) and has a close affinity cytogenetically with A-genome species but is distinct from other D isozyme groups (Grant et al. 1984a; Singh et al. 1988). Although the D4 isozyme group is morphologically distinct from the A-genome species, it does have long and narrow leaves and a longer pod length that is a characteristic feature of A-genome species and that distinguishes it from other diploid *G. tomentella* accessions (Singh et al. 1998b). The histone H3-D gene sequence also grouped D4 isozyme accessions with A-genome species (Brown et al. 2002), and the D4 isozyme group *G. tomentella* was classified as *Glycine syndetika* B.E. Pfeil and Craven (Pfeil et al. 2006). The A_6-genome symbol was assigned to PI441000 (Singh and Chung 2007; Singh et al. 2007b). No viable hybrid plants were produced in crosses between accessions of the D5 and D1, D2, D3, and D4 isozyme groups (Fig. 2.12).

2.4.2 Origin of Polyploid Complexes of *Glycine tabacina* and *G. tomentella*

Of the 26 species of the subgenus *Glycine*, *G. hirticaulis*, *G. tabacina*, and *G. tomentella* contain $2n = 40$ and $2n = 80$ chromosomes (Table 2.3). Furthermore, *G. tomentella* consists of aneudiploid ($2n = 38$) and aneutetraploid ($2n = 78$) accessions. Tetraploid *G. hirticaulis* has restricted geographical distribution in the Northern Territory of Australia. *Glycine tomentella* accessions with $2n = 80$ chromosomes are also distributed in Taiwan, and it was designated as. *G. dolichocarpa* Tateishi and Ohashi (Tateishi and Ohashi 1992) and species status was verified by seed protein patterns (Hsieh et al. 2001). On the other hand, aneutetraploid ($2n = 78$) *G. tomentella* is found in Australia and Papua New Guinea and tetraploid ($2n = 80$) *G. tomentella* is distributed in Australia, Papua New Guinea, Philippines, Timor Island of Indonesia, and

Taiwan. Aneutetraploid and tetraploid tomentellas are allopolyploid and exhibit diploid-like meiosis (Singh and Hymowitz 1985a) and complexes of multiple origins (Putievsky and Broué 1979; Singh et al. 1987b, 1989, 1992b; Singh and Hymowitz 1985b; Doyle and Brown 1989; Kollipara et al. 1994; Doyle et al. 1999, 2004; Hsing et al. 2001; Brown et al. 2002; Rauscher et al. 2004). Classification of the polyploid *G. tabacina* and *G. tomentella* accessions into discretely defined, reproductively isolated groups using various methods (morphological, cytogenetic, biochemical, and molecular) helps in better understanding the origin of the species complex (Doyle et al. 1990c; Kollipara et al. 1994).

2.4.2.1 *Glycine tabacina* (2*n* = 80)

Diploid *G. tabacina* is indigenous to Australia. A tetraploid ($2n = 80$) cytotype is found sympatrically with diploids in Australia, but 80-chromosome tabacinas are distributed also in the islands of the south Pacific (New Caledonia, Vanuatu, Fiji, Tonga, Niue) and west central Pacific (Taiwan, Ryuku, Marianas) (Singh et al. 1992b). Morphological observations (Costanza and Hymowitz 1987), cytogenetic investigation (Singh et al. 1987b, 1992b), and molecular studies (Doyle et al. 1990a, b, 1999) have shown two distinct groups in the 80-chromosome *G. tabacina*. It is an allopolyploid complex of multiple origins. One group contains adventitious roots, while the other group has long and narrow leaves (like A-genome species) and has no adventitious roots. All the intraspecific F_1 hybrids within each group showed normal meiosis and seed fertility. However, F_1 hybrids between two groups were sterile owing to disturbed meiosis. At metaphase-I, a model chromosome association of 40I + 20II was recorded (Singh et al. 1987b), indicating that both groups have one genome in common and differ for the second genome.

Singh et al. (1992a, b) proposed, based on cytogenetics, that the 80-chromosome *G. tabacina* without adventitious roots is a complex, probably synthesized from A-genome (*G. canescens*, *G. clandestina*, *G. argyrea*, *G. syndetika*) and *G. tabacina* with adventitious roots evolved through segmental allopolyploidy from B-genome (*G. latifolia*, *G. microphylla*, *G. tabacina, G. stenophita*) or by allopolyploidy having one of the B-genome species and second unknown genome donor species. Doyle et al. (2000) included *G. stenophita* into B-genome group; however, cytogenetic relationship with other B-genome species has not been determined.

Doyle et al. (1999) suggested, based on sequencing of histone H3-D locus, the multiple origins with gene exchange among lineage increase the genetic base of a polyploid and help better colonization of polyploid *G. tabacina* relative to its diploid progenitors. Hybridization is unlikely in a highly self-inbreeder in the nature; however, F_1 hybrids among B-genome species are completely fertile (Putievsky and Broué 1979; Newell and Hymowitz 1983; Grant et al. 1984b; Singh and Hymowitz 1985c). Since B-genome species are sympatric (Doyle et al. 1999), adventitious root trait is controlled by recessive gene (Singh et al. 1988) and Bowman Birk Inhibitor (BBI) is present in A-genome species including 80-chromosome *G. tabacina* without adventitious roots but absent in B-genome species and 80-chromosome *G. tabacina* with adventitious roots (Kollipara et al. 1997).

2.4.2.2 *Glycine tomentella* (2*n* = 78, 80)

Diploid-like meiosis, isozyme banding patterns among the accessions and meiotic pairing in intraspecific and interspecific F_1 hybrids, wide geographical distribution, and aggressive and vigorous growth habit suggest that 78- and 80-chromosome tomentellas are of allopolyploid origin and are polyploid complexes (Singh and Hymowitz 1985a).

Genomic complexes within species can be determined by obtaining intraspecific hybrids involving parental accessions of diverse morphology, cytology, and geographical origins. Meiotic pairing and molecular results in intraspecific plants of 78-chromosome tomentellas have revealed three complexes (designated based on isozyme; T1, T5, and T6; Doyle and Brown 1985). Hybrids within groups showed normal chromosome pairing. All hybrids between genomic complexes showed one common genome (EE-genome; 38-chromosome *G. tomentella*), and this was verified by molecular methods (Kollipara et al. 1994). This suggests that some complexes have one genome common and differ for the second genome. T1 group aneutetraploid tomentella predominates and is distributed in Queensland, and one accession in Papua New Guinea. T5 group is found in New South Wales, while T6 group is found in Western Australia. Thus, these isozyme groups are geographically isolated and may have originated independently with an aneudiploid ($2n = 38$; E-genome) as common genome donor.

Based on isozyme banding patterns, Doyle and Brown (1985) separated 80-chromosome *G. tomentella* accessions into three (T2, T3, T4) groups. They did not examine accessions from Indonesia. Kollipara et al. (1994) assigned T7 group to accessions from Indonesia. Cytogenetics, total seed protein profiles, protease inhibitor activity band profiles, immunostained banding patterns, and RFLP analysis clearly identified four distinct groups (T2, T3, T4, and T7) in 80-chromosome *G. tomentella* (Kollipara et al. 1994). Meiotic chromosome pairing at diakinesis-metaphase-I in the F_1 hybrids within isozyme groups was normal and the plants were completely fertile, but hybrids between groups were sterile. All F_1 hybrids between T2 × T3, T3 × T4, and T2 × T7 were sterile; presumably as a consequence of disturbed meiotic chromosome pairing. Chromosome pairing results suggest one common genome in the T2, T3, T4, and T7 groups. Accessions from Indonesia (T7) did not show complete genome affinity with T3 group (accessions from Cooktown, Qld, Port Moresby, PNG, and Gratto Creek, Western Australia). However, maximum chromosome association of 30II + 20I at metaphase-I was recorded, and it produced few cleistogamous pods and mature seeds. This suggests that accessions of T7 probably carry the same genome and geographical isolation played a major role in their divergence. An independent origin cannot be proposed because *Glycine* species with diploid cytotypes have not been identified in Indonesia.

A way to ascertain the ancestors of 78-and 80-chromosome tomentellas is to synthesize amphidiploids from their putative parental species (Singh et al. 1998a). Aneuallotetraploid (DDEE, AAEE; $2n = 78$) and allotetraploid (AADD; $2n = 80$) were produced by somatic chromosome doubling of $2n = 39$ and $2n = 40$ F_1 hybrids. Meiotic chromosome pairing in the synthesized amphiploid was diploid-like and produced normal pods and seeds. Synthesized amphidiploids were hybridized with accessions of tomentellas of T1, T2 ($2n = 78$), and T2 ($2n = 80$) isozyme groups. Meiotic pairing was normal and fertile (Singh et al. 1989), and molecular results verified the cytogenetic results (Kollipara et al. 1994). Rauscher et al. (2004) reported that the genome donors of T2 *G. tomentella* are diploid *G. tomentella* of D3 and D4 isozyme groups. The D4 (PI44100) isozyme showed genomic affinity with *G. clandestina* (PI505161) (Singh et al. 1987). Pfeil et al. (2006) taxonomically classified PI441001 as *G. syndetika,* and Singh et al. (2007) assigned it genome symbol A_6 (Table 2.3). Cytogenetics and molecular studies of Kollipara et al. (1994) support *G. canescens* or any A-genome species including *G. syndetika* as possible genome donor to T2 tomentella.

Kollipara et al. (1994) concluded "both aneutetraploid (T1, T5, T6) and tetraploid accessions (T2, T3, T4, and T7) may have originated by independent events. Reconstruction of hypothesized ancestors through synthesis of artificial hybrids appears to provide an excellent way to analyze polyploid complexes such as those of *G. tomentella*. However, it is also important to understand the diversity among diploid donors of these polyploids." Rauscher et al. (2004) verified this statement by molecular studies.

2.4.3 Chromosomal Aberrations-Structural Aberrations

Chromosomal structural changes such as deficiencies, duplications, interchanges, and inversions have not been systematically produced, identified, and used in physical genetic mapping in soybean. An interchange of spontaneous origin in soybean (Sadanaga and Grindeland (1984) has been used to located the w_1 (white flower) locus on the satellite chromosome (chromosome13). Palmer et al. (1987) surveyed 56 *G. soja* accessions from China and the Russia that also included PI81762 studied by Singh and Hymowitz (1988). They concluded that these accessions have a single similar or identical interchange. Singh and Hymowitz (1988) examined an interspecific F_1 hybrid of soybean × PI81762 and observed one quadrivalent that was always associated with the nucleolus. Mahama et al. (1999) identified six of the possible 20 reciprocal translocation lines. Mahama and Palmer (2003) identified that classical linkage groups (CLGs) six and eight are the same by reciprocal translocations.

Inversions (paracentric and pericentric) have neither been produced nor used in physical mapping of the soybean genome. Study has been limited to identifying a paracentric inversion in soybean × *G. soja* hybrid (Ahmad et al. 1977; Palmer et al. 2000). Wild perennial *Glycine* species with similar genomes are differentiated by a paracentric inversion (Singh 2003) as the majority of the sporocytes showed normal pairing at metaphase-I (Fig. 2.9a), but at anaphase-I, a chromatin bridge and an acentric fragment were observed (Fig. 2.9b).

2.4.4 Chromosomal Aberrations—Numerical Changes

2.4.4.1 Autopolyploidy

Darlington and Wylie (1955) proposed × = 10 as the basic chromosome number of the genus *Glycine*. Haploid ($2n = 20$) (Crane et al. 1982), triploid ($2n = 60$) (Chen and Palmer 1985), and tetraploid ($2n = 80$) (Sen and Vidyabhusan 1960) soybean plants have been reported. Haploid and triploid are completely sterile, and tetraploid soybean produces few one or two large seeded pods. Tetraploid soybean has no commercial value. Tetraploid × diploid crosses in natural population have failed to produce an autotriploid (Sadanaga and Grindeland 1981), an excellent source for producing primary trisomics (Singh 1998a). However, Chen and Palmer (1985) identified autotriploids from the progenies of male sterile lines, but the derived autotriploid was not used to produce primary trisomics. Xu et al. (2000b) found a hypertriploid ($2n = 3x + 1 = 61$) plant from a cross T31 [a homozygous recessive glabrous (*pp*)] × T190-47-3 (an unidentified primary trisomic). The hypertriploid plant produced 98 selfed seed, and the chromosome numbers of the plants ranged from $2n = 50$ to 69. Chromosome number in hypertriploid × diploid seeds ranged from $2n = 44$ to 56. These lines were not used in producing primary trisomics either.

2.4.4.2 Aneuploidy—Trisomics

Primary Trisomics

Primary trisomics in soybean (an individual with normal chromosome complements plus an extra complete chromosome; $2n = 2x + 1 = 41$) have been isolated from the progenies of asynaptic and desynaptic mutants (Palmer 1976; Gwyn et al. 1985; Xu et al. 2000c). Gwyn et al. (1985) examined four primary ($2n = 41$) trisomics (Tri A, B, C, D) and observed that they were similar to the diploid ($2n = 40$). Singh and Hymowitz (1991) identified, by using pachytene chromosome, Tri A as Triplo 5, Tri C as Triplo 1, Tri D as Triplo 4, and Tri S (Skorupska et al. 1989) as Triplo 13. Triplo 13 contains 3 satellite chromosomes. Ahmad et al. (1992) identified four new primary trisomics (Triplo 2, 3, 10, and 14), and Xu et al. (2000c) isolated 12 additional primaries from 37 aneuploid lines ($2n = 41, 42, 43$) and tentatively identified 20 simple primary trisomics that were designated, based on the length of pachytene chromosome, as Triplo 1 (contains the longest chromosome) to Triplo 20 (the shortest chromosome). These aneuploid lines originated from the progenies of asynaptic and desynaptic mutants that were supplied by the late Reid Palmer (USDA/ARS, Iowa State University, Ames, IA, 50011-1010).

At metaphase-I of meiosis, a majority of the microsporocytes in primary trisomics exhibit 1III + 19II or 20II + 1I. Average female transmission of 20 soybean primary trisomics was

42% with a range of 27 (Triplo 20) to 59% (Triplo 9). Female transmission rate has been estimated from the hybrid population (Xu et al. 2000c). This may be the reason for high female transmission rate of the extra chromosome in primary trisomics of soybean; heterozygosity often favors the higher female transmission rate (Singh 2003).

Primary trisomics of soybean have been used to associate a few classical genetic markers. Three marker genes *Eu1* (seed urease), *Lx1* (lipoxygenase-1), and *P2* (puberulent) were located on chromosome 5, 13, and 20, respectively (Xu et al. 2000c). Zou et al. (2003a, b) associated yellow leaf mutant *y10* with chromosome 3 by primary trisomics method. Griffin et al. (1989) located *fr1* gene on CLG 12 which was separated from *Ep* gene with a distance of 41.4 ± 0.8 cM, and Jin et al. (1999) associated CLG 12 with molecular linkage group (MLG) K. Root fluorescence gene (*fr1*) showed trisomic ratio with Triplo 9 (216:14) and disomics ratio with Triplo 1 (23:5), 2 (21:7), 3 (25:4), 4 (32:6); 5 (23:6), 6 (19:5), 12 (27:8), and 15 (39:10). This clearly demonstrates that the *fr1* is located on chromosome 9 which contains MLG K and CLG 12.

Hedges and Palmer (1991) used five primary trisomics (A, B, C, D, and S) to locate several isozyme loci. *Dia1* (diaphorase) showed trisomics ratio with trisomic D which is chromosome 4 of Singh and Hymowitz (1988, 1991). Thus, chromosome 4 has a genetic marker because none of the MLGs has been associated with chromosome 4 (Zou et al. 2003b).

Gardner et al. (2001) associated *Rps1-K* (*Phytophthora sojae*) with chromosome 3 by primary trisomics because this gene is dominant and showed 2:1 (trisomics ratio) with Triplo 3 and disomic ratio (3:1) with 9 other primary trisomics. MLG N is associated with chromosome 3. This demonstrates that primary trisomics are an excellent cytogenetic tool to associate genes and linkage maps physically to the chromosomes. Seeds of primary trisomics are unavailable and are likely perished because they have not been deposited in the USDA soybean germplasm collection (https://npgsweb.ars-grin.gov/gringlobal/site.aspx?id=24).

Monosomics

In soybean, monosomics ($2n - 1 = 39$; an individual lacking one chromosome is called monosomic) have been isolated from the progenies of synaptic mutant (Skorupska and Palmer 1987) and Triplo 3 and Triplo 6 (Xu et al. 2000a). Skorupska and Palmer (1987) isolated two plants with $2n = 39$ and 92 seedlings with $2n = 40$ from 1380 seeds; only 94 seeds germinated. Morphologically, mono-3 was smaller with reduced vigor, while mono-6 was similar to the disomics. Female transmission in mono-3 was 6.5%, while mono-6 was not transmitted among 105 plants. The transmission rate in monosomics is sporadic in soybean. It was concluded that monosomics in soybean are viable and fertile and can be produced; however, no systematic effort is being made to isolate monosomics in this economically important crop.

Tetrasomics

In soybean, tetrasomics ($2n + 2 = 42$; an individual carrying two extra homologous chromosomes in addition to its normal somatic chromosome complement is called tetrasomics) are identified in low frequencies from the selfed progenies of primaries ($2n = 41$) (Singh and Chung 2007). Tetrasomic 13 plants were weak and died prematurely. Gwyn and Palmer (1989) observed, based on morphological measurement, that tetrasomics and double trisomics ($2n + 1 + 1$) could be distinguished accurately from their disomics sibs. Tetrasomics mostly breed true, and occasionally related trisomics ($2n = 41$) and diploids ($2n = 40$) are identified. Most primary trisomics plants are produced from tetrasomics × disomics crosses. Tetrasomics in the soybean are unique cytogenetic stock because they are not viable in diploid crops such as maize, barley, rice, and tomato. It is sad that primary trisomics and tetrasomics stock of soybean are unavailable and probably they have been lost forever.

2.4.5 Chromosome Mapping

Chromosome, genetic, and cytogenetic maps in the model economically important crops such as rice, maize, barley, wheat, and tomato were developed first, and molecular maps followed. By contrast, several molecular maps have been developed first in the soybean and these maps were not associated with the chromosomes by primary trisomics.

2.4.5.1 Chromosome Map

Although the precise chromosome number of soybean was determined in 1925 (Karpechenko 1925), several chromosome maps have been developed. However, none of the papers convincingly identified individual somatic metaphase chromosomes because the chromosomes are symmetrical and only a pair of nucleolus organizer chromosome is identified in one of the best chromosome spreads. Singh and Hymowitz (1988) constructed a chromosome map of soybean by using pachynema chromosomes (Figs. 2.7 and 2.8). This pioneering research has set the stage to produce all possible primary trisomics in the soybean (Xu et al. 2000c).

2.4.5.2 Classical Linkage Groups

A genetic linkage map with 20 linkage groups, designated as CLGs of soybean, has been proposed (Palmer et al. 2004). Each of the classical linkage groups 2, 3, 12, 13, 15, 16, 18, 20, (21?) has two qualitative trait loci. Thus, the genetic linkage map of soybean is not saturated with classical markers as compared to other economically important crops. Mahama and Palmer (2003) associated CLGs 6 and 8 on the same linkage groups using translocation stocks. Translocation tester sets involving 20 soybean chromosomes have not been produced so far.

2.4.5.3 Cytogenetic Map

By using SSR markers from 20 MLGs and primary trisomics, Zou et al. (2003b) associated 11 MLGs with the 11 chromosomes and they failed to associate nine MLGs with the remaining chromosomes. It is possible that the trisomic set is not complete. Segregation distortion is common using primary trisomics and SSR markers, and this may be due to the preferential selection of gametes containing certain genotypes (Zou et al. 2006).

2.5 Wide Hybridization

2.5.1 Genetic Resources

Harlan and de Wet (1971) developed the concept of three gene pools—primary (GP-1), secondary (GP-2), and tertiary (GP-3) based on the success rate of hybridization among/between species. The clear understanding of taxonomic and evolutionary relationships between a cultigen and its wild relatives is a prerequisite for the exploitation of the primary, secondary, and tertiary gene pools.

2.5.1.1 Soybean GP-1

Soybean GP-1 consists of biological species that can be crossed to produce vigorous hybrids that exhibit normal meiotic chromosome pairing and possess total seed fertility. Gene segregation is normal, and gene exchange is generally easy. ARS GRIN maintains 20073 accessions of *G. max* and 1181 accessions of *G. soja* as of May 24, 2017 (https://npgsweb.ars-grin.gov/gringlobal/site.aspx?id=24).

2.5.1.2 Soybean GP-2

GP-2 species can hybridize with GP-1 easily, and F_1 plants exhibit at least some seed fertility (Harlan and de Wet 1971). *Glycine max* is without GP-2 because no known species has such a relationship with soybean. It is possible that species in the soybean GP-2 do exist in Southeast Asia where the *Glycine* genus may have originated. However, it is merely a speculation and extensive plant exploration in this part of the world is required to validate this assumption.

2.5.1.3 Soybean GP-3

GP-3 is the third outer limit of potential genetic resource. Hybrids between GP-1 and GP-3 are lethal, or completely sterile, and gene transfer is not possible or requires radical techniques (Harlan and de Wet 1971). Based on this definition, GP-3 includes the 26 wild perennial species of the subgenus *Glycine*. These species are indigenous to Australia, and various surrounding islands and are geographically isolated from *G. max* and *G. soja*. Table 2.3 shows the *Glycine* species and their $2n$ chromosome numbers, nuclear genomes, and geographical distributions. The USDA Soybean Germplasm Collection, Urbana, Illinois, as of May 24, 2017, maintains 1006 accessions of the 19 wild perennial species (http://www.ars-grin.gov/cgi-bin/npgs/html/site_holding.pl?SOY).

2.5.2 Intersubgeneric Hybridization

The twenty-six wild perennial species of the genus *Glycine* subgenus *Glycine* have not been exploited in soybean breeding programs. These species are extremely diverse morphologically, cytologically, and genomically, grow in very diverse climatic and soil conditions, and have a wide geographical distribution (Singh and Hymowitz 1999). Wild perennial *Glycine* species have great potential for soybean improvement. They are a rich source of agronomically useful genes and alleles (Chung and Singh 2008). However, extensive and reproducible screening for abiotic and biotic stresses of 26 wild perennial *Glycine* species are lacking. Schoen et al. (1992) studied resistance of *G. tomentella* to 3 Australian isolates of soybean leaf rust and found PI 441001 ($2n = 78$), used in this study, resistant to all three races. It has been demonstrated that resistance to soybean rust in PI441001 has a chemical basis; a chemical inhibits the growth of fungus spores (Damla et al. 2008).

Ladizinsky et al. (1979b) initiated producing intersubgeneric hybrids between soybean and five wild perennial species of the subgenus *Glycine* and concluded "*there would be little chance that these wild species could be exploited for breeding purposes.*" Since 1979, several researchers have attempted to hybridize wild perennial *Glycine* species with the soybean, but only a few sterile intersubgeneric F_1 hybrid combinations have been achieved by using embryo rescue method (Table 2.6). Pod abortion in intersubgeneric hybridization is a post-hybridization problem (Singh and Hymowitz 1987). Pod retention can be achieved by spraying a growth hormone mixture (100 mg GA_3 + 25 mg NAA + 5 mg Kinetin/L distilled water) once a day for 19 days. The developing pods should be removed 19–21 days post–pollination, and the immature seeds should be extracted aseptically and cultured onto a seed maturation medium (Singh 2007). Thus far, only Singh et al. (1990, 1993) have successfully produced backcross-derived fertile progenies from the soybean and a wild perennial, *G. tomentella* ($2n = 78$). Monosomic alien addition lines (MAALs) and modified diploid ($2n = 40$) lines are being isolated and identified (Singh et al. 1998a; Singh 2007; Singh and Nelson 2014, 2015).

A schematic diagram to produce reciprocal intersubgeneric hybrid between soybean and *G. tomentella* is shown in Fig. 2.13. The F_1 hybrid plants between soybean cv. Dwight ($2n = 40$) and *G. tomentella,* PI441001 ($2n = 78$) were rescued through embryo culture after 1 year. F_1 plants were vigorous, sterile, and contained $2n = 59$ chromosomes. Chromosomes were doubled (amphidiploid; $2n = 118$) by 0.1% colchicine treatment (Singh 1998a). An amphidiploid plant produced one (mostly) and 2-seeded pods and was backcrossed to Dwight (recurrent parent), and many BC_1 plants with $2n = 79$ chromosomes (40 chromosomes from Dwight and 39 chromosomes from *G. tomentella*) were produced through embryogenesis. All BC_1 plants were totally self-sterile. Backcrossed to Dwight, pod abortion was common but 40 mature seeds were harvested that produced 24 BC_2 plants. Chromosome number in BC_2 plants ranged from $2n = 55$ (40 chromosomes from Dwight + 15 chromosomes from *G. tomentella*) to 59 (40 chromosomes from Dwight + 19 chromosomes from *G. tomentella*). These plants are known as hypotriploids, and all plants were morphologically distinct in ways depending upon the presence of *G. tomentella* chromosome combinations. Amphidiploid plants ($2n = 118$) were induced by colchicine treatment. Amphidiploid × Dwight (BC_1) was

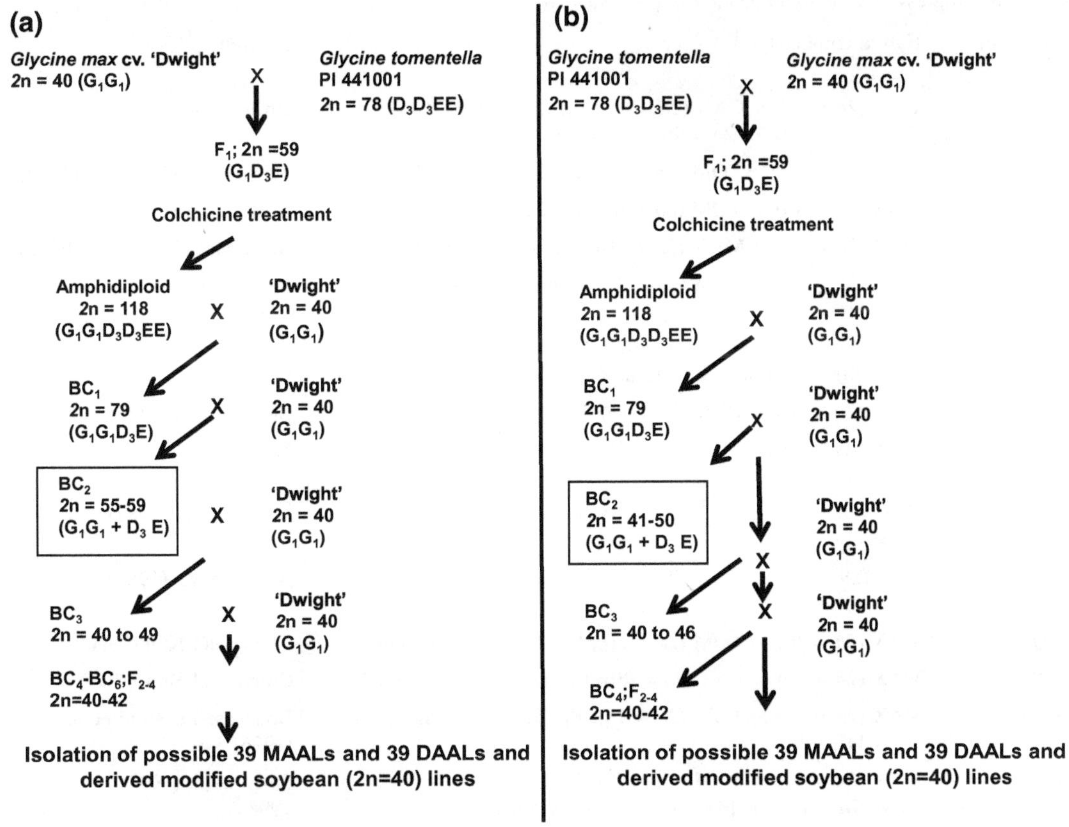

Fig. 2.13 A cytogenetic diagrammatic scheme to produce fertile soybean lines from *G. max* cv. Dwight ($2n = 40$) × *G. tomentella* ($2n = 78$) (**a**) and reciprocal hybrid (**b**). Note, chromosome numbers in BC_2 plants

developed through immature seed culture, and BC_1 plant contained $2n = 79$ chromosomes. Chromosome number in BC_2F_1 plants ranged from $2n = 55$ to 60. All BC_2 plants were self-sterile and so were backcrossed to Dwight. Chromosome number in BC_3 plants ranged from $2n = 41$ to 49. Plants with higher than $2n = 42$ chromosomes were self-sterile and so were back-crossed to Dwight. At the end, we expect to isolate plants $2n = 41$ chromosomes and hope to categorize them into the 39 possible monosomic alien addition lines (MAALs). Progenies of $2n = 41$ chromosome plants segregated about 70% $2n = 40$ chromosome plants, 29% $2n = 41$ chromosome plants, and 1% $2n = 42$ (disomic alien addition lines; DAALs) plants. Transfer of plants from test tube to the field required about 4 years. The modified diploid ($2n = 40$) lines are being screened for resistance to pests and pathogens. This study has broken the crossability barrier and sets the stage for the exploitation of perennial, wild *Glycine* germplasm so-called a weed from Australia to broaden the genetic base of the cultivated soybean. Materials are distributed to public soybean breeders through material transfer agreement (MTA) (Table 2.6).

Wide hybridization in soybean calls for perseverance, commitment, dedication, patience, and skill. The initial process requires tissue culture expertise, and the post-BC_1 period needs knowledge of cytogenetics and precise record-keeping ability. The handling of soybean chromosomes is required because misidentification of plants from $2n = 40$ to 41 (Sadanaga and Grindeland 1981) may occur if the chromosome spreads are not excellent.

Table 2.6 Progress of wide hybridization in the genus *Glycine*

Number of attempts	Hybrid combinations	References
1	TOM ($2n = 38$) × CAN ($2n = 40$); F_1; $2n = 39$ = CT ($2n = 78$) × MAX ($2n = 40$); F_1 sterile	Broué et al. (1982)
2	MAX ($2n = 40$) × TOM ($2n = 78$); F_1; $2n = 59$; sterile	Newell and Hymowitz (1982)
3	MAX ($2n = 40$) × TOM ($2n = 80$); F_1; $2n = 60$; sterile	Newell and Hymowitz (1982)
4	TOM ($2n = 78$) × MAX ($2n = 40$); F_1; $2n = 59$; sterile	Singh and Hymowitz (1985d)
5	MAX ($2n = 40$) × TOM ($2n = 80$); F_1 embryo ($2n = 64$); no F_1 plant	Sakai and Kaizuma (1985)
6	ARG ($2n = 40$) × CAN ($2n = 40$); F_1; $2n = 40$ × MAX ($2n = 40$) = CT ($2n = 80$); sterile	Grant et al. (1986)
7	MAX ($2n = 40$) × CLA ($2n = 40$); F_1; $2n = 40$; sterile	Singh et al. (1987a)
8	MAX ($2n = 40$) × TOM ($2n = 78$); F_1; $2n = 59$; sterile = CT ($2n = 118$)	Newell et al. (1987)
9	TOM ($2n = 78$) × MAX ($2n = 40$); F_1; $2n = 59$; sterile = CT ($2n = 118$)	Newell et al. (1987)
10	CAN ($2n = 40$) × MAX ($2n = 40$); F_1; $2n = 40$; sterile = CT ($2n = 80$)	Newell et al. (1987)
11	MAX ($2n = 40$) × TOM ($2n = 80$); F_1; $2n$ = Not determined	Chung and Kim (1990)
12	MAX ($2n = 40$) × LAT ($2n = 40$); F_1; $2n$ = Not determined	Chung and Kim (1991)
13	MAX ($2n = 40$) × TOM ($2n = 78$); F_1; $2n = 59$; sterile = CT ($2n = 118$)	Bodanese-Zanettini et al. (1996)
14	MAX ($2n = 40$) × TOM ($2n = 78$); F_1; $2n = 59$; CT = ($2n = 118$) × MAX (BC_1-BC_6); MAALs	Singh et al. (1990, 1993, 1998b)
15	TOM ($2n = 78$) × Max ($2n = 40$); F1; $2n = 59$; CT = ($2n = 118$) × MAX (BC1-BC4); MAALs	Singh and Nelson (2014)
16	MAX ($2n = 40$) × TOM ($2n = 78$); F_1; $2n = 59$; CT = ($2n = 118$) × MAX (BC_1-BC_6); MAALs	Singh and Nelson (2015), Singh et al. 2007a)

TOM *G. tomentella*, *CAN G. canescens*, *MAX G. max*, *ARG G. argyrea*, *LAT G. latifolia*, *CT* Colchicine treatment

References

Ahmad QN, Britten EJ, Byth DE (1977) Inversion bridges and meiotic behavior in species hybrids of soybeans. J Hered 68:360–364

Ahmad QN, Britten EJ, Byth DE (1979) Inversion heterozygosity in the hybrid soybean × *Glycine soja*. J Hered 70:358–364

Ahmad QN, Britten EJ, Byth DE (1983) A quantitative method of karyotypic analysis applied to the soybean, *Glycine max*. Cytologia 48:879–892

Ahmad QN, Britten EJ, Byth DE (1984) The karyotype of *Glycine soja* and its relationship to that of the soybean, *Glycine max*. Cytologia 49:645–658

Ahmad F, Singh RJ, Hymowitz T (1992) Cytological evidence for four new primary trisomics in soybean. J Hered 83:221–224

Bodanese-Zanettini MH, Lauxen MS, Richter SNC, Cavalli-Molina S, Lange CE, Wang PJ, Hu CY (1996) Wide hybridization between Brazilian soybean cultivars and wild perennial relatives. Theor Appl Genet 93:703–709

Broich SL, Palmer RG (1980) A cluster analysis of wild and domesticated soybean phenotypes. Euphytica 29:23–32

Broué P, Marshall DR, Müller WJ (1977) Biosystematics of subgenus *Glycine* (Verdc.): Isoenzymatic data. Aust J Bot 25:555–566

Broué P, Douglass J, Grace JP, Marshall DR (1982) Interspecific hybridization of soybeans and perennial Glycine species indigenous to Australia via embryo culture. Euphytica 31:715–724

Brown AHD, Doyle JL, Grace JP, Doyle JJ (2002) Molecular phylogenetic relationships within and among diploid races of *Glycine tomentella* (Leguminosae). Aust Syst Bot 15:37–47

Carlson JB, Lersten NR (2004) Reproductive morphology. In: Boerma HR, Specht J E (eds) *Soybeans: improvement, production, and uses*, 3rd edn, *Agron.*

Monogra. 16. American Society of Agronomy, Inc./ Crop Science Society of America, Inc./ Soil Science Society of America, Inc, pp 59–95
Chen LF, Palmer RG (1985) Cytological studies of triploids and their progeny from male-sterile (*ms1*) soybean. Theor Appl Genet 71:400–407
Chung GH, Kim JH (1990) Production of interspecific hybrids between *Glycine max* and *G. tomentella* through embryo culture. Euphytica 48:97–101
Chung GH, Kim KS (1991) Obtaining intersubgeneric hybridization between *Glycine max* and *G. latifolia* through embryo culture. Korean J Plant Tissue Cult 18:39–45
Chung G, Singh RJ (2008) Broadening the genetic base of soybean: a multidisciplinary approach. Crit Rev Plant Sci 27:295–341
Clarindo WR, de Carvalho CR, Alves BMG (2007) Mitotic evidence for the tetraploid nature of *Glycine max* provided by high quality karyograms. Plant Syst Evol 265:101–107
Costanza SH, Hymowitz T (1987) Adventitious roots in *Glycine* subg. *Glycine (Leguminosae)*: morphological and taxonomic indicators of the B genome. Plant Syst Evol 158:37–46
Crane CF, Beversdorf WD, Bingham ET (1982) Chromosome pairing and association at meiosis in haploid soybean (*Glycine max*). Can J Genet Cytol 24: 293–300
Damla B, DeLucia EH, Zangerl AR, Singh RJ (2008) Plant-derived biofungicide against soybean rust disease. U.S. Provosional Application no. 61/028,459
Darlington CD, Wylie AP (1955) Chromosome atlas of flowering plants. George Allen and Unwin Ltd., London
Doyle JJ (1988) 5S ribosomal gene variation in the soybean and its progenitor. Theor Appl Genet 75:621–624
Doyle JJ, Beachy RN (1985) Ribosomal gene variation in soybean (*Glycine*) and its relatives. Theor Appl Genet 70:369–376
Doyle MJ, Brown AHD (1985) Numerical analysis of isozyme variation in *Glycine tomentella*. Biochem Syst Ecol 13:413–419
Doyle JJ, Brown AHD (1989) 5S nuclear ribosomal gene variation in the *Glycine tomentella* polyploid complex (Leguminosae). Syst Bot 14:398–407
Doyle MJ, Grant JE, Brown AHD (1986) Reproductive isolation between isozyme groups of *Glycine tomentella* (Leguminosae), and spontaneous doubling in their hybrids. Aust J Bot 34:523–535
Doyle JJ, Doyle JL, Brown AHD (1990a) Analysis of a polyploid complex in *Glycine* with chloroplast and nuclear DNA. Aust Syst Bot 3:125–136
Doyle JJ, Doyle JL, Brown AHD (1990b) Chloroplast DNA phylogenetic affinity of newly described species in *Glycine* (Leguminosae: Phaseoleae). Syst Bot 15:466–471
Doyle J, Doyle JL, Grace JP, Brown AHD (1990c) Reproductively isolated polyploid races of *Glycine tabacina* (leguminosae) had different chloroplast genome donors. Syst Bot 15:173–181
Doyle JJ, Doyle JL, Brown AHD (1999) Origins, colonization, and lineage recombination in a widespread perennial soybean polyploid complex. Proc Natl Acad Sci USA 96:10741–10745
Doyle JJ, Doyle JL, Brown AHD, Pfeil BE (2000) Confirmation of shared and divergent genomes in the *Glycinea tabacin* polyploid complex (Leguminosae) using histone H3-D sequences. Syst Bot 25:437–448
Doyle JJ, Doyle JL, Brown AHD, Palmer RG (2002) Genomes, multiple origins, and lineage recombination in the *Glycine tomentella* (Leguminosae) polyploid complex: histone H3-D gene sequences. Evolution 56:1388–1402
Doyle JJ, Doyle JL, Rauscher JT, Brown AHD (2004) Evolution of the perennial soybean polyploid complex (*Glycine* subgenus *Glycine*): a study of contrasts. Biol J Linnn Soc 82:583–597
Findley SD, Cannon S, Varala K, Du J, Ma J, Hudson ME, Birchler JA, Stacey G (2010) A fluorescence in situ hybridization system for karyotyping soybean. Genetics 185:727–744
Fukuda Y (1933) Cyto-genetical studies on the wild and cultivated Manchurian soy beans (*Glycine* L.). Jpn J Bot 6:489–506
Gardner ME, Hymowitz T, Xu SJ, Hartman GL (2001) Physical map location of the *Rps-1-k* allele in soybean. Crop Sci 41:1435–1438
Grant JE, Brown AHD, Grace JP (1984a) Cytological and isozyme diversity in *Glycine tomentella* Hayata (Leguminosae). Aust J Bot 32:665–677
Grant JE, Grace JP, Brown AHD, Putievsky E (1984b) Interspecific hybridization in *Glycine* Willd. subgenus *Glycine* (Leguminosae). Aust J Bot 32:655–663
Grant JE, Pullen R, Brown AHD, Grace JP, Gresshoff PM (1986) Cytogenetic affinity between the new species Glycine argyrea and its congeners. J Hered 77:423–426
Griffin JD, Broich SL, Delannay X, Palmer RG (1989) The loci *Fr1* and *EP* define soybean linkage group 12. Crop Sci 29:80–82
Griffor MC, Vodkin LO, Singh RJ, Hymowitz T (1991) Fluorescent *in situ* hybridization to soybean metaphase chromosomes. Plant Mol Biol 17:101–109
Gwyn JJ, Palmer RG (1989) Morphological discrimination among some aneuploids of soybean (*Glycine max* [L.] Merr.): 2. Double trisomics, tetrasomics. J Hered 80:209–213
Gwyn JJ, Palmer RG, Sadanaga K (1985) Morphological discrimination among some aneuploids in soybean (*Glycine max* (L.) Merr.). I. Trisomics. Can J Genet Cytol 27:608–613
Harlan JR, de Wet JMJ (1971) Toward a rational classification of cultivated plants. Taxon 20:509–517
Hedges BR, Palmer RG (1991) Tests of linkage of isozyme loci with five primary trisomics in soybean, *Glycine max* (L.) Merr. J Hered 82:494–496
Hermann FJ (1962) A revision of the genus *Glycine* and its immediate allies. United States Department of

Agriculture, Agricultural Research Service. Technical bulletin no. 1268, p 82
Hsieh JS, Hsieh KL, Tsai YC, Hsing YI (2001) Each species of *Glycine* collected in Taiwan has a unique seed protein pattern. Euphytica 118:67–73
Hsing YIC, Hsieh JS, Peng CI, Chou CH, Chiang TY (2001) Systematic status of the *Glycine tomentella* and *G. tabacina* species complexes (Fabaceae) based on ITS sequences of nuclear ribosomal DNA. J Plant Res 114:435–442
Hymowitz T, Singh RJ, Larkin RP (1990) Long-distance dispersal: The case for the allopolyploid *Glycine tabacina* (labill.) benth. and *G. tomentella* Hayata in the West- Central Pacific. Micronesica 23:5–13
Jin W, Palmer RG, Horner HT, Shoemaker RC (1999) *Fr1* (root fluorescence) locus is located in a segregation distortion region on linkage group K of soybean genetic map. J Hered 90:553–556
Karasawa K (1936) Crossing experiments with *Glycine soja* and *G. ussuriensis*. Jpn J Bot 8:113–118
Karpechenko GD (1925) On the chromosomes of Phaseolinae. Bull Appl Bot Genet Plant Breed Leningr 14 (2):143–148 (In Russian with English summary)
Kollipara KP, Singh RJ, Hymowitz T (1993) Genomic diversity in aneudiploid ($2n = 38$) and diploid ($2n = 40$) *Glycine tomentella* revealed by cytogenetic and biochemical methods. Genome 36:391–396
Kollipara KP, Singh RJ, Hymowitz T (1994) Genomic diversity and multiple origins of tetraploid ($2n = 78$, 80) *Glycine tomentella*. Genome 37:448–459
Kollipara KP, Singh RJ, Hymowitz T (1995) Genomic relationships in the genus *Glycine* (Fabaceae: Phaseoleae): use of a monoclonal antibody to the soybean Bowman-Birk inhibitor as a genome marker. Amer J Bot 82:1104–1111
Kollipara KP, Singh RJ, Hymowitz T (1997) Phylogenetic and genomic relationships in the genus Glycine Willd. based on sequences from the ITS region of nuclear rDNA. Genome 40:57–68
Krishnan P, Sapra VT, Soliman KM, Zipf A (2001) FISH mapping of the 5S and 18S-28S rDNA loci in different species of *Glycine*. J Hered 92:295–300
Lackey JA (1977) *Neonotonia*, a new generic name to include *Glycine wightii* (Arnott) Verdcourt (Legumonosae, Papilionoideae). Phytologia 37:209–212
Ladizinsky G, Newell CA, Hymowitz T (1979a) Giemsa staining of soybean chromosomes. J Hered 70:415–416
Ladizinsky G, Newell CA, Hymowitz T (1979b) Wide crosses in soybean: prospects and limitations. Euphytica 28:421–423
Lersten NR, Carlson JB (2004) Vegetative morphology. In: Boerma HR, Specht JE (eds) *Soybeans: improvement, production, and uses*, 3rd edn. American Society of Agronomy, Inc./ Crop Science Society of America, Inc./ Soil Science Society of America, Inc., Agron Monogra 16, pp 15–57
Lynch AJJ (1994) The identification and distribution of *Glycine latrobeana* (meissn.) Benth. In Tasmania. Proc Royal Soc Tasman 128:17–20
Mahama AA, Palmer RG (2003) Translocation breakpoints in soybean classical genetic linkage groups 6 and 8. Crop Sci 43:1602–1609
Mahama AA, Deaderick LM, Sadanaga K, Newhouse KE, Palmer RG (1999) Cytogenetic analysis of translocations in soybean. J Hered 90:648–653
Newell CA, Hymowitz T (1980) A taxonomic revision in the genus *Glycine* subgenus *Glycine* (Leguminosae). Brittonia 32:63–69
Newell CA, Hymowitz T (1982) Successful wide hybridization between the soybean and a wild perennial relative, G. tomentella Hatyata. Crop Sci 22:1062–1065
Newell CA, Hymowitz T (1983) Hybridization in the genus *Glycine* subgenus *Glycine* Willd. (Leguminosae, Papilionoideae). Am J Bot 70:334–348
Newell CA, Delannay X, Edge ME (1987) Interspecific hybrids between the soybean and wild perennial relatives. J Hered 78:301–306
Ohmido N, Sato S, Tabata S, Fukui K (2007) Chromosome maps of legumes. Chromosome Res 15:97–103
Palmer RG (1976) Chromosome transmission and morphology of three primary trisomics in soybean (*Glycine max*). Can J Genet Cytol 18:131–140
Palmer RG, Heer H (1973) A root tip squash technique for soybean chromosomes. Crop Sci 13:389–391
Palmer RG, Newhouse KE, Graybosch RA, Delannay X (1987) Chromosome structure of the wild soybean. J Hered 78:243–247
Palmer RG, Sun H, Zhao LM (2000) Genetics and cytology of chromosome inversions in soybean germplasm. Crop Sci 40:683–687
Palmer RG, Pfeiffer TW, Buss GR, Kilen TC (2004) Qualitative genetics. In: Boerma HR, Specht JF, (eds) Soybean: improvement, production, and uses, 3rd edn, vol 16. American Society of Agronomy, Inc., Crop Science Society of America, Inc., Soil Science Society of America, Inc., Madison, Wisconsin, pp 137–233
Pfeil BE, Craven LA (2002) New taxa in *Glycine* (Fabaceae: Phaseolae) from north-western Australia. Aust Syst Bot 15:565–573
Pfeil BE, Craven LA, Brown AHD, Murray BG, Doyle JJ (2006) Three new species of northern Australian *Glycine* (Fabaceae, Phaseolae), *G. gracei*, *G. montis-douglas*, and *G. syndetika*. Aust Syst Bot 19:245–258
Pfeil BE, Tindale MD, Craven LA (2001) A review of the Glycine clandestina species complex (Fabaceae: Phaseolae) reveals two new species. Aust Syst Bot 14:891–900
Putievsky E, Broué P (1979) Cytogenetics of hybrids among perennial species of *Glycine* subgenus *Glycine*. Aust J Bot 27:713–723
Rauscher JT, Doyle JJ, Brown AHD (2004) Multiple origins and nrDNA internal transcribed spacer homeologue evolution in the *Glycine tomentella* (Leguminosae) allopolyploid complex. Genetics 166:987–998
Sadanaga K, Grindeland R (1981) Natural cross-pollination in diploid and autotetraploid soybeans. Crop Sci 21:503–506

Sadanaga K, Grindeland RL (1984) Locating the *w1* locus on the satellite chromosome in soybean. Crop Sci 24:147–151

Sakai B (1951) Karyotype analysis in Leguninous plants I. La Kromosoma 11:425–429 (In Japanese with English summary)

Sakai T, Kaizuma N (1985) Hybrid embryo formation in an intersubgeneric cross of soybean (Glycine max Merrill) with a wild relative (G. tomentella Hayata). Jpn J Breed 35:363–374

Schoen DJ, Burdon JJ, Brown AHD (1992) Resistance of Glycine tomentella to soybean leaf rust Phakopsora pachyrhizi in relation to ploidy level and geographical distribution. Theor Appl Genet 83:827–832

Sen NK, Vidyabhusan RV (1960) Tetraploid soybeans. Euphytica 9:317–322

Singh RJ (2003) Plant cytogenetics, 2nd edn. CRC Press Inc, Boca Raton

Singh RJ (2007) Methods for producing fertile crosses between wild and domestic soybean species. United States Patent Pub, No. US2007/0261139A1

Singh RJ (2017) Practical manual on plant cytogenetics. CRC press Inc, Boca Raton

Singh RJ, Chung GH (2007) Cytogenetics of soybean: progress and prospectives. Nucleus 50:403–425

Singh RJ, Hymowitz T (1985a) Diploid-like meiotic behavior in synthesized amphiploids of the genus *Glycine* Willd. subgenus *Glycine*. Can J Genet Cytol 27:655–660

Singh RJ, Hymowitz T (1985b) The genomic relationships among six wild perennial species of the genus *Glycine* subgenus *Glycine* Willd. Theor Appl Genet 71:221–230

Singh RJ, Hymowitz T (1985c) Intra-and interspecific hybridization in the genus *Glycine*, subgenus *Glycine* Willd: chromosome pairing and genomic relationships. Z Pflanzenzüchtg 95:289–310

Singh RJ, Hymowitz T (1985d) An intersubgeneric hybrid between Glycine tomentella Hayata and the soybean, G. max (L.) Merr Euphytica 34:187–192

Singh RJ, Kollipara KP Hymowitz T (1987a) Intersubgeneric hybridization of soybeans with a wild perennial species, Glycine clandestina. Wendl Theor Appl Genet 74:391–396

Singh RJ, Hymowitz T (1987b) Intersubgeneric crossability in the genus *Glycine* Willd. Plant Breed 98:171–173

Singh RJ, Hymowitz T (1988) The genomic relationships between *Glycine max* (L.) Merr. and *G. soja* Sieb. and Zucc. as revealed by pachytene chromosome analysis. Theor Appl Genet 76:705–711

Singh RJ, Hymowitz T (1991) Identification of four primary trisomics of soybean by pachytene chromosome analysis. J Hered 82:75–77

Singh RJ, Hymowitz T (1999) Soybean genetic resources and crop improvement. Genome 42:605–616

Singh RJ, Nelson RL (2014) Methodology for creating alloplasmic soybean lines by using *Glycine tomentella* as a maternal parent. Plant Breed 133:624–631

Singh RJ, Nelson RL (2015) Intersubgeneric hybridization between *Glycine max* and *G tomentella*: production of F_1, amphidiploid, BC_1, BC_2, BC_3, and fertile soybean plants. Theor Appl Genet 128:1117–1136

Singh RJ, Kollipara KP, Hymowitz T (1987b) Polyploid complexes of *Glycine tabacina* (Labill.) Benth. and *G. tomentella* Hayata revealed by cytogenetic analysis. Genome 29:490–497

Singh RJ, Kollipara KP, Hymowitz T (1988) Further data on the genomic relationships among wild perennial species ($2n$ = 40) of the genus *Glycine* Willd. Genome 30:166–176

Singh RJ, Kollipara KP, Hymowitz T (1989) Ancestors of 80- and 78-chromosome *Glycine tomentella* Hayata (Leguminosae). Genome 32:796–801

Singh RJ, Kollipara KP, Hymowitz T (1990) Backcrossed-derived progeny from soybean and *Glycine tomentella* Hayata intersubgeneric hybrids. Crop Sci 30:871–874

Singh RJ, Kollipara KP, Ahmad F, Hymowitz T (1992a) Putative diploid ancestors of 80-chromosome *G. tabacina*. Genome 35:140–146

Singh RJ, Kollipara KP, Hymowitz T (1992b) Genomic relationships among diploid wild perennial species of the genus Glycine Willd. subgenus *Glycine* revealed by crossability, meiotic chromosome pairing and seed protein electrophoresis. Theor Appl Genet 85:276–282

Singh RJ, Kollipara KP, Hymowitz T (1993) Backcross (BC_2–BC_4)-derived fertile plants from *Glycine max* and *G. tomentella* intersubgeneric hybrids. Crop Sci 33:1002–1007

Singh RJ, Kollipara KP, Hymowitz T (1998a) The genomes of *Glycine canescens* F. J. Herm., and *G. tomentella* Hayata of Western Australia and their phylogenetic relationships in the genus Glycine Willd. Genome 41:669–679

Singh RJ, Kollipara KP, Hymowitz T (1998b) Monosomic alien addition lines derived from *Glycine max* (L.) Merr. and *G. tomentella* Hayata: production, characterization, and breeding behavior. Crop Sci 38:1483–1489

Singh RJ, Kim HH, Hymowitz T (2001) Distribution of rDNA loci in the genus *Glycine* Willd. Theor Appl Genet 103:212–218

Singh RJ, Chung GH, Nelson RL (2007a) Landmark research in legumes. Genome 50:525–537

Singh RJ, Nelson RL, Chung GH (2007b) Soybean (*Glycine max* (L.) Merr.). In: Singh RJ (ed) Genetic resources, chromosome engineering, and crop improvement volume 4 oilseed crops. CRC Press, Boca Raton, pp 13–50

Skorupska H, Palmer RG (1987) Monosomics from synaptic KS mutant. Soybean Genet Newsl 14:174–178

Skorupska H, Palmer RG (1989) Genetics and cytology of *ms6* male-sterile soybean. J Hered 80:304–310

Skorupska H, Albertsen MC, Langholz KD, Palmer RG (1989) Detection of ribosomal RNA genes in soybean, *Glycine max* (L.) Merr., by *in situ* hybridization. Genome 32:1091–1095

Song QJ, Marek LF, Shoemaker RC, Lark KG, Concibido VC, Delannay X, Specht JF, Cregan PB (2004) A new integrated genetic linkage map of the soybean. Theor Appl Genet 109:122–128

Tateishi Y, Ohashi H (1992) Taxonomic studies on *Glycine* of Taiwan. J Jpn Bot 67:127–147

Thseng FS, Tsai SJ, Abe J, Wu ST (1999) *Glycine formosana* Hosokawa in Taiwan: pod morphology, allozyme, and DNA polymorphism. Bot Bull Acad Stn 40:251–257

Tindale MD (1984) Two new Eastern Australian species of *Glycine* Willd. (Fabaceae). Brunonia 7:207–213

Tindale MD (1986a) A new North Queensland species of *Glycine* Willd. (Fabaceae). Brunonia 9:99–103

Tindale MD (1986b) Taxonomic notes on three Australian and Norfolk Island species of *Glycine* Willd. (Fabaceae:Phaseolae) including the choice of a neotype for *G. clandestina* Wendl. Brunonia 9:179–191

Tindale MD, Craven LA (1988) Three new species of *Glycine* (Fabaceae: Phaseolae) from North-western Australia, with notes on amphicarpy in the genus. Aust Syst Bot 1:399–410

Tindale MD, Craven LA (1993) *Glycine pindanica* (Fabaceae, Phaseolae), a new species from West Kimberly, Western Australia. Aust Syst Bot 6:371–376

Veatch C (1934) Chromosomes of the soy bean. Bot Gaz 96:189

Verdcourt B (1966) A proposal concerning Glycine L. Taxon 15:34–36

Xia Z, Tsubokura T, Hoshi M, Hanawa M, Yano C, Okamura K, Ahmed TA, Anai T, Watanabe S, Hayashi M, Kawai T, Hossain KG, Masaki H, Asai K, Yamanaka N, Kubo N, Kadowaki K, Nagamura Y, Yano M, Sasaki T, Harada K (2007) An integrated high-density linkage map of soybean with RFLP, SSR, STS, and AFLP markers using a single F_2 population. DNA Res 14:257–269

Xu SJ, Singh RJ, Hymowitz T (2000a) Monosomics in soybean: origin, identification, cytology, and breeding behavior. Crop Sci 40:985–989

Xu SJ, Singh RJ, Kollipara KP, Hymowitz T (2000b) Hypertriploid in soybean: origin, identification, cytology, and breeding behavior. Crop Sci 40:72–77

Xu SJ, Singh RJ, Kollipara KP, Hymowitz T (2000c) Primary trisomics in soybean: origin, identification, breeding behavior, and use in linkage mapping. Crop Sci 40:1543–1551

Yanagisawa T, Tano S, Fukui K, Harada K (1991) Marker chromosomes commonly observed in the genus *Glycine*. Theor Appl Genet 81:606–612

Zhu T, Shi L, Doyle JJ, Keim P (1995) A single nuclear locus phylogeny of soybean based on DNA sequence. Theor Appl Genet 90:991–999

Zou JJ, Singh RJ, Hymowitz T (2003a) Association of the yellow leaf (*y10*) mutant to soybean chromosome 3. J Hered 94:352–354

Zou JJ, Singh RJ, Lee J, Xu SJ, Cregan PB, Hymowitz T (2003b) Assignment of molecular linkage groups to soybean chromosomes by primary trisomics. Theor Appl Genet 107:745–750

Zou JJ, Singh RJ, Lee J, Xu SJ, Hymowitz T (2006) SSR markers exhibit trisomic segregation distortion in soybean. Crop Sci 46:1456–1461

Classical and Molecular Genetic Mapping

3

Qijian Song and Perry B. Cregan

Abstract

A brief history of classical and molecular genetic mapping in soybean [*Glycine max* (L.) Merr.] is followed by a description of the most current version of the soybean genetic linkage map, which is primarily based on simple sequence repeat (SSR) or microsatellite markers as well as single nucleotide polymorphism (SNP) markers. Like many plant and animal species, the first molecular map of soybean was based on restriction fragment length polymorphism (RFLP) markers. Because of the relatively low level of sequence diversity in soybean, the first RFLP maps were constructed in populations derived from crosses of cultivated (*G. max*) × wild soybean [*G. soja* (Seib. et Zucc.)]. Random amplified polymorphic DNA (RAPD) and amplified fragment length polymorphism loci were briefly used in map construction, but soybean was the first plant species in which SSR markers were employed and SSRs soon became the marker of choice in map development. SNP markers are now being widely used by plant and animal geneticists, and the most recent molecular genetic map of soybean incorporates a large number of SNPs from genic and nongenic regions.

3.1 Introduction

Classical genetic mapping in soybean (*Glycine max* (L.) Merr.) was initially based on a relatively small number of qualitative traits most of which were related to pigmentation, morphological and physiological traits, response to disease and insect pests, and other easily observed traits. A number of seed isozymes were also among the traits that were used in classical genetic mapping. As an advance with unparalleled importance to the analysis of

Q. Song (✉) · P.B. Cregan
Soybean Genomics and Improvement Laboratory, U. S. Department of Agriculture, Agricultural Research Service, Beltsville, MD 20705, USA
e-mail: Qijian.Song@ars.usda.gov

P.B. Cregan
e-mail: pbcregan@verizon.net

H.T. Nguyen and M.K. Bhattacharyya (eds.), *The Soybean Genome*, Compendium of Plant Genomes, DOI 10.1007/978-3-319-64198-0_3

genetic diversity and genetic map development in plant, animal, and other species was the work of Botstein et al. (1980) that described the use of restriction fragment length polymorphisms (RFLPs) in the construction of molecular genetic maps. The availability of abundant DNA markers provided geneticists the tools for studying a wide range of plant and animal species for the analysis of genetic diversity and for the creation of relatively dense molecular genetic maps. The subsequent development of the polymerase chain reaction (PCR) (Mullis et al. 1986) provided a method to very specifically and easily analyze DNA sequence polymorphisms and was the basis of the development of a number of DNA marker technologies. One of these was PCR-based microsatellite or simple sequence repeat (SSR) markers (Litt and Luty 1989; Tautz 1989; Weber and May 1989). Subsequently, random amplified polymorphic DNA (RAPD) markers (Williams et al. 1990) or arbitrarily primed PCR (AP-PCR) markers (Welsh and McClelland 1990) that were based on PCR amplification using a single arbitrary primer and amplification fragment length polymorphism (AFLP) (Vos et al. 1995) markers were developed. As more high-throughput and less expensive DNA sequence analysis systems became available, single nucleotide polymorphism (SNP) DNA markers have become widely used. Rapid analysis of thousands of SNPs in large numbers of individuals using systems such as the Infinium® HD Assay (Illumina, Inc. San Diego, CA) and next-generation sequence analysis have resulted in the widespread use of SNP markers in genetic analysis and molecular genetic map development.

3.2 Classical Genetic Maps

Classical genetic maps are created via genetic mapping of a range of morphological traits that include classical loci controlling easily observed traits such as flowering and maturity, plant morphology, pigmentation, fertility and sterility, disease and insect resistance, physiological traits, and seed composition traits. Isozyme variants in soybean seed or germinating seed were also among the traits used in classical genetic mapping. The first genetic maps of soybean were described in a series of reports by Weiss (1970a, b, c, d, e) that established a set of seven small linkage groups based upon previous reports, as well as, additional tests of linkage between a number of classical loci controlling plant and seed traits. The seven linkage groups contained two or three loci each with a total of 17 loci. A subsequent summary of classical soybean linkage groups by Palmer and Kilen (1987) reported the addition of three loci, two of which were isozyme loci, to the classical Linkage Group 1 reported by Weiss (1970a) while Linkage Group 2 though 7 remained as defined by Weiss (1970b, c, d). Palmer and Kilen (1987) also reported the addition of six new classical linkage groups that contained from two to four loci. The resulting classical genetic linkage map with 13 linkage groups contained a total of 35 morphological and isozyme loci. In a subsequent report, Palmer and Shoemaker (1998) provided an updated classical linkage map with 20 classical linkage groups (CLG01–CLG15 and CLG17–CLG21) with a total of 67 classical and isozyme loci. Devine (2003) defined CLG16 that contained two loci. The most recent summary of classical linkage groups in soybean was provided by Palmer et al. (2004) and contained 20 linkage groups (CLG01–CLG05 and CLG07–CLG21) with a total of 72 classical and isozyme loci ranging from 2 to 9 loci per linkage group.

3.3 RFLP-Based Genetic Maps

Keim et al. (1990) was the first report of an RFLP-based genetic map of soybean. This map was created using a mapping population derived from a cross of cultivated soybean (A81-356022) crossed with a wild soybean (*G. soja* Seib. and Zucc.) PI468916. Apuya et al. (1988) had previously reported the relatively low level of RFLP polymorphism in cultivated soybean as compared to other species such as maize (*Zea maize*

L.) with high levels of RFLP allelic diversity (Burr et al. 1983; Helentjaris et al. 1986). Thus, progeny from the cultivated × wild soybean cross were used with the anticipation of greater sequence diversity and RFLP polymorphism than would be expected in progeny from a cross of two cultivated soybean genotypes. The resulting genetic map contained a total of 127 RFLP loci and three classical trait loci and was composed of 26 linkage groups with a total length of 1200 cM (Keim et al. 1990). Subsequent additions to this map by Diers et al. (1992), Shoemaker and Olson (1993), and Shoemaker and Specht (1995) added more than 200 RFLP loci and resulted in a map 2473 cM in length with 25 linkage groups (Table 3.1). Researchers at the E.I. Dupont Corporation (Rafalski and Tingey 1993) also used a population derived from a *G. max* × *G. soja* cross to create a genetic map with more than 600 RFLP loci. Despite the relatively low level of genetic polymorphism in cultivated soybean, Lark et al. (1993) created a mapping population from the cross of the cultivars Minsoy and Noir 1 and successfully mapped 125 RFLP loci. This was followed by the development of a Minsoy × Noir 1 mapping population with 284 recombinant inbred lines (RILs) (Mansur et al. 1996), which provided an excellent tool for soybean genetic mapping studies. Another molecular genetic map based upon a population of 190 F_2 progeny from a cross of *G. max* cultivar Misuzudaizu with Moshidou Gong 503, an experimental line that is an intermediate between cultivated soybean and wild soybean, was created by Yamanaka et al. (2001) and was predominately constructed using RFLP markers. The map contained 401 RFLP markers to which Xia et al. (2007) subsequently added more than 100 additional RFLP markers as well as other marker types. In addition, a map based upon 300 RILs from a cross of the *G. max* genotypes BSR301 × PI437654 was created, which contained 165 RFLP loci along with a large number of other marker types (Keim et al. 1997).

3.4 Genetic Maps Based on RAPD and AFLP Markers

RAPD (Williams et al. 1990) or AP-PCR markers (Welsh and McClelland 1990) are PCR based, require no prior knowledge of DNA sequence, and are analyzed based on the presence or absence of an amplicon using agarose gel electrophoresis. Because of their simplicity, RAPDs provided an appealing alternative to RFLP and Southern blot analysis, however, while soybean geneticists have used RAPDs extensively in germplasm classification (Brown-Guedira et al. 2000; Li and Nelson 2002; Thompson et al. 1998), RAPD markers were not widely used in the development of soybean genetic maps. The genetic maps with the largest number of RAPD loci were those developed by Ferreira et al. (2000) and Lightfoot et al. (2005) and contained 106 and 88 RAPD loci, respectively. In neither case were RAPD markers the predominant marker type. There were no maps created with larger numbers of RAPD loci. In contrast, extensive maps based primarily on AFLP markers were constructed by Keim et al. (1997) and Xia et al. (2007). A total of 650 AFLP loci were mapped in a 42 RIL subset of a 300 RIL BSR 101 × PI437654 population developed by Keim et al. (1997). The map was anchored with 165 RFLP and 25 RAPD markers based upon the analysis of the 300 RILs and was populated with the AFLP markers analyzed in the 42 RIL subset to create a map with 840 molecular markers. They noted some clustering of AFLP markers but indicated that the AFLPs were well distributed relative to the RFLP and RAPD markers. Xia et al. (2007) updated the RFLP-based genetic map created by Yamanaka et al. (2001) with the addition of 318 AFLP markers, plus well over 400 markers of other types. Xia et al. (2007) noted that the RFLP loci were generally distributed evenly among the linkage groups while the AFLP markers appeared to be 'moderately' clustered.

Table 3.1 Summary of soybean molecular genetic linkage maps and numbers of various types of DNA markers positioned in each

Mapping population and literature citation	Population size and type	Linkage groups	Map length	DNA Marker type							Classical & Isozyme markers
				Total	RFLP	RAPD	AFLP	SSR	STS	SNP	
		No.	cM		No.						
USDA/Iowa St. (A81-356022 × G. soja PI468.916)											
Keim et al. (1990)	60 F_2 lines	26	1200	130	127						3
Diers et al. (1992)	60 F_2 lines	31	2147	241	234						7
Shoemaker and Olson (1993)	60 F_2 lines	23	3371	370	351	10					7
Shoemaker and Specht (1995)	60 F_2 lines	25	2473	368	358	10					7
Cregan et al. (1999)	59 F_2 lines	23	3003	1004	501	10		486			7
E. I. DuPont Bonus × G. soja PI 81762	68 F_2 lines										
Rafalski and Tingey (1993)		21	2678	600+	600+	4		1			
Univ. of Utah (Minsoy × Noir 1)											
Lark et al. (1993)	69 F_3 families	31	1550	132	125						7
Mansur et al. (1996)	284 RILs	35	1981	277	226			44			7
Cregan et al. (1999)	240 RILs	22	2413	633	209			412			12
Univ. of Nebraska (Clark × Harosoy isolines)	57 F_2 lines										
Shoemaker and Specht (1995)		26	1004	100	80	4					16
Akkaya et al. (1995)		29	1486	118	80	4		34			15
Cregan et al. (1999)		28	2787	523	95	57	11	339			21
BSR101 × PI 437654											
Keim et al. (1997)	300 RILs	28	3441	840	165	25	650				
Ferreira et al. (2000)	330 RILs	35	3275	356	250	106					
Changnong 4 × Xinmin 6	88 RILs										
Liu et al. (2000)		22	3713	237	100	62	42	33			
Misuzudaizu × Moshidou Gong 503											

(continued)

Table 3.1 (continued)

Mapping population and literature citation	Population size and type	Linkage groups	Map length	DNA Marker type							Classical & Isozyme markers
				Total	RFLP	RAPD	AFLP	SSR	STS	SNP	
		No.	cM		No.						
Yamanaka et al. (2001)	190 F_2 plants	20	2909	498	401	1		96			5
Xia et al. (2007)	190 F_2 plants	20	3080	1277	509	1	318	318	126		5
Hisano et al. (2007)	94 RILs	20	2700	935	105			829			1
Kefeng 1 × Nannong 1138-2	184 RILs										
Wu et al. (2001)		24	2321	792	196	18	486	87			4
Noir 1 × BARC-2	149 F_2 plants										
Matthews et al. (2001)		35	1400	180	39	17	25	105			4
Essex × Forrest	100 RILs										
Lightfoot et al. (2005)		20	2823	335	41	88		206			
Forrest × Williams 82											
Wu et al. (2011)	1025 RILs	20	2724	990				474		516	
Williams 82 × G. soja PI479752											
Hyten et al. (2010b)	444 RILs	20	2537	1790						1790	
Song et al. (2016)	1083 RILs	20	2446	21,478						21,478	
Charleston × Donguong594	147 RILs										
Qi et al. (2014)		20	2294	5308						5308	
Essex × Williams 82	922 RILs										
Song et al. (2016)		20	2648	11,922						11,922	
Consensus Maps											
Song et al. (2004)	5 populations	20	2524	1849	709	73	6	1015			34
Choi et al. (2007)	3 populations	20	2550	2793	634	4		983		1157	15
Hwang et al. (2009)	3 populations	20	2443	1810				1810			
Hyten et al. (2010a)	5 populations	20	2296	5485	664			1006		3792	23

3.5 Simple Sequence Repeat (SSR) or Microsatellite Markers

The discovery, development, and use of microsatellite or SSR markers in humans (Litt and Luty 1989; Tautz 1989; Weber and May 1989) led researchers to explore the use of SSRs in soybean. Indeed, the first demonstration of SSR allelic variation and heritability in a plant species was reported in soybean (Akkaya et al. 1992; Morgante and Olivieri 1993). Akkaya et al. (1992) found as many as eight SSR alleles at one locus in a group of 38 *G. max* and five *G. soja* genotypes. Further reports of SSR allelic variation in soybean (Maughan et al. 1995; Morgante et al. 1994; Rongwen et al. 1995; Song et al. 2010) detected very high levels of allelic variation including one locus with 26 alleles among a group of 91 cultivated and five wild soybean genotypes (Rongwen et al. 1995). The abundance and variation of SSRs in the soybean genome have been examined. After screening for a number of factors including locus specificity using e-PCR based on the whole genome DNA sequence of Williams 82 (Schmutz et al 2010), Song et al (2010) developed a soybean candidate SSR database, BARCSOYSSR_1.0, containing 33,065 SSR DNA markers. Evaluation of 1034 primer sets by amplifying DNAs of seven diverse soybean genotypes and one wild soybean genotype showed that a total of 978 (94.6%) of the primer sets amplified a single discrete PCR product and 798 (77.2%) amplified polymorphic amplicons as determined by 4.5% agarose gel electrophoresis. The first report describing the mapping of SSR loci in soybean (Akkaya et al. 1995) found no evidence of clustering of SSR loci. The high level of polymorphism combined with their random distribution on the genetic map as well as their single locus nature and analysis via PCR suggested that SSRs were an excellent complement to RFLP markers for use in soybean molecular biology, genetics, and plant breeding research.

3.6 An SSR-Based Soybean Genome Map

With the support of the United Soybean Board, a project was initiated in 1995 to develop and map a large set of soybean SSR markers. A collaborative project between workers at the USDA, Beltsville, MD and Ames, IA; the Univ. of Nebraska and the Univ. of Utah as well as Bio-Genetic Services Inc. of Brookings, SD, resulted in the development and mapping of more than 600 SSR loci in one, two, and occasionally three different mapping populations (Cregan et al. 1999). One population was the USDA/Iowa St. *G. max* × *G. soja* population (Keim et al. 1990) and the second, the expanded 240 RIL University of Utah population developed from the cultivars Minsoy and Noir 1 (Mansur et al. 1996). The third population was the University of Nebraska Clark × Harosoy isoline population consisting of 57 F_2-derived lines (Shoemaker and Specht 1995). A total of 187 loci were mapped in each of the three populations, while many more loci were common to any two of the populations. Thus, because of the single locus nature of the SSRs, it was a simple matter to align homologous linkage groups and to create the first soybean linkage map with 20 linkage groups, which were assumed to correspond to the 20 soybean chromosomes. The 20 sets of aligned linkage groups contained a total of 1423 unique marker loci including 606 SSRs, 689 RFLP, 26 classical loci, and 10 isozyme loci. The consensus soybean linkage map with an average of more than 30 SSR markers per linkage group (Cregan et al. 1999) was useful for the alignment of linkage groups in preexisting or newly created linkage maps with the consensus linkage groups in the SSR-based genetic map. A small number of SSRs could be used to associate linkage groups with the corresponding linkage groups on the SSR-based consensus map. Wu et al. (2001) mapped 792 markers in a population of 201 RILs from a cross of the genotypes Kefeng

1 × Nannong 1138-2 and aligned their linkage groups with the consensus linkage groups via the analysis of 87 SSRs common to the Cregan et al. (1999) map. Similarly, the population developed by Yamanaka et al. (2001) with 190 F_2 plants from a cross of Misuzudaizu × Moshidou Gong 503 was used to map 401 RFLP loci along with 96 SSRs. The resulting map consisted of 20 major linkage groups with a total length of 2908.7 cM. Matthews et al. (2001) successfully positioned cDNA and genomic clones on consensus linkage groups in the SSR-based genome map using a small number of SSR loci to align corresponding linkage groups. An updated version of the SSR-based linkage map was published by Song et al. (2004) and provided a consensus genetic map. This map included 420 additional SSR markers that were mapped in one or more of the three populations included in the Cregan et al. (1999) map as well as in two additional RIL populations from the Univ. of Utah; Minsoy × Archer and Archer × Noir 1. The resulting consensus map of the five populations was created using JoinMap (Van Ooijen and Voorrips 2001) software. This integrated genetic map spanned 2523.6 cM of Kosambi map distance across 20 linkage groups and contained 1849 markers, including 1015 SSRs, 709 RFLPs, 73 RAPDs, six AFLPs, 24 classical traits, and ten isozymes. In the same F_2 population used by Yamanaka et al. (2001), Xia et al. (2007) mapped an additional 222 SSR and 108 RFLP markers, as well as 318 AFLP markers. Hisano et al. (2007) mapped 829 SSR markers including markers derived from expressed sequence tags (ESTs) and SSR markers from Cregan et al. (1999) and Song et al. (2004) along with 105 RFLP loci using a population of 94 RILs developed from Misuzudaizu × Moshidou Gong 503. Similarly, Hwang et al. (2009) constructed an integrated genetic map with 1810 SSR or sequence-tagged site markers, which were mapped in one or more of the three RIL populations: Misuzudaizu × Moshidou Gong 503, Jack × Fukuyutaka, and/or Peking × Akita. The map contained 1074 EST-derived SSRs, 595 SSR markers reported by Cregan et al. (1999) and Song et al. (2004), as well as, 141 additional SSR and sequence-tagged site (STS) markers. JoinMap analysis resulted in a genetic map with 20 linkage groups spanning 2442.9 cM of Kosambi map distance. As a result of the large number of markers common to the maps developed by Cregan et al. (1999) and Song et al. (2004), Hwang et al. (2009) demonstrated the clear correspondence of the 20 linkage groups they reported to those reported by Cregan et al. (1999) and Song et al. (2004). Wu et al. (2011) constructed an integrated linkage map containing 474 SSRs markers including 145 newly developed SSR markers and 323 SSR markers reported by Cregan et al. (1999) and Song et al. (2004) in the Forrest × Williams 82 RIL population; the map was constructed for integration of the Forrest physical map with the Williams 82 physical map.

3.7 Integration of Classical and Molecular Linkage Groups

Beginning with the work of Shoemaker and Specht (1995), classical genetic linkage groups began to be integrated into genetic maps based on molecular genetic markers. They created a molecular genetic map of a population derived from the cross of two *G. max* genotypes (near isogenic lines of the cultivars Clark and Harosoy) that differed at seven isozyme loci as well as 13 classical loci consisting of seven pigmentation loci and six morphological loci. The population was analyzed for the classical and isozyme loci as well as with RFLP and RAPD markers that had previously been mapped by Shoemaker and Olson (1993) in the USDA/Iowa St. A81-356022 × *G. soja* PI468916 population. Thus, the common markers allowed identification of homologous linkage groups in the two populations. The presence of the isozyme and classical markers in the map derived from the Clark × Harosoy progeny allowed five of the 20 classical linkage groups to be unambiguously associated with molecular linkage groups (Table 3.2). Further association of classical and molecular linkage groups was based on the previously described molecular genetic map created via the mapping of more than 600 SSR loci in one, two, and occasionally three different mapping populations along with a large number of RFLP markers (Cregan et al. 1999). These maps included a total of 10 isozyme and 26 classical loci and resulted in the association of 13 additional CLG of the 20 CLG, defined by Palmer and Shoemaker (1998), with a molecular linkage group. GLG20 was associated with either molecular linkage group D2 or H, but it was not clear which. CLG 6, with the *Df2* and *Y11* classical loci, was not associated with a molecular linkage group. Subsequently, Mahama et al. (2002) reported that the *Df2* and *Y11* markers of CLG06 were linked with the classical markers on CLG08 and the two linkage groups were consolidated and referred to as CLG08. Devine (2003) defined a new classical linkage group that was referred to as CLG16 and contained the *Lf2* and *Pd2* loci and was not associated with any molecular linkage group. CLG 16 was ultimately associated with molecular linkage group B1 by Seversike et al. (2008). Thus, with the exception of CLG 20 all of the classical linkage groups have been unambiguously associated with a molecular linkage group and with a specific chromosome (Table 3.2).

3.8 Single Nucleotide Polymorphisms (SNPs)

Because of a lack of definition of the soybean gene-space, which was thought to be clustered in approximately 25% of the genome, Mudge et al. (2004) and Stacey et al. (2004) suggested that

Table 3.2 Twenty soybean chromosomes and the associated molecular and classical linkage groups

Chromosome number	Molecular linkage group[a]	Classical linkage group (CLG)	Reference associating CLG with a molecular linkage group
Chromosome 1	D1a	CLG03	Cregan et al. (1999)
Chromosome 2	D1b	CLG11	Cregan et al. (1999)
Chromosome 3	N	CLG10	Cregan et al. (1999)
Chromosome 4	C1	CLG21	Cregan et al. (1999)
Chromosome 5	A1		
Chromosome 6	C2	CLG01	Shoemaker and Specht (1995)
Chromosome 7	M		
Chromosome 8	A2	CLG07 & CLG09	Cregan et al. (1999)
Chromosome 9	K	CLG02 & CLG 12	CLG02 assigned by Shoemaker and Specht (1995), CLG12 assigned by Cregan et al. (1999)
Chromosome 10	O	CLG15	Cregan et al. (1999)
Chromosome 11	B1	CLG16	Seversike et al. (2008)
Chromosome 12	H	CLG20?[b]	Cregan et al. (1999)
Chromosome 13	F	CLG08 & CLG13	Cregan et al. (1999)
Chromosome 14	B2	CLG17	Cregan et al. (1999)
Chromosome 15	E	CLG14	Shoemaker and Specht (1995)
Chromosome 16	J	CLG19	Cregan et al. (1999)
Chromosome 17	D2	CLG20?[b]	Cregan et al. (1999)
Chromosome 18	G	CLG18	Shoemaker and Specht (1995)
Chromosome 19	L	CLG05	Shoemaker and Specht (1995)
Chromosome 20	I	CLG04	Cregan et al. (1999)

[a]Molecular linkage groups assigned to chromosomes by Schmutz et al. (2010)
[b]CLG20 has not been unambiguously assigned to a specific molecular linkage group

2000–3000 cDNA sequences be placed on the physical map. Alternatively, coding sequences could be genetically mapped onto the existing SSR-based map. Such a genetic map would indicate the positions of coding sequences and would also provide information on the relative positions of genes with existing SSR and RFLP markers. The mapping of SSRs from EST sequence is one means of positioning genes on the genetic map. This approach has worked successfully in a number of species including *Medicago truncatula* (Eujayl et al. 2004), wheat (*Triticum aestivum* L.) (Gao et al. 2004; Yu et al. 2004a), and rice (*Oryza sativa* L.) (Yu et al. 2004b) for the genetic mapping of the ESTs. In contrast, only a small proportion of soybean ESTs contain polymorphic SSRs as indicated by Song et al. (2004) who reported the successful development and mapping of only 24 polymorphic SSR markers from more than 136,000 soybean ESTs. However, the discovery of SNPs (which include single base changes and insertion/deletions) in genic sequence would provide a source of markers to expedite the positioning of genes on the genetic map. SNPs are now the marker of choice in human genetics studies and in plant species including *Arabidopsis thaliana* (Horton et al. 2012; Cao et al. 2011; Jander et al. 2002; Nordborg et al. 2005; Schmid et al. 2003, 2005), maize (*Zea mays* L.) (Ganal et al. 2011; Ching et al. 2002; Cook et al. 2012; Tenaillon et al. 2001), rice (Huang et al. 2012; McNally et al. 2009; Feltus et al. 2004; Nasu et al. 2002), barley (*Hordeum vulgare* L.) (Mascher et al. 2013; Rostoks et al. 2005), sorghum [*Sorghum bicolor* (L.) Moench.] (Morris et al. 2013), foxtail millet (*Setaria italica*) (Jia et al. 2013), sunflower (*Helianthus annuus* L.) (Bachlava et al. 2012), and poplar (*Populus trichocarpa* Torr. & Gray) (Geraldes et al. 2011; Tuskan et al. 2006).

3.9 SNPs in Soybean

Zhu et al. (2003) reported the frequency of SNPs in a diverse set of 25 soybean genotypes via the analysis of more than 76 kbp of coding sequence, untranslated regions (UTRs), introns, and genomic DNA in close proximity to coding sequence. The frequency of SNPs was reported in terms of nucleotide diversity (θ) (Watterson 1975). θ was equal to 0.00053 in 28.7 kbp of coding sequence (an average of approximately 0.5 SNPs per kbp corrected for sample size) and was more than twofold higher in UTRs, introns, and genomic DNA ($\theta = 0.00111$). The level of sequence diversity in soybean is low in comparison with species such as maize. For example, Wright et al. (2005) reported nucleotide diversity of $\theta = 0.00627$ in modern maize inbreds. The report by Hyten et al. (2006) indicated that the frequency of sequence variants in soybean is low due to historical genetic bottlenecks and low sequence diversity in soybean's wild ancestor, *G. soja*. Nonetheless, the nucleotide diversity in soybean is very similar to that in humans where most estimates indicate nucleotide diversity (θ) values of less than 0.001 (Sachidanandam et al. 2001; Cargill et al. 1999; Halushka et al. 1999). Indeed, a calculation from Zhu et al. (2003) who reported the discovery of 280 SNPs in 76.4 kbp of sequence indicated there are likely in excess of 4 million SNPs in cultivated soybean. While this estimate was based mostly on coding sequence and may be biased, discoveries of >6 million SNPs (Song et al. 2013; Li et al. 2013; Lam et al. 2010) via re-sequencing diverse sets of *G. max* and *G. soja* accessions clearly verified previous conclusions that there were an adequate number of sequence variants for successful SNP discovery in soybean.

3.10 Soybean SNP-Based Maps

The first SNP-based genetic linkage map of the soybean genome (Choi et al. 2007) was constructed via the JoinMap (van Ooijen and Voorrips 2001) analysis of three mapping populations: the University of Utah Minsoy × Noir 1 and Minsoy × Archer RIL populations and a population of 77 RILs derived from a cross of the cultivar Evans × *G. max* PI 209332 (Concibido et al. 1996). This map was built upon the SSR framework map defined by Song et al. (2004)

with the objective of discovering and mapping SNPs in, or in close proximity to, soybean genes. SNPs were discovered via the re-sequencing of STS developed by the analysis of 9459 polymerase chain reaction primer sets, more than 8500 of which were designed to soybean unigene sequences reported by Vodkin et al. (2004). The analysis of these primer sets resulted in the development of 4240 STS which were re-sequenced in a set of six diverse *G. max* genotypes: Minsoy, Noir 1, Archer, Evans, Peking, and PI209332. In the resulting 2.44 Mbp of aligned sequence, a total of 5551 SNPs were discovered, including 4712 single base changes and 839 indels for an average nucleotide diversity of $\theta = 0.000997$ (an average of almost 1 SNP per kbp adjusted for the size of the population analyzed). At least one SNP was discovered in 2032 of the 4240 STS. A total of 1157 SNPs derived from 1141 different genes were positioned in the 20 soybean linkage groups. The SNP detection assays were conducted using single base extension on either the Sequenom MassARRAY™ platform (Griffin et al. 1999) or the Luminex flow cytometer (Chen et al. 2000). The analysis of the observed genetic distances between adjacent genes *versus* the theoretical distribution based upon the assumption of a random distribution of genes across the 20 soybean linkage groups clearly indicated that genes were clustered. One interesting result of the transcript mapping on the SSR framework was the opportunity to better understand the relationship of SSRs and genes. Marek et al. (2001) had reported that end-sequences of soybean bacterial artificial chromosome (BAC) clones identified with *Pst*1-derived RFLP probes had 50% more gene-like sequences and 45% less repetitive sequence than end-sequences of BAC clones identified with SSR markers. This suggested that in relation to SSRs, RFLPs were more closely associated with gene-rich regions. However, an analysis of the relationship of SSR and RFLP loci and mapped transcripts on the new map of Choi et al. (2007) detected no difference in the proximity of SSRs and RFLP loci to the closest flanking genic sequences. This result was similar to that reported by Morgante et al. (2002) who analyzed DNA sequence data from *Arabidopsis thaliana*, rice, soybean, maize, and wheat. They concluded that the frequency of SSRs was significantly higher in transcribed regions and that SSRs are associated with low-copy portions of plant genomes rather than with regions of repetitive DNA.

The SNP/SSR-based transcript map of the soybean genome (Choi et al. 2007) more than doubled the number of available sequence-based markers and provided an important resource to soybean geneticists for quantitative trait locus discovery and map-based cloning, as well as to soybean breeders who depend upon marker assisted selection in cultivar improvement. This map represented the first step in the development of a much denser SNP-based soybean linkage map based upon the use of the Illumina Inc. (San Diego, CA) GoldenGate SNP detection assay. Via the design and testing of a 384-SNP GoldenGate assay, Hyten et al. (2008) reported that the GoldenGate assay functioned extremely well in the complex soybean genome. Hyten et al. (2010a) designed two additional GoldenGate assays that each contained 1536 SNPs that were identified following the procedures described by Choi et al. (2007). These were used to create the Soybean Consensus Map 4.0 that contained 3792 SNPs including 2651 new SNPs segregating in one or more of the Minsoy × Noir 1, Minsoy × Archer, or Evans × Peking mapping populations in addition to the SNPs previously mapped by Choi et al. (2007). An integrated genetic linkage map spanning a genetic distance of 2296.4 cM was constructed by compositing the SNP locus data of the three populations with the SNP locus data from Evans × PI 209332 and A81-356022 × PI 468916 mapping populations used by Choi et al (2007). In addition to the SNP markers, the linkage map contained a total of 1700 other markers including SSRs and RFLPs, which were previously mapped by Cregan et al. (1999) and Song et al. (2004). The Soybean Consensus Map 4.0 was used to anchor 95.6% of the Williams 82 soybean whole-genome sequence developed by the Joint Genome Institute, US Department of Energy (Schmutz et al. 2010), but the marker density was not sufficient to orient 66% of the sequence scaffolds. Thus, it was necessary to discover and map additional SNP markers to properly orient scaffolds and to map additional sequence scaffolds. Hyten et al. (2010b) used second-generation sequence analysis to identify SNPs targeted to the Williams 82 preliminary 6.5 × scaffold assembly that would anchor additional scaffolds and assist in orienting unoriented scaffolds. A set of 1536 Illumina GoldenGate SNPs was selected and used to genotype 444 F_5-derived RILs from a cross of Williams 82 and *G. soja* PI479752. A portion of the RIL population was also genotyped with a set of 1536 Illumina GoldenGate SNPs that had previously been used by Hyten et al. (2010a) in generating the Soybean Consensus Map 4.0. JoinMap 3.0 software (Van Ooijen and Voorrips 2001) was used to create a higher resolution genetic map by combining the mapping data from the Soybean Consensus Map 4.0 with the data from the Williams 82 × PI479752.RIL genotyping. The result was the Williams 82 whole-genome sequence (Schmutz et al. 2010) with 20 chromosomes comprised of 950 megabases in 397 anchored sequence scaffolds all but 20 of which were unambiguously oriented. Similarly, a linkage map with 516 SNPs was reported in the Forrest × Williams 82 RIL population (Wu et al. 2001).

Sun et al (2013) described a method of high-throughput SNP discovery and genotyping that was referred to as Specific Locus Amplification Fragment sequencing or SLAF-seq. SLAF-seq is implemented by the creation and second-generation sequence analysis of reduced representation libraries (RRL) of the parents and individuals of a mapping population. The restriction enzymes used for the RRL construction are first selected via the sequence analysis of the RRLs to determine the choice of restriction enzymes and the restriction fragment size range that provides a desirable number of unique fragments that facilitate the discovery of a sufficiently large number of SNPs in the resulting paired-end sequence analysis of the library. Sun et al. (2013) conducted a pilot study of SLAF-seq with soybean using the restriction enzymes *Hae*III and *Mse*I for RLL construction with 380–450 bp size-selected fragments. Based upon the analysis of the soybean whole-genome sequence, the expected number of restriction fragments (SLAFs) in the 380–450 size range was 76,970. The sequence analysis of the resulting library found 61,056 SLAFs or 79.3% of the predicted fragments. While slightly more than 25% of the SLAFs were of repetitive sequence, there were nonetheless a large number of SLAFs that were suggested to be single copy fragments. Qi et al. (2014) used the SLAF-seq approach to create a high-density soybean genetic map based upon the analysis of 147 RILs from a cross of 'Charleston' × 'Donguong594'. A total of 12,577 SLAFs showed polymorphism and after elimination of low-quality SLAFs and those with segregation distortion, 5308 markers were suitable for linkage map construction. The resulting linkage map with 20 linkage groups was 2294.43 cM in length with an average distance of 0.43 cM between adjacent markers. Markers were well distributed across the 20 chromosomes with relatively few gaps between markers that exceeded 5 cM.

With the use of second-generation sequencing technology coupled with the availability of the soybean whole-genome sequence (Schmutz et al. 2010), large numbers of sequence variants can be efficiently identified via the alignment of short sequence reads from diverse genotypes to the whole-genome sequence. Song et al. (2013) obtained a total of 498,921,777 reads, 35–45 bp in length, from DNA of reduced representation libraries from six cultivated and two wild soybean genotypes. These reads were mapped to the soybean whole-genome sequence for SNP identification. Of 60,800 SNPs selected for Illumina Infinium BeadChip design, 50,701 were targeted to euchromatic regions and 10,000 to heterochromatic regions of the 20 soybean chromosomes. Of the 60,800 SNPs, a total of 52,041 passed Illumina's manufacturing phase to produce the SoySNP50K iSelect BeadChip (Song et al. 2013). The SoySNP50K iSelect BeadChip was used to genotype 1083 RILs from a Williams 82 × *G. soja* PI479752 population and 922 RILs from Essex × Williams 82 population (Song et al. 2016). A total of 21,478 loci were mapped in the Williams 82 × PI479752 and 11,922 loci in the Essex × Williams 82 population, and the total genetic map length was 2446 and 2648 cM in the Williams 82 × PI479752 and Essex × Williams 82, respectively. Of the 27,431 SNPs that were mapped in one or both of the mapping populations, a total of 5969 SNPs were common to the two populations and the number of SNPs per chromosome on the two linkage maps ranged from 938 to 2481. The linkage maps were used to re-position or re-orient or anchor scaffolds of the soybean whole-genome sequence Glyma1.01 assembly (Schmutz et al. 2010) and to build the Wm82.a2.v1 assembly of soybean, which is available at the Department of Energy, Joint Genome Institute site: http://www.phytozome.org/.

References

Akkaya MS, Bhagwat AA, Cregan PB (1992) Length polymorphisms of simple sequence repeat DNA in soybean. Genetics 132:1131–1139

Akkaya MS, Shoemaker RC, Specht JE, Bhagwat AA, Cregan PB (1995) Integration of simple sequence repeat DNA markers into a soybean linkage map. Crop Sci 35:1439–1445

Apuya NR, Frazier BL, Keim P, Roth EJ, Lark KG (1988) Restriction fragment length polymorphisms as genetic markers in soybean, *Glycine max* (L) Merr. Theor Appl Genet 75:889–901

Bachlava E, Taylor CA, Tang S, Bowers JE, Mandel JR, Burke JM, Knapp SJ (2012) SNP discovery and development of a high-density genotyping array for sunflower. PLoS ONE 7(1):e29814. doi:10.1371/journalpone0029814

Botstein D, White RL, Skolnick M, Davis RW (1980) Construction of a genetic linkage map in man using restriction fragment length polymorphisms. Am J Hum Genet 32:314–331

Brown-Guedira GL, Thompson JA, Nelson RL, Warburton ML (2000) Evaluation of genetic diversity of soybean introductions and North American ancestors using RAPD and SSR markers. Crop Sci 40:815–823

Burr B, Evola SV, Burr FA, Beckmann JS (1983) The application of restriction fragment length polymorphism to plant breeding. In: Setlow JK, Hollander A (eds) Genetic engineering principles and methods, vol 5. Plenum Press, New York, London, pp 45–59

Cao J, Schneeberger K, Ossowski S, Günther T, Bender S, Fitz J, Koenig D, Lanz C, Stegle O, Lippert C (2011) Whole-genome sequencing of multiple *Arabidopsis thaliana* populations. Nat Genet 43:956–963

Cargill M, Altshuler D, Ireland J, Sklar P, Ardlie K, Patil N, Shaw N, Lane CR, Lim EP, Kalyanaraman N, Nemesh J, Ziaugra L, Friedland L, Rolfe A, Warrington J, Lipshutz R, Daley GQ, Lander ES (1999) Characterization of single-nucleotide polymorphisms in coding regions of human genes. Nat Genet 22:231–238

Chen J, Iannone MA, Li MS, Taylor JD, Rivers P, Nelsen AJ, Slentz-Kesler KA, Roses A, Weiner MP (2000) A microsphere-based assay for multiplexed single nucleotide polymorphism analysis using single base chain extension. Genome Res 10:549–557

Ching A, Caldwell KS, Jung M, Dolan M, Smith OS, Tingey S, Morgante M, Rafalski AJ (2002) SNP frequency, haplotype structure and linkage disequilibrium in elite maize inbred lines. BMC Genet 3:19

Choi IY, Hyten DL, Matukumalli LK, Song QJ, Chaky JM, Quigley CV, Chase K, Lark KG, Reiter RS, Yoon MS, Hwang EY, Yi SI, Young ND, Shoemaker RC, van Tassell CP, Specht JE, Cregan PB (2007) A soybean transcript map: gene distribution, haplotype and SNP analysis. Genetics 176:685–686

Concibido VC, Denny RL, Lange DA, Orf JH, Young ND (1996) RFLP mapping and marker-assisted selection of soybean cyst nematode resistance in PI 209332. Crop Sci 36:1643–1650

Cook JP, McMullen MD, Holland JB, Tian F, Bradbury P, Ross-Ibarra J, Buckler ES, Flint-Garcia SA (2012) Genetic architecture of maize kernel composition in the nested association mapping and inbred association panels. Plant Physiol 158:824–834

Cregan PB, Jarvik T, Bush AL, Shoemaker RC, Lark KG, Kahler AL, Kaya N, VanToai TT, Lohnes DG, Chung J (1999) An integrated genetic linkage map of the soybean genome. Crop Sci 39:1464–1490

Devine TE (2003) The *Pd2* and *Lf2* loci define soybean linkage group 16. Crop Sci 43:2028–2030

Diers BW, Keim P, Fehr WR, Shoemaker RC (1992) RFLP analysis of soybean seed protein and oil content. Theor Appl Genet 83:608–612

Eujayl I, Sledge MK, Wang L, May GD, Chekhovskiy K, Zwonitzer JC, Mian MA (2004) *Medicago truncatula* EST-SSRs reveal cross-species genetic markers for *Medicago* spp. Theor Appl Genet 108:414–422

Feltus FA, Wan J, Schulze SR, Estill JC, Jiang N, Paterson AH (2004) An SNP resource for rice genetics and breeding based on subspecies *indica* and *japonica* genome alignments. Genome Res 14:1812–1819

Ferreira AR, Foutz KR, Keim P (2000) Soybean genetic map of RAPD markers assigned to an existing scaffold RFLP map. J Hered 91:392–396

Ganal MW, Durstewitz G, Polley A, Bérard A, Buckler ES, Charcosset A, Clarke JD, Graner EM, Hansen M, Joets J (2011) A large maize (*Zea mays* L) SNP genotyping array: development and germplasm genotyping, and genetic mapping to compare with the B73 reference genome. PLoS ONE 6:e28334

Gao LF, Jing RL, Huo NX, Li Y, Li XP, Zhou RH, Chang XP, Tang JF, Ma ZY, Jia JZ (2004) One hundred and one new microsatellite loci derived from ESTs (EST-SSRs) in bread wheat. Theor Appl Genet 108:1392–1400

Geraldes A, Pang J, Thiessen N, Cezard T, Moore R, Zhao Y, Tam A, Wang S, Friedmann M, Birol I (2011) SNP discovery in black cottonwood (*Populus trichocarpa*) by population transcriptome resequencing. Mol Ecol Resour 11:81–92

Griffin TJ, Hall JG, Prudent JR, Smith LM (1999) Direct genetic analysis by matrix-assisted laser desorption/ionization mass spectrometry. Proc Natl Acad Sci USA 96:6301–6306

Halushka MK, Fan JB, Bentley K, Hsie L, Shen N, Weder A, Cooper R, Lipshutz R, Chakravarti A (1999) Patterns of single-nucleotide polymorphisms in candidate genes for blood-pressure homeostasis. Nat Genet 22:239–247

Helentjaris T, Slocum M, Wright S, Schaefer A, Nienhuis J (1986) Construction of genetic maps in maize and tomato using restriction fragment length polymorphisms. Theor Appl Genet 72:761–769

Hisano H, Sato S, Isobe S, Sasamoto S, Wada T, Matsuno A, Fujishiro T, Yamada M, Nakayama S, Nakamura Y (2007) Characterization of the soybean genome using EST-derived microsatellite markers. DNA Res 14:271–281

Horton MW, Hancock AM, Huang YS, Toomajian C, Atwell S, Auton A, Muliyati NW, Plat A, Sperone FG, Vilhjálmsson BJ (2012) Genome-wide patterns of genetic variation in worldwide *Arabidopsis thaliana* accessions from the RegMap panel. Nat Genet 44:212–216

Huang X, Zhao Y, Wei X, Li C, Wang A, Zhao Q, Li W, Guo Y, Deng L, Zhu C (2012) Genome-wide association study of flowering time and grain yield traits in a worldwide collection of rice germplasm. Nat Genet 44:32–39

Hwang TY, Sayama T, Takahashi M, Takada Y, Nakamoto Y, Funatsuki H, Hisano H, Sasamoto S, Sato S, Tabata S, Kono I, Hoshi M, Hanawa M, Yano C, Xia Z, Harada K, Kitamura K (2009) High-density integrated linkage map based on SSR markers in soybean. DNA Res 16:213–225

Hyten DL, Song Q, Zhu Y, Choi IY, Nelson RL, Costa JM, Specht JE, Shoemaker RC, Cregan PB (2006) Impacts of genetic bottlenecks on soybean genome diversity. Proc Natl Acad Sci USA 103:16666–16671

Hyten DL, Song Q, Choi IY, Yoon MS, Specht JE, Matukumalli LK, Nelson RL, Shoemaker RC, Young ND, Cregan PB (2008) High-throughput genotyping with the GoldenGate assay in the complex genome of soybean. Theor Appl Genet 116:945–952

Hyten DL, Choi IY, Song QJ, Specht JE, Carter TE, Shoemaker RC, Hwang EY, Matukumalli LK, Cregan PB (2010a) A high density integrated genetic linkage map of soybean and the development of a 1536 universal soy linkage panel for quantitative trait locus mapping. Crop Sci 50:960–968

Hyten DL, Cannon SB, Song QJ, Weeks N, Fickus EW, Shoemaker RC, Specht JE, Farmer AD, May GD, Cregan PB (2010b) High-throughput SNP discovery through deep resequencing of a reduced representation library to anchor and orient scaffolds in the soybean whole genome sequence. BMC Genomics 11:38. doi:10.1186/1471-2164-11-38

Jander G, Norris SR, Rounsley SD, Bush DF, Levin IM, Last RL (2002) Arabidopsis map-based cloning in the post-genome era. Plant Physiol 129:440–450

Jia G, Huang X, Zhi H, Zhao Y, Zhao Q, Li W, Chai Y, Yang L, Liu K, Lu H, Zhu C, Lu Y, Zhou C, Fan D, Weng Q, Guo Y, Huang T, Zhang L, Lu T, Feng Q, Hao H, Liu H, Lu P, Zhang N, Li Y, Guo E, Wang S, Wang S, Liu J, Zhang W, Chen G, Zhang B, Li W, Wang Y, Li H, Zhao B, Li J, Diao X, Han B (2013) A haplotype map of genomic variations and genome-wide association studies of agronomic traits in foxtail millet (*Setaria italica*). Nat Genet 45:957–961

Keim P, Diers BW, Olson TC, Shoemaker RC (1990) RFLP mapping in soybean: association between marker loci and variation in quantitative traits. Genetics 126:735–742

Keim P, Schupp JM, Travis SE, Clayton K, Zhu T, Shi L, Ferreira A, Webb DM (1997) A high-density soybean genetic map based on AFLP markers. Crop Sci 37:537–543

Lam HM, Xu X, Liu X, Chen W, Yang G, Wong FL, Li MW, He W, Qin N, Wang B (2010) Resequencing of 31 wild and cultivated soybean genomes identifies patterns of genetic diversity and selection. Nat Genet 42:1053–1059

Lark KG, Weisemann JM, Matthews BF, Palmer R, Chase K, Macalma T (1993) A genetic map of soybean (*Glycine max* L) using an intraspecific cross of two cultivars: 'Minsoy' and 'Noir 1'. Theor Appl Genet 86:901–906

Li YH, Zhao SC, Ma JX, Li D, Yan L, Li J, Qi XT, Guo XS, Zhang L, He WM, Chang RZ, Liang QS, Guo Y, Ye C, Wang XB, Tao Y, Guan RX, Wang JY, Liu YL, Jin LG, Zhang XG, Liu ZX, Zhang LJ, Chen J, Wang KJ, Nielsen R, Li RQ, Chen PY, Li WB, Reif JC, Purugganan M, Wang J, Zhang MC, Wang J, Qui LJ (2013) Molecular footprints of domestication and improvement in soybean revealed by whole genome re-sequencing. BMC Genomics 14:579

Li Z, Nelson RL (2002) RAPD marker diversity among cultivated and wild soybean accessions from four Chinese provinces. Crop Sci 42:1737–1744

Lightfoot DA, Njiti VN, Gibson PT, Kassem MA, Iqbal JM, Meksem K (2005) Registration of the Essex × Forrest recombinant inbred line mapping population. Crop Sci 45:1678–1681

Litt M, Luty JA (1989) A hypervariable microsatellite revealed by in vitro amplification of a dinucleotide repeat within the cardiac muscle actin gene. Am J Hum Genet 44:397–401

Liu F, Dong JJ, Zou JJ, Chen SY, Zhuang BC (2000) Soybean germplasm diversity and genetic variation detected by microsatellite markers. Acta Genet Sin 27:628–633

Mahama AA, Lewers KS, Palmer RG (2002) Genetic linkage in soybean: classical genetic linkage groups 6 and 8. Crop Sci 42:1459–1464

Mansur LM, Orf JH, Chase K, Jarvik T, Cregan PB, Lark KG (1996) Genetic mapping of agronomic traits using recombinant inbred lines of soybean. Crop Sci 36:1327–1336

Marek LF, Mudge J, Darnielle L, Grant D, Hanson N, Paz M, Huihuang Y, Denny R, Larson K, Foster-Hartnett D, Cooper A, Danesh D, Larsen D, Schmidt T, Staggs R, Crow JA, Retzel E, Young ND, Shoemaker RC (2001) Soybean genomic survey: BAC-end sequences near RFLP and SSR markers. Genome 44:572–581

Mascher M, Wu S, Amand PS, Stein N, Poland J (2013) Application of genotyping-by-sequencing on semiconductor sequencing platforms: a comparison of genetic and reference-based marker ordering in barley. PLoS ONE 8:e76925

Matthews BF, Devine TE, Weisemann JM, Beard HS, Lewers KS, MacDonald MH, Park YB, Maiti R, Lin JJ, Kuo J (2001) Incorporation of sequenced cDNA and genomic markers into the soybean genetic map. Crop Sci 41:516–521

Maughan PJ, Saghai Maroof MA, Buss GR (1995) Microsatellite and amplified sequence length polymorphisms in cultivated and wild soybean. Genome 38:715–723

McNally KL, Childs KL, Bohnert R, Davidson RM, Zhao K, Ulat VJ, Zeller G, Clark RM, Hoen DR, Bureau TE (2009) Genomewide SNP variation reveals relationships among landraces and modern varieties of rice. Proc Nat Acad Sci USA 106:12273–12278

Morgante M, Olivieri AM (1993) PCR-amplified microsatellites as markers in plant genetics. Plant J 3:175–182

Morgante M, Hanafey M, Powell W (2002) Microsatellites are preferentially associated with nonrepetitive DNA in plant genomes. Nat Genet 30:194–200

Morgante M, Rafalski A, Biddle P, Tingey S, Olivieri AM (1994) Genetic mapping and variability of seven soybean simple sequence repeat loci. Genome 37:763–769

Morris GP, Ramu P, Deshpandeb SP, Hash CT, Shah T, Upadhyaya HD, Riera-Lizarazu O, Brown PJ, Acharya CB, Mitchell SE, Harriman J, Glaubitz JC, Buckler ES, Kresovich S (2013) Population genomic and genome-wide association studies of agroclimatic traits in sorghum. Proc Natl Acad Sci USA 110:453–458

Mudge J, Huihuang Y, Denny RL, How DK, Danesh D, Marek LF, Retzel E, Shoemaker RC, Young ND (2004) Soybean bacterial artificial chromosome contigs anchored with RFLPs: insights into genome duplication and gene clustering. Genome 47:361–372

Mullis K, Faloona F, Scharf S, Saiki R, Horn G, Erlich H (1986) Specific enzymatic amplification of DNA in vitro: the polymerase chain reaction. Cold Spring Harb Symp Quant Biol 51(Pt 1):263–273

Nasu S, Suzuki J, Ohta R, Hasegawa K, Yui R, Kitazawa N, Monna L, Minobe Y (2002) Search for and analysis of single nucleotide polymorphisms (SNPs) in rice (*Oryza sativa*, *Oryza rufipogon*) and establishment of SNP markers. DNA Res 9:163–171

Nordborg M, Hu TT, Ishino Y, Jhaveri J, Toomajian C, Zheng H, Bakker E, Calabrese P, Gladstone J, Goyal R, Jakobsson M, Kim S, Morozov Y, Padhukasahasram B, Plagnol V, Rosenberg NA, Shah C, Wall JD, Wang J, Zhao K, Kalbfleisch T, Schulz V, Kreitman M, Bergelson J (2005) The pattern of polymorphism in *Arabidopsis thaliana*. PLoS Biol 3: e196

Palmer RG, Kilen TC (1987) Qualitative genetics and cytogenetics In: Wilcox JR (ed) Soybeans: improvement, production, and uses, 2nd edn. Agron Monogr 16 ASA, CSSA and SSSA, Madison, WI, pp 135–209

Palmer RG, Pfeiffer TW, Buss GR, Kilen TC (2004) Qualitative genetics In: Boerma HR, Specht JE (eds) Soybeans, improvement, production and uses, 3rd edn. Agron Monogr 16 ASA, CSSA and SSSA, Madison, WI pp 137–233

Palmer RG, Shoemaker RC (1998) Soybean genetics. In: Hrustic MVM, Jockovic D (eds) Soybean institute of field and vegetable crops. Novi Sad, Yugoslavia, pp 45–82

Qi Z, Hunag L, Zhu R, Xin D, Liu C, Han X, Jiang H, Hong W, Hu G, Zheng H, Chen Q (2014) A high-density genetic map for soybean based on specific length amplified fragment sequencing. PLoS ONE 9(8):e104871

Rafalski A, Tingey S (1993) RFLP map of soybean (*Glycine max*). In: O'Brien SJ (ed) Genetic maps: locus maps of complex genomes. Cold Spring Harbor Laboratory Press, New York, pp 1-6149–1-6156

Rongwen J, Akkaya MS, Bhagwat AA, Lavi U, Cregan PB (1995) The use of microsatellite DNA markers for soybean genotype identification. Theor Appl Genet 90:43–48

Rostoks N, Mudie S, Cardle L, Russell J, Ramsay L, Booth A, Svensson JT, Wanamaker SI, Walia H, Rodriguez EM, Hedley PE, Liu H, Morris J, Close TJ, Marshall DF, Waugh R (2005) Genome-wide SNP discovery and linkage analysis in barley based on genes responsive to abiotic stress. Mol Genet Genomics 274:515–527

Sachidanandam R, Weissman D, Schmidt SC, Kakol JM, Stein LD, Marth G, Sherry S, Mullikin JC, Mortimore BJ, Willey DL, Hunt SE, Cole CG, Coggill PC, Rice CM, Ning Z, Rogers J, Bentley DR, Kwok PY, Mardis ER, Yeh RT, Schultz B, Cook L, Davenport R, Dante M, Fulton L, Hillier L, Waterston RH, McPherson JD, Gilman B, Schaffner S, Van Etten WJ, Reich D, Higgins J, Daly MJ, Blumenstiel B, Baldwin J, Stange-Thomann N, Zody MC, Linton L, Lander ES, Altshuler D (2001) A map of human genome sequence variation containing 142 million single nucleotide polymorphisms. Nature 409:928–933

Schmid KJ, Ramos-Onsins S, Ringys-Beckstein H, Weisshaar B, Mitchell-Olds T (2005) A multilocus sequence survey in *Arabidopsis thaliana* reveals a genome-wide departure from a neutral model of DNA sequence polymorphism. Genetics 169:1601–1615

Schmid KJ, Sorensen TR, Stracke R, Torjek O, Altmann T, Mitchell-Olds T, Weisshaar B (2003) Large-scale identification and analysis of genome-wide single-nucleotide polymorphisms for mapping in *Arabidopsis thaliana*. Genome Res 13:1250–1257

Schmutz J, Cannon SB, Schlueter J, Ma J, Mitros T, Nelson W, Hyten DL, Song QJ, Thelen JJ, Cheng J, Xu D, Hellsten U, May GD, Yu Y, Sakurai T, Umezawa T, Bhattacharyya MK, Sandhu D, Valliyodan B, Lindquist E, Peto M, Grant D, Shu S, Goodstein D, Barry K, Futrell-Griggs M, Abernathy B, Du J, Tian Z, Zhu L, Gill N, Joshi T, Libault M, Sethuraman A, Zhang XC, Shinozaki K, Nguyen HT, Wing RA, Cregan P, Specht J, Grimwood J, Rokhsar D, Stacey G, Shoemaker RC, Jackson SA (2010) Genome sequence of the palaeopolyploid soybean. Nature 463:178–183

Seversike TM, Ray JD, Shultz JL, Purcell LC (2008) Soybean molecular linkage group B1 corresponds to classical linkage group 16 based on map location of the *lf2* gene. Theor Appl Genet 117:143–147

Shoemaker RC, Olson TC (1993) Molecular linkage map of soybean (*Glycine max* L Merr). In: O'Brien SJ (ed) Genetic Maps: locus maps of complex genomes. Cold Spring Harbor Laboratory Press, New York, pp 1-6131–1-6138

Shoemaker RC, Specht JE (1995) Integration of the soybean molecular and classical genetic linkage groups. Crop Sci 35:436–446

Song QJ, Marek LF, Shoemaker RC, Lark KG, Concibido VC, Delannay X, Specht JE, Cregan PB (2004) A new integrated genetic linkage map of the soybean. Theor Appl Genet 109:122–128

Song QJ, Jia GF, Zhu Y, Grant D, Nelson RT, Hwang EY, Hyten DL, Cregan PB (2010) Abundance of SSR motifs and development of candidate polymorphic SSR markers (BARCSOYSSR_1 0) in soybean. Crop Sci 50:1950–1960

Song QJ, Hyten DL, Jia GF, Quigley CV, Fickus EW, Nelson RL, Cregan PB (2013) Development and evaluation of SoySNP50K, a high-density genotyping array for soybean. PLoS ONE 8:e54985. doi:10.1371/journal.pone.0054985

Song QJ, Jenkins J, Jia GF, Hyten DL, Pantalone V, Jackson SA, Schmutz J, Cregan PB (2016) Construction of high resolution genetic linkage maps to improve the soybean genome sequence assembly Glyma101. BMC Genomics 17(1):33

Stacey G, Vodkin L, Parrott WA, Shoemaker RC (2004) National Science Foundation-sponsored workshop report draft plan for soybean genomics. Plant Physiol 135:59–70

Sun X, Liu D, Zhang X, Li W, Liu H, Hong W, Jiang C, Guan N, Ma C, Zeng H, Xu C, Song J, Huang L, Wang C, Shi J, Wang R, Zheng X, Lu C, Wang X, Zheng H (2013) SLAF-seq: an efficient method of large-scale *De novo* SNP discovery and genotyping using high throughput sequencing. PLoS ONE 8(3): e58700. doi:10.1371/journalpone0058700

Tautz D (1989) Hypervariability of simple sequences as a general source for polymorphic DNA markers. Nucleic Acids Res 17:6463–6471

Tenaillon MI, Sawkins MC, Long AD, Gaut RL, Doebley JF, Gaut BS (2001) Patterns of DNA sequence polymorphism along chromosome 1 of maize (*Zea mays* ssp *mays* L). Proc Natl Acad Sci USA 98:9161–9166

Thompson JA, Nelson RL, Vodkin LO (1998) Identification of diverse soybean germplasm using RAPD markers. Crop Sci 38:1348–1355

Tuskan GA, Difazio S, Jansson S, Bohlmann J, Grigoriev I, Hellsten U, Putnam N, Ralph S, Rombauts S, Salamov A, Schein J, Sterck L, Aerts A, Bhalerao RR, Bhalerao RP, Blaudez D, Boerjan W, Brun A, Brunner A, Busov V, Campbell M, Carlson J, Chalot M, Chapman J, Chen GL, Cooper D, Coutinho PM, Couturier J, Covert S, Cronk Q, Cunningham R, Davis J, Degroeve S, Dejardin A, Depamphilis C, Detter J, Dirks B, Dubchak I, Duplessis S, Ehlting J, Ellis B, Gendler K, Goodstein D, Gribskov M, Grimwood J, Groover A, Gunter L, Hamberger B, Heinze B, Helariutta Y, Henrissat B, Holligan D, Holt R, Huang W, Islam-Faridi N, Jones S, Jones-Rhoades M, Jorgensen R, Joshi C, Kangasjarvi J, Karlsson J, Kelleher C, Kirkpatrick R, Kirst M, Kohler A, Kalluri U, Larimer F, Leebens-Mack J, Leple JC, Locascio P, Lou Y, Lucas S, Martin F, Montanini B, Napoli C, Nelson DR, Nelson C, Nieminen K, Nilsson O, Pereda V, Peter G, Philippe R, Pilate G, Poliakov A, Razumovskaya J, Richardson P, Rinaldi C, Ritland K, Rouze P, Ryaboy D, Schmutz J, Schrade J, Segerman B, Shin H, Siddiqui A, Sterky F, Terry A, Tsai CJ, Uberbacher E, Unneberg P, Vahala J, Wall K, Wessler S, Yang G, Yin T, Douglas C, Marra M, Sandberg G, Van de Peer Y, Rokhsar D (2006) The genome of black cottonwood, *Populus trichocarpa* (Torr & Gray). Science 313:1596–1604

Van Ooijen JW, Voorrips RE (2001) JoinMap 30 software. Plant Research International, Wageningen

Vodkin LO, Khanna A, Shealy R, Clough SJ, Gonzalez DO, Philip R, Zabala G, Thibaud-Nissen F, Sidarous M, Stromvik MV, Shoop E, Schmidt C, Retzel E, Erpelding J, Shoemaker RC, Rodriguez-Huete AM, Polacco JC, Coryell V, Keim P, Gong G, Liu L, Pardinas J, Schweitzer P (2004) Microarrays for global expression constructed with a low redundancy set of 27,500 sequenced cDNAs representing an array of developmental stages and physiological conditions of the soybean plant. BMC Genom 5:73

Vos P, Hogers R, Bleeker M, Reijans M, van de Lee T, Hornes M, Frijters A, Pot J, Peleman J, Kuiper M, Zabeau M (1995) AFLP: a new technique for DNA fingerprinting. Nucleic Acids Res 23:4407–4414

Watterson GA (1975) On the number of segregating sites in genetical models without recombination. Theor Popul Biol 7:256–276

Weber JL, May PE (1989) Abundant class of human DNA polymorphisms which can be typed using the polymerase chain reaction. Am J Hum Genet 44:388–396

Welsh J, McClelland M (1990) Fingerprinting genomes using PCR with arbitrary primers. Nucleic Acids Res 18:7213–7218

Weiss MG (1970a) Genetic linkage in soybean Linkage group I. Crop Sci 10:69–72

Weiss MG (1970b) Genetic linkage in soybean Linkage groups II and III. Crop Sci 10:300–303

Weiss MG (1970c) Genetic linkage in soybean Linkage group IV. Crop Sci 10:368–370

Weiss MG (1970d) Genetic linkage in soybean Linkage groups V and VI. Crop Sci 10:469–470

Weiss MG (1970e) Genetic linkage in soybean Linkage group VII. Crop Sci 10:627–629

Williams JG, Kubelik AR, Livak KJ, Rafalski JA, Tingey SV (1990) DNA polymorphisms amplified by arbitrary primers are useful as genetic markers. Nucleic Acids Res 18:6531–6535

Wright SI, Bi IV, Schroeder SG, Yamasaki M, Doebley JF, McMullen MD, Gaut BS (2005) The effects of artificial selection on the maize genome. Science 308:1310–1314

Wu X, Vuong TD, Leroy JA, Shannon JG, Sleper DA, Nguyen HT (2011) Selection of a core set of RILs from Forrest x Williams 82 to develop a framework map in soyeban. Theor Appl Genet 122(6):1179–1187

Wu XL, He CY, Wang YJ, Zhang ZY, Dongfang Y, Zhang JS, Chen SY, Gai JY (2001) Construction and analysis of a genetic linkage map of soybean. Yi Chuan Xue Bao 28:1051–1061

Xia Z, Tsubokura Y, Hoshi M, Hanawa M, Yano C, Okamura K, Ahmed TA, Anai T, Watanabe S, Hayashi M (2007) An integrated high-density linkage map of soybean with RFLP, SSR, STS, and AFLP markers using a single F_2 population. DNA Res 14:257–269

Yamanaka N, Ninomiya S, Hoshi M, Tsubokura Y, Yano M, Nagamura Y, Sasaki T, Harada K (2001) An informative linkage map of soybean reveals QTLs for flowering time, leaflet morphology and regions of segregation distortion. DNA Res 8:61–72

Yu JK, Dake TM, Singh S, Benscher D, Li W, Gill B, Sorrells ME (2004a) Development and mapping of EST-derived simple sequence repeat markers for hexaploid wheat. Genome 47:805–818

Yu JK, La Rota M, Kantety RV, Sorrells ME (2004b) EST derived SSR markers for comparative mapping in wheat and rice. Mol Gen Genomics 271:742–751

Zhu YL, Song QJ, Hyten DL, Van Tassell CP, Matukumalli LK, Grimm DR, Hyatt SM, Fickus EW, Young ND, Cregan PB (2003) Single-nucleotide polymorphisms in soybean. Genetics 163:1123–1134

Structural Variation and the Soybean Genome

4

Justin E. Anderson and Robert M. Stupar

Abstract

Genomic structural variation is an important component to genetic diversity in soybean (*Glycine max*). These large-scale genomic differences are now known as the underlying genetic mechanisms for a number of important phenotypic traits. Identifying structural variants across numerous individuals at higher resolution is increasingly possible with a number of improving genome analyses platforms. Understanding where these polymorphisms occur and why some are maintained over evolutionary time has important biological and agronomic implications. This chapter elaborates on detecting, describing, and developing structural variation in the soybean genome and how to incorporate these polymorphisms in ongoing soybean research and improvement.

4.1 Introduction

Plants contain more types of genetic diversity than an "assembled genome" leads one to believe. When interested in genetics and genomics, many researchers in the recent past have focused on single-nucleotide polymorphisms (SNPs). This is true in soybean, where the modern sequencing technologies and the release of the soybean genome assembly (Schmutz et al. 2010) have facilitated the detection of SNPs across numerous cultivars and accessions (Lam et al. 2010; Wu et al. 2010; Hyten et al. 2010; Chung et al. 2014; Qiu et al. 2014; Zhou et al. 2015). After implementing appropriate filtering steps, a list of thousands to millions of SNPs distributed across the genome can be developed. The convenience and predictability of this process has allowed researchers from all realms of genetics to participate in the genomics era. While hugely beneficial and impactful, this framework has generally ignored the larger scale genomic variants.

Structural variation, an inexact term used to describe relatively large genomic sequence variants, is known to affect substantial portions of

J.E. Anderson · R.M. Stupar (✉)
Department of Agronomy and Plant Genetics, University of Minnesota, 1991 Upper Buford Circle, 411 Borlaug Hall, St. Paul, MN 55108-6026, USA
e-mail: stup0004@umn.edu

J.E. Anderson
e-mail: ande9112@umn.edu

H.T. Nguyen and M.K. Bhattacharyya (eds.), *The Soybean Genome*, Compendium of Plant Genomes, DOI 10.1007/978-3-319-64198-0_4

many plant genomes (Żmieńko et al. 2014). A structural variant (SV) can vary widely in size, and the threshold for defining a SV event is largely dependent on the platform use for detection (e.g., array-based platforms can usually only detect variants larger than one or two kb, while sequence-based platforms may have higher resolution). In addition to variation in size, SVs also vary in type, including deletions, duplications, inversions, and translocations. Broadly, these polymorphisms can be described as either nucleotide content variation or genomic context rearrangements. Some recently published SV profiles in plants include apple (*Malus domestica*) (Boocock et al. 2015), Arabidopsis (*Arabidopsis thaliana*) (Santuari et al. 2010; Cao et al. 2011), barley (*Hordeum vulgare*) (Munoz-Amatriain et al. 2013), cucumber (*Cucumis* sativas) (Zhang et al. 2015), maize (*Zea mays*) (Swanson-Wagner et al. 2010; Chia et al. 2012; Hirsch et al. 2014), rice (*Oriza sativa*) (Yu et al. 2013; Schatz et al. 2014), sorghum (*Sorghum bicolor*) (Zheng et al. 2011), and soybean (*Glycine max*) (Lam et al. 2010; McHale et al. 2012; Anderson et al. 2014; Zhou et al. 2015).

SV formation is not fully understood in plants but is likely attributable to homologous recombination (HR) or non-homologous recombination repair mechanisms as revealed in human studies (Hastings et al. 2009). HR requires long stretches (hundreds of base pairs) of high sequence similarity. HR between regions of the genome that are not alleles, known as nonallelic homologous recombination (NAHR), is one mechanism of SV formation. Repair pathways following DNA double-strand breakage, such as non-homologous end joining (NHEJ), single-strand annealing, or microhomology-mediated end joining (MMEJ), can also result in gene deletions. NHEJ-based repair involves less than 5 bp of sequence homology, while MMEJ involves 5–25 bp (Hastings et al. 2009). Fork stalling and template switching is a replication-based mechanism proposed to result in deletions or duplications, but also potentially lead to complex SV events (Gu et al. 2008). SVs can also result from other biological processes including T-DNA insertion (Kyndt et al. 2015) or transposon activity (Lisch 2013). While the frequency of each is unclear, estimates in barley (Munoz-Amatriain et al. 2013) and cucumber (Zhang et al. 2015) suggest deletions are most frequently attributable to double-strand break repair mechanisms.

These large-scale genetic polymorphisms are associated with a number of fine-mapped phenotypic traits. Specific examples within the plant community include glyphosate resistance in Palmer amaranth (*Amaranthus palmeri*) (Gaines et al. 2010, 2011), boron tolerance and winter hardiness in barley (*Hordeum vulgare*) (Sutton et al. 2007; Knox et al. 2010), dwarfism and flowering time in wheat (*Triticum* spp.) (Pearce et al. 2011; Díaz et al. 2012; Li et al. 2012), female gamete fitness in potato (*Solanum tuberosum*) (Iovene et al. 2013), submergence tolerance and grain size in rice (*Oriza sativa*) (Xu et al. 2006; Wang et al. 2015b), reproductive morphology in cucumber (*Cucumis sativas*) (Zhang et al. 2015), and aluminum tolerance and glume formation in maize (*Zea mays*) (Han et al. 2012; Wingen et al. 2012; Maron et al. 2013). See Żmieńko et al. (2014) for a comprehensive review in plants.

SVs can modify gene expression in a number of ways (Fig. 4.1). Whole-gene deletions result in no DNA template for transcription and therefore an absence of that particular mRNA and protein. Gene duplications increase the amount of DNA template and may lead to additional transcription (higher mRNA expression) and downstream translation products (more protein). Gene dosage is therefore affected directly by the nucleotide content in these cases. Alternatively, genomic context can also affect gene dosage. For example, a rearrangement SV, such as an inversion or translocation, could move a gene from heterochromatin to euchromatin (or euchromatin to heterochromatin) resulting in transcriptional alterations due to the new genomic location.

SV events that only partially overlap a gene can have unique consequences. These events have less predictable effects on overall expression but generally result in a compromised transcript. For example, deleting a single internal exon might not affect a gene's transcription but may produce a non-functional protein missing an important domain. Deleting the first exon or

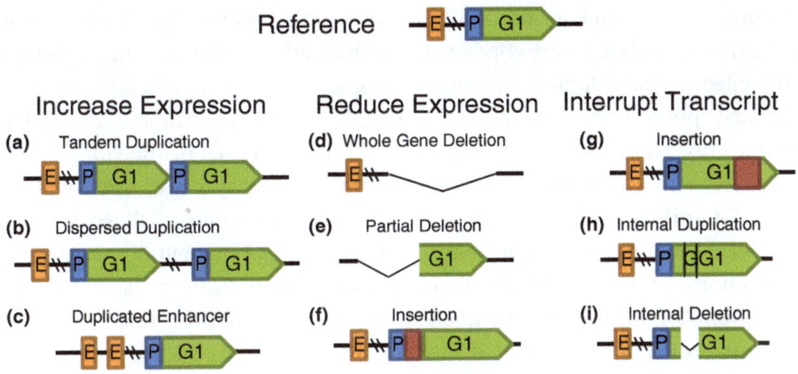

Fig. 4.1 Potential effects of a SV on gene content and transcript production. **a–c** Expression can increase as a result of tandem or dispersed whole-gene duplication or duplication of an enhancer region. **d–f** Expression can decrease through whole-gene deletion, partial deletion, or interruption of gene promoter region. **g–i** Internal changes can lead to interrupted genes and altered transcripts. SV detection with CGH or read depth variance is most likely to detect large-scale changes (**a–e**) and unable to detect rearrangements or insertions (**f, g**). This figure is modeled after a figure in Żmieńko et al. (2014)

promoter region could be sufficient to turn off expression of a gene entirely. A SV break point overlapping coding sequences might even result in novel transcript formation from incidental alignment of exons. This is just a sampling of the types of disruptive scenarios a SV can cause. In addition to modifying expression, a SV such as transposon insertions (Yao et al. 2002) or inversions also can inhibit local recombination. This can limit the ability to introgress traits from a specific locus as well as have long-term evolutionary implications.

Ancient polyploidization events are also an important factor in the exploration of soybean SVs. Often referred to as a palaeopolyploid, the soybean genome has evidence of two relatively recent whole-genome duplication events, occurring around 59 million years ago (mya) and 13 mya, respectively (Schmutz et al. 2010; Severin et al. 2011). Whole-genome duplications (WGD) are often followed by a period of fractionation where rearrangements occur and copies of genes are deleted. Interestingly, even after millions of years the soybean genome has retained a large portion of its genes still present in multiple copies. Examining the context of this genetic redundancy is influential in shaping hypotheses surrounding SVs.

4.2 Functional Structural Variants in Soybean

In soybean, the known examples of SVs that influence phenotypic variation have been discovered in fine-mapping experiments. Perhaps the first such association was identified for a soybean seed coat color trait. In soybean production, it is common to find spontaneous black seed coat mutants in yellow seed coat varieties. Todd and Vodkin (1996) investigated this issue and found mutations in a cluster of chalcone synthase genes underlying this phenotype (Todd and Vodkin 1996). This family contained three genes, *CHS1*, *CHS3*, and *CHS4*. A duplication of *CHS1* (termed *dCHS1*) was associated with the yellow seed coat, yellow hilum varieties. Spontaneous black seed coats in the offspring of full yellow seed varieties no longer had detectable full *dCHS1* duplication. The authors also observed a reversion mutation from a yellow seed, colored hilum to a fully black seed. This spontaneous reversion was the result of a deletion in the CHS4 promoter region in seven out of ten cultivars. Bacterial artificial chromosome sequencing found these gene family members occurred in multiple clusters within a confined area on chromosome 8, likely contributing to the frequent and recurring SVs (Tuteja and Vodkin 2008). Gene transcription analyses were conducted to understand the effects of structural variation on gene regulation and phenotype. Unexpectedly, both spontaneous reversion mutation types, resulting in black seed coats, were associated with increased total *CHS* family mRNA. The duplication in *dCHS1* reduced transcription in all family members and deletions in the *CHS4* promoter increased transcription of all family members. Todd and Vodkin suspected this was likely due to an RNAi-like system of family-wide silencing. The advent of siRNA sequencing confirmed the presence of natural RNAi targeting this gene family to explain the unique SV effects on transcription (Tuteja et al. 2009).

Recently, the *Rhg1* soybean cyst nematode resistance quantitative trait locus (QTL) (Cook et al. 2012) was also characterized as a functional SV in the soybean genome. After many attempts at cloning this QTL, researchers discovered the resistance locus is a 31.2-kb segment encompassing five genes and arranged as a tandem duplication of varying copy numbers among accessions. The authors reported that three of the five genes within this segment are required for enhanced resistance, and haplotypes with more copies exhibited greater levels of resistance. Unlike the *CHS* example above, haplotypes with additional copies of the *Rhg1* segment exhibited greater transcription of these genes. Furthermore, silencing any one of these three genes reduced soybean cyst nematode resistance. Conversely, simultaneous overexpression of these three genes in a susceptible line conferred enhanced resistance.

A larger screen of soybean germplasm followed these initial findings and identified a wider variety of SV states at the *Rhg1* locus (Cook et al. 2014). Two types of resistant classes were identified: the three copy class, and a high copy class ranging from seven to ten copies. Phenotypic screens further confirmed a relationship between copy number and resistance level. Additionally, the three copy number genotypes had higher methylation than one-copy genotypes, as has been observed previously with duplicated genes (Rodin and Riggs 2003). The relationship between methylation and copy number variation is not yet clear in this situation.

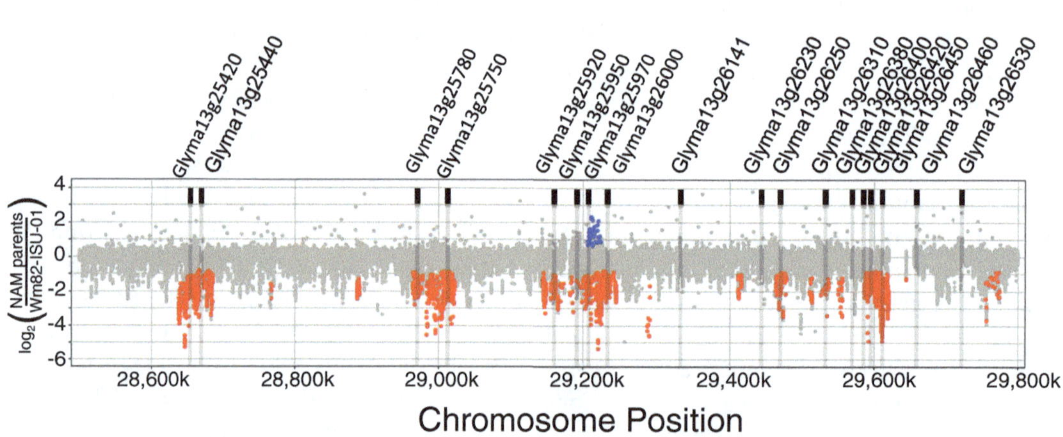

Fig. 4.2 Structural variation in an R-gene cluster of soybean chromosome 13 detected with CGH in 41 diverse soybean lines (Anderson et al. 2014). Plotted points are the $\log_2$ ratio of each genotype as compared to the Williams82-ISU-01 reference for each probe. Colored points denote putative duplications (*blue*) and putative deletions (*red*). Labeled across the top are the location of NBS–LRR genes according to the soybean genome assembly version Glyma.v1.a1.1 (Schmutz et al. 2010). Multiple forms of disease resistance are mapped to this region including resistance to *Phytophthora sojae* (Rps3a, Rps3b, Rps3c, and Rps8) (Gordon et al. 2006), *soybean mosaic virus* (Rsv1—one of the R-genes near Glyma13g25440) (Hayes et al. 2004; Zhang et al. 2012), *peanut mottle virus* (Rpv1), *Pseudomonas syringae* (Rpg1-b) (Ashfield et al. 2007), and *Aphis glycines* (Soybean Aphid, Rag2) (Kim et al. 2010a)

Attempts to map genes resistant to a range of fungal and viral diseases (R-genes) frequently localize to regions of enhanced SVs. One particularly active R-gene cluster in soybean is on chromosome 13 (Fig. 4.2). *Rsv1*, resistance to soybean mosaic virus (SMV), is a cloned single member of this cluster of nucleotide binding site–leucine-rich repeat (NBS–LRR) genes (Hayes et al. 2004). The *Rsv1* gene is responsible for resistance to many strains of SMV. Additional members of this R-gene cluster are also implicated in unique resistant and necrotic reactions to other SMV strains, depending on their presence or absence (Zhang et al. 2012). Furthermore, this locus exhibits higher total NBS–LRR genes in accessions with higher levels of SMV resistance.

The *Rpp4* locus is another soybean disease resistance locus that exhibits a relationship between gene content variation in NBS–LRR genes and disease resistance levels. *Rpp4* is one of few natural sources of resistance to Asian soybean rust. After the resistance locus was fine-mapped to a small space on chromosome 18, Meyer et al. discovered variation in the number of NBS–LRR type genes in this region (Meyer et al. 2009). Specifically, the susceptible reference type, Williams 82, had a cluster of three R-genes, while the resistant cultivar had five R-genes. Within this five gene cluster, *Rpp4C-4* was expressed in the resistant cultivar and the other members were nearly undetectable. Susceptible cultivars simply do not have this *Rpp4C-4* gene.

These four putatively functional SVs exemplify the complexity of this type of genetic polymorphism. Gene duplication, as in the case of *Rhg1*, might increase expression. Furthermore, a gene deletion will typically reduce/eliminate a gene's expression, as was the case for *Rpp4* and *Rsv1*. These examples are intuitive. However, a duplication could also initiate a feedback loop, thus reducing expression of a gene or its entire family, as in the case of *dCHS1* with yellow seed coat and hilum. In this case, a deletion might knock out a gene or gene family regulator and increase expression, as in the case of deleting part of *CHS4* resulting in increased chalcone synthase family expression and a black seed coat.

4.3 Genome Scans for SV

While the discovery of functional SVs has relied on fine-mapping of specific loci, some soybean researchers have used cytogenetics to identify large SV events and genomics methods to catalog SV events genomewide (reviewed by Chung and Singh 2008). Cytogeneticists have long been capable of documenting large chromosomal abnormalities. Microscopy-based studies can detect aneuploidy, polyploidization, large inversions, and rearrangements. In plants, the first cytological observations of these large-scale events were performed in maize (McClintock 1931). However, such observations tend to be limited by chromosomes amenable to visualization under a microscope and rearrangements large enough for visual detection.

Genome analysis platforms are now allowing SV detection at a much finer scale in a wider range of species. These studies often use array-based techniques, next-generation sequencing, or a combination of both. Comparative genome hybridization (CGH) arrays are the prevailing array-based technology for detecting SVs in soybean. This technique utilizes preset probes designed to bind a specific region of DNA. Probes can be designed at adjacent locations along each chromosome, allowing for a genomewide view of SVs. With this method, two genotypes can be labeled with separate fluorescent dyes and co-hybridized to the probe array, producing a comparative fluorescent readout that indicates the relative DNA copy abundance for each genotype at each probe location. Like all array technology, this method has background technical variation that can cause low signal-to-noise ratios for a subset of probes. Furthermore, probes designed to match a reference sequence will hybridize more efficiently to that sequence than a genotype containing substitutions or small indels in the probe-binding region. The number of probes developed, and therefore their spacing, limits the size of SV that can be detected (Gresham et al. 2008). Nonetheless, this technique has proved highly valuable in detecting SVs in a wide assortment of species including yeast (Dunham et al. 2002), humans (Iafrate et al. 2004; Sebat et al. 2004), and soybeans (Haun et al. 2011).

The other common genomewide SV scanning platform utilizes next-generation sequencing. Unlike CGH, every nucleotide of the genome could theoretically be assayed. There are four main approaches to detect SVs with whole-genome sequencing (Tattini et al. 2015). The most widespread technique is based on read depth variation (RDV), wherein sequence reads from a given genotype are mapped to a genome reference sequence and quantified as the number of reads mapped per genomic interval (e.g., reads per gene). RDV analysis would therefore predict that genomic regions in which few or no reads are mapped are putatively deleted, whereas regions with disproportionally high read-mapping coverage are duplicated. Generally, there is some necessary scaling to account for the non-normal distribution of read mapping.

In addition to RDV, data from paired-end reads can also be helpful in SV detection and characterization. Read pairs can orient SVs, such as detecting an inversion or determining whether a duplication is tandemly located or dispersed to a new location. Orientation and genomic location are something CGH simply cannot answer. Read pairs can also bridge deletion gaps, validating RDV-detected deletions.

Split read mapping, where part of a read is masked during mapping, can also be used to detect SVs and increase resolution down to a single base pair at a break point. CREST (Wang et al. 2011) and BreakDancer (Chen et al. 2009) are examples of read pair or split read-based algorithms used to detect SVs in soybean (Bolon et al. 2014; Qiu et al. 2014).

The final next-generation sequencing-based approach incorporates de novo assembly, requiring much higher levels of sequence coverage. Detecting structural variation through de novo assembly first requires the development of scaffolds and then aligning these to a previously assembled genome for comparison. SVs can then be assessed based on large-scale differences between the assembled scaffolds and the reference genome (Li et al. 2014). An alternative approach begins by mapping reads to the reference genome, and any unmapped reads are used to assemble scaffolds. This approach provides novel assemblies for genomic regions not found in the reference genome.

The use of next-generation sequencing also has limitations (Sims et al. 2014). Plant genomes generally have a high degree of repetitive elements, making read mapping unclear or inaccurate in many genomic regions. Mistakes or misassembles in the reference genome could accidentally be interpreted as a SV. Furthermore, nucleotides that do not exist in the reference genome but are present in other lines are difficult to incorporate and often ignored.

Long read sequencing technologies, now increasingly available, might alleviate some of the deficiencies of both CGH and next-generation sequencing. Current long read technologies now produce single reads many kilobases in length. Single-molecule sequencing, such as with PacBio, can be very helpful in reference genome assembly, particularly across highly repetitive regions (Huddleston et al. 2014). Sequencing across long genomic regions also greatly facilitates accurate SV detection (Wang et al. 2015a). While this technology is comparatively expensive and has a higher error rate in calling nucleotides (Wang et al. 2015a), techniques are being developed to account for and correct these errors, and the long read technology will undoubtedly appear in soybean SV publications in the near future.

4.4 Understanding the Limitations of a Reference Genome

The largest limiting factor for all of the aforementioned approaches is the biases associated with a single reference genome sequence. Improvements in the reference genome will help to anchor scaffolds, bridge gaps, and confirm orientation of segments. However, even a perfect reference genome assembly will not solve all of the problems. One issue is caused by genetic heterogeneity among individuals within any given soybean cultivar. Small amounts of residual intra-cultivar variation are not likely a problem for farmers, but can cause major issues in genomics. Intra-cultivar variation is primarily a consequence of the plant breeding process. Most soybean breeding strategies require only a limited number of single-seed descent generations following an initial cross, which is then followed by bulk harvesting for seed increase and evaluation. Any remaining heterozygous regions at the time of the last single-seed descent generation will be free to segregate and differentially fix among sub-lineages of the population (which will eventually become the cultivar). The soybean reference cultivar, Williams 82, has a number of documented regions of genomic variation among such sub-lineages (Haun et al. 2011). Documentation of the cultivar release specifies that the final inbred was a combination of four separate BC_6F_3 families (Bernard and Cremeens 1988). This variation is not specific to Williams 82. Nearly all soybean cultivars are likely to have some level of intra-cultivar variation. Additionally, mutation is an ongoing process, wherein novel substitutions and SVs continuously arise, resulting in slight differences between the individual plants of study and the reference genome (Ossowski et al. 2010).

Even if a perfectly inbred, mutation-free, reference genome was assembled, analysis platforms based on it would still not detect all forms of variation between genotypes. For example, a gene absent in the reference but present in a different cultivar would not be detected in most current genomewide scans. Of the soybean SV examples discussed, both *Rpp4* and *Rsv1* would not be detected based on strict comparisons of their source lines against the Williams 82 reference, as these genes do not exist in Williams 82. Studies of whole-genome biology are attempting to address this limitation by developing species-wide genome catalogs known as a pan-genome. The idea of a pan-genome comes from the bacterial community, where scientists detected gene content variation between isolates (Tettelin et al. 2005). A few key patterns arose. Firstly, a subset of genes is found in all isolates, termed the "core" genome. It is presumed that all (or nearly all) individuals in the entire species have these "core" genes. Alternatively, those genes found in some but not all individuals in a species were termed the "dispensable" genes.

These pan-genome ideas of "core" and "dispensable" genome components can be applied to any species exhibiting SVs, including plants (Morgante et al. 2007). It is tempting to consider the "core" genome as a list of the essential genes but this would be an over simplification. The core genome is defined by observing naturally occurring variation, and therefore genes conserved across the entire species range. However, from a molecular biology perspective, essential genes are generally those that are necessary for survival in a laboratory or specific controlled conditions. Therefore, essential genes are likely part of the core genome but all genes in the core might not be essential for plant survival under laboratory conditions (Klein et al. 2012).

Genes initially classified as "dispensable" might be beneficial under certain environmental conditions (Marroni et al. 2014). A specific R-gene, for example, might be necessary to survive under a certain disease pressure, but entirely dispensable in the absence of this pressure. Genetic redundancy might also be misclassified as dispensable. An example of this occurs in Arabidopsis where following a dispersed gene duplication, divergent evolution resulted in separate lineages each carrying only one functional copy of an essential gene (Bikard et al. 2009). Since not all individuals have a copy of this gene at the same location, this essential gene is considered dispensable.

A recent publication with de novo assembly of seven geographically diverse *Glycine soja* accessions is the first pan-genome analysis in soybean (Li et al. 2014). The authors estimated around 80% of the genome was present in all samples, making up the core genome. According to their results, thousands of genes in the pan-genome do not occur in the current *Glycine max* reference genome. Even with a limited sample size of seven genotypes, the *G. soja* pan-genome is estimated to contain 30.2 Mb more than any single individual's assembly (Li et al. 2014). Development of a complete soybean pan-genome would require de novo assembly of many more individuals. This level of assembly isn't currently feasible in most plant species. Instead, studies of SVs often focus only on gene space. Analyzing gene space can produce a genic pan-genome, or similarly a pan-transcriptome, surmised from transcriptome data. For example, transcriptome data from a wide array of individuals and a bulked tissue type was recently used to infer a maize pan-genome (Hirsch et al. 2014).

4.5 Diversity and Structural Variation Within and Between *Glycine soja* and *Glycine max*

The pan-genome study of *G. soja* is one of several publications that have assayed the genomic diversity in soybean's wild relative (Table 4.1). Researchers are interested in *G. soja* because modern soybean lost much of its genetic diversity in the domestication and improvement process (Hyten et al. 2006). As a close relative, *G. soja* has a similar genome to soybean making it amenable to crossing or genome scans for SVs.

The first SV observed between a *G. soja* and *G. max* comparison was an inversion (Ahmad et al. 1979). Since then, a number of additional inversions have been also detected within some *G. soja* individuals (Palmer et al. 2000; Kim et al. 2010b; Qiu et al. 2014). These inversions are segregating in *G.* soja and do not represent fixed differences between the species (Palmer et al. 2000). Inversions, like those observed, naturally maintained in the wild are often associated with adaptation to clinal variation or even speciation (Kirkpatrick and Barton 2006). Inversions are inherently negative due to the deleterious meiotic consequences of unequal crossing over. In order for an inversion to be maintained and at detectable frequencies, it must include a beneficial pair of alleles (Kirkpatrick 2010). If two beneficial alleles are included in an inversion, then recombination cannot separate them and the benefit of having both alleles outweighs the deleterious meiotic consequences. This concept has been discussed in the population genetics community and documented in other crop wild relatives (Fang et al. 2012). In addition to the previously discussed SV detection methods, inversions can also be discovered as regions of highly elevated linkage disequilibrium. Other factors also affect linkage disequilibrium, such as reduced recombination in soybean's large pericentromeric regions (Song et al. 2013), suggesting any putative inversions should be validated using an additional technique.

Translocations, though rare, have been discovered in soybean individuals (Mahama et al. 1999). Through a combination of mapping populations and cytology using fluorescence in situ hybridization, recent studies have characterized the chromosomes involved and approximate break points of the seven known translocation events in soybean (Findley et al. 2010, 2011). These individual events were derived from a range of backgrounds including: *G. soja, Glycine gracilis* (close relative of *G. soja* and *G. max*), fast neutron irradiated *G. max* populations, and a spontaneous translocation in a *G. max* cultivar cross. The presence of these large translocations in heterozygous individuals can produce a single chain or ring of multiple chromosomes pairing in meiosis potentially resulting in pollen or ovule sterility. One of these translocations, occurring frequently in *G. soja* accessions from northern China, might explain the occasional semi-sterility found in *G. soja* by *G. max* crosses (Findley et al. 2010). Smaller translocation events that could be detected through the use of paired-end or de novo sequence assembly and comparison likely also occur but have yet to be explored in soybean.

Recent genome scans of *G. soja* and *G. max* have detected widespread nucleotide content SVs within and between these species (Table 4.1). These include SVs segregating in *G. max* and *G. soja* as well as those only present in *G. soja* (Li et al. 2014; Zhou et al. 2015). Many of these studies were focused on detecting SNPs and indels then followed with a scan for SVs based on RDV. These re-sequencing studies often analyzed *G. soja* accessions, *G. max* landraces, and/or cultivars in order to detect QTLs related to the domestication or improvement process. More deletions are discovered than duplications, or other types of SVs, as these are most easily

Table 4.1 Genomewide SV genotyping studies in soybean and *G. soja* (years 2010–2015)

SV detection method	Publication	*G. soja*/landraces/elite lines in SV scan	Coverage depth	Deleted	Duplicated	Deleted and duplicated[a]	Novel[b]
RDV and de novo unmapped	Kim et al. (2010b)	1/0/0	43×	32.4 Mb	–	–	8.3 Mb
De novo	Lam et al. (2010)	1/0/0	80×	856 genes	–	–	–
CGH and RDV	McHale et al. (2012)	0/0/4	–	672 SV genes		–	
RDV	Li et al. (2013)	8/8/9 plus previous data	3.38×	22.3 Mb	–	–	–
RDV and de novo unmapped	Chung et al. (2014)	6/4/6	>14×	1737 genes	–	–	343 genes with plant homologues
De novo	Qiu et al. (2014)	1/0/0	55×	–	–	–	10 Mb
BreakDancer		0/1/0	41×	8.7 Mb	–	–	–
CGH and RVD	Anderson et al. (2014)	0/0/41	>2×	1200 genes	223 genes	105 genes	–
De novo pan-genome	Li et al. (2014)	7/0/0	>83	1179 genes	726 genes	73 genes	2.3–3.9 Mb/line
RDV	Zhou et al. (2015)	62/130/110	>11×	73.6 Mb	15.14 Mb	–	–

[a]'Deleted and Duplicated' refers to genes found deleted in a subset of individuals and duplicated in other individuals in a single study
[b]The column "Novel" refers to genomic sequence or genes not included in the *Glycine max* reference genome

detected with RDV or CGH. Some disparity between these studies is linked to the number of genotypes, the depth of sequencing, and the parameters used. Based on this collection of diverse studies, in soybean elite lines, landraces, and wild relatives SVs effect up to nearly 10% of the genome and around 3% of genes. Similarly, an Arabidopsis study involving 80 lines observed RDV in around 2% of the reference genome (Cao et al. 2011). The rates of SVs in maize are much higher with estimates up to 30% or more (Chia et al. 2012).

4.6 Evolving R-gene Clusters

SVs often occurs in genes functionally annotated as biotic stress response (McHale et al. 2012; Li et al. 2014; Anderson et al. 2014). One major family is the NBS–LRR genes, the same family responsible for most of the cloned disease resistance genes in plants (Dangl and Jones 2001; McHale et al. 2006). As demonstrated in Fig. 4.2, these R-genes tend to occur in clusters. These clusters average nearly five NBS–LRR genes per locus (Shao et al. 2014). R-gene clustering is a pattern occurring in a wide variety of plant species studied (Michelmore and Meyers 1998) that develops from tandem and segmental duplication (Leister 2004). The locally repetitive structure of R-gene clusters can lead to additional SVs through gene conversion and unequal crossing over. Rapid changes in disease-resistant gene content, especially in these gene clusters, are likely an important component to evolving disease resistance (Michelmore and Meyers 1998).

NBS–LRR type R-genes generally act to directly or indirectly recognize pathogen effector proteins and trigger a defense response (Jones and Dangl 2006). This gene-for-gene interaction model between plant and pathogen results in a constantly evolving arms race (Flor 1971; Takken and Rep 2010; Ravensdale et al. 2011). Genomic studies of plant pathogens, such as *Phytophthora sojae,* have discovered SVs in their avirulence genes as well (Qutob et al. 2009). In this evolutionary arms race, gene deletion and duplication appear to be an important evolutionary mechanism for both plants and pathogens. One might ask why R-gene clusters aren't constantly expanding to defend against all pathogens. In the presence of a pathogen, a specific R-gene might be essential, but without this selective force the gene may be dispensable or even have a fitness costs (Tian et al. 2003; Bomblies and Weigel 2007). To explore this hypothesized fitness cost, one group created a pair of near-isogenic lines in Arabidopsis where the only variant was the gene responsible for resistance. Under non-inoculated conditions, the genotype without the R-gene was found to yield 9% more (Tian et al. 2003).

4.7 Evolutionary Dynamics of Structural Variation

Gene duplications are less frequently discovered than gene deletions in genome scans, but their implications are no less significant. Gene duplication has long been implicated as the route to new function (Ohno 1970). Initially, gene duplication simply results in genetic redundancy. This idea of redundancy suggests both of these paralogs are contributing to the same gene function. While potentially beneficial (Liu et al. 2008), this redundancy is often unnecessary and potentially disruptive leading to silencing, subfunctionalization, or neofunctionalization over time (Lynch and Conery 2000). Subfunctionalization is when both members of the pair accumulate degenerative mutations to a point where combined they serve the same function as the ancestral gene (Lynch and Force 2000). Neofunctionalization is when one of the duplicated pair develops into a role unrelated to its previous evolutionary function. New gene duplications arise at higher rates than nucleotide substitution rates, but these new events are often under purifying selection (Katju and Bergthorsson 2013). This was evidenced in soybean by the excess of rare variants found in the frequency distribution of duplications in the soybean nested association mapping parents (SoyNAM) (Anderson et al. 2014). Studies in other species have also noted this purifying selection (Epstein et al. 2014). If the duplicated genes affect the stoichiometry in a biochemical pathway, then increased dosage can be deleterious, as suggested in the gene balance hypothesis (Birchler and Veitia 2012). A more thorough discussion of the population genetic implications of gene duplication is reviewed in Katju and Bergthorsson (2013).

Deletions, unlike duplications under purifying selection, are often neutral. For example, the frequency of deletions found in the SoyNAM parents resembles a simulated neutral model (Anderson et al. 2014). This finding is a bit counterintuitive. On the surface, it suggests genes can be lost without negative consequences. While certainly not true for all genes, genetic redundancy might imply many of these copies aren't necessary. Deletions removing pseudogenes, annotated as genes, would also be inconsequential. Current studies of fast neutron mutagenesis lines suggest large chromosomal regions can be deleted and still result in wild-type looking soybean plants (Bolon et al. 2011, 2014).

4.8 Whole-Genome Duplication

The history of WGD is an important factor when discussing SVs in soybean. All plant species, and many other organisms, are now believed to have undergone WGD at least once in their history. As mentioned earlier, the soybean genome has evidence of recent WGDs occurring approximately 59 and 13 mya (Schmutz et al. 2010). In the legume family, many individuals share the more ancient event, while the 13 mya event is exclusive to the genus *Glycine* (Shoemaker et al. 2006;

Severin et al. 2011). One of these WGD events appears to be an allopolyploidization, as evidenced by two types of centromeric repeats (Gill et al. 2009). Since WGD, the soybean genome has gone through a process of unbiased fractionation, where genes present and expressed today are relatively equally derived from both ancestral genomes (Garsmeur et al. 2014). Through fractionation and diploidization, the soybean genome still maintains 60–70% of its genes as paralogs (Schmutz et al. 2010; Anderson et al. 2014). One might expect these gene duplicates act as a buffer of genetic redundancy. If this were the case, there should be an enrichment for SVs in the duplicated subset of genes among soybean genotypes, but instead the opposite is observed (Anderson et al. 2014). SVs, and especially deletions, in the SoyNAM parents were less likely to overlap WGD-derived paralogs. This same pattern is found in mammals and other vertebrates, where SVs infrequently overlap preserved WGD-derived paralogs (Makino et al. 2013). Subfunctionalization, observed in many of these duplicated genes (Roulin et al. 2013), wherein both copies are now necessary, is one possible explanation for the preferential maintenance. The gene balance hypothesis also suggests SVs in WGD-derived paralogs would be deleterious because they affect the stoichiometry of biochemical pathways (Birchler and Veitia 2012). Therefore, the apparent genetic redundancy found in the soybean genome does not necessarily imply full functional redundancy.

4.9 Mapping Using SVs as a Marker

The number and dispersion of SVs makes these polymorphisms also useful as markers in mapping experiments (Wang et al. 2014; Shen et al. 2015). This was recently implemented in a soybean domestication and improvement study (Zhou et al. 2015). Using re-sequencing data, the authors called both SNPs and RDVs in 302 phenotyped lines. With these markers, they scanned for signals of selection during domestication and improvement and conducted genomewide association study (GWAS) on a number of different phenotypes. The use of SVs in GWAS was successful at detecting previously known functional structural variants. When assaying for seed coat color, they detected a strong signal on chromosome 8, where variation in the chalcone synthase family is known to affect seed hilum color. When assessing soybean cyst nematode resistance, they associated major resistance with the *Rhg1* SV locus on chromosome 18. Interestingly, a GWAS for plant height with SVs detected four significant loci on chromosome 12, including one overlapping a strong selection signal during domestication. Incorporating SVs with SNPs in association mapping can improve resolution and even aid in the detection of causative genetic variants for complex phenotypes (Stranger et al. 2007).

4.10 Inducing SVs

Mutagenic irradiation, such as fast neutrons (FN) or X-rays, can induce large-scale SVs and unique phenotypes. Using CGH and next-generation sequencing, sufficiently large SVs can be easily detected in mutants. In order to develop novel soybean phenotypes for breeding and gene function applications, the soybean community has recently developed two large FN irradiated populations. The resulting unique phenotypes and detected SVs for a subset of mutant lines are now publicly available on Soybase.org/mutants. FN-induced SVs have been associated with a number of interesting traits, including hyper-nodulation (Men et al. 2002), dwarfism (Hwang et al. 2014), seed protein and oil content, and short petioles (Bolon et al. 2011, 2014). Associating detected SVs with the unique phenotypes in these populations is an ongoing process. This FN mutant population database also serves as a community resource for reverse genetic studies.

The patterns of SVs induced by FN mutagenesis suggest a highly malleable soybean genome (Bolon et al. 2014). Mutagenesis-induced SVs can affect many more genes per locus than the SVs observed in diverse germplasm scans of natural variation. Among 264 FN mutant lines assayed to date (Bolon et al. 2014), more than

40% of the soybean genes have been identified within at least one duplicated segment, 9% of the genes have been found within at least one homozygous deletion, and 19% have been found within at least one hemizygous deletion. Much like the SVs observed in the SoyNAM, FN-induced SVs were enriched for genes without a retained paralog from the last WGD. These findings further enforce the WGD-derived paralogs might play essential roles in biological processes.

The advent of genome editing technologies makes targeted SV induction possible (Voytas 2013). Zinc Finger Nucleases (ZFN), TALENs, and CRISPR/Cas9 have all been demonstrated to work in soybean (Curtin et al. 2011; Haun et al. 2014; Jacobs et al. 2015; Michno et al. 2015). Using one of these technologies to simultaneously target two separate loci on one chromosome can induce a SV. In Arabidopsis, simultaneous ZFNs induced large deletions and inversions (Qi et al. 2013), and in rice large deletions were induced with the CRISPR/Cas9 system (Zhou et al. 2014). The recent publication of successful gene editing in soybean (Li et al. 2015) further expands the potential for novel gene insertion, arrangement, or other SVs. These new technologies will likely be instrumental tools in future studies of gene function, genome evolution, and the development of new phenotypic traits.

4.11 Conclusions

Advancements in genome scanning technologies have facilitated the accurate and precise detection of SVs in plants genomes. Recognizing where SVs occur and how they are maintained will continue to improve our understanding of their role in adaptation and crop improvement. Structural variation is found throughout the soybean genome, exhibiting substantial enrichment in biotic defense response genes. Future research will likely uncover additional functional SVs underlying important phenotypic traits. The accelerated use and advancement of next-generation sequencing platforms is rapidly enabling this process. Since the initial submission of this manuscript in December 2015, there have been several new publications that have used re-sequencing data to investigate genomewide structural variation in soybean. This includes re-sequencing of lines representing standing variation in the soybean germplasm and lines representing variation associated with genetic transformation and mutagenesis (e.g., Anderson et al. 2016; Maldonado dos Santos et al. 2016; Valliyodan et al. 2016). Undoubtedly, numerous other such studies will be forthcoming, promising higher resolution/detection of SVs and identifying connections to phenotypic variation.

Tapping into the currently underexplored genetic diversity in the soybean germplasm is important in the search for agronomically essential traits. Furthermore, inducing a SV de novo through traditional or biotechnology-aided mutagenesis will be useful for generating novel phenotypic variation to enable mutation breeding and studies of gene function, genome evolution, and the limits of the soybean genome.

Acknowledgements The authors are grateful to Candice Hirsch, Peter Morrell, Chad Myers, and Nevin Young for reviewing the manuscript and contributing helpful suggestions. The authors are also grateful to the US Department of Agriculture, United Soybean Board, and the Minnesota Soybean Research and Promotion Council for project support. We apologize if any relevant publications were overlooked in this review.

Conflict of interest The authors declare that they do not have any conflict of interest.

References

Ahmad QN, Britten EJ, Byth DE (1979) Inversion heterozygosity in the hybrid soybean × *Glycine soja*: evidence from a pachytene loop configuration and other meiotic irregularities. J Hered 70:358–364

Anderson JE, Kantar MB, Kono TY, Fu F, Stec AO et al (2014) A roadmap for functional structural variants in the soybean genome. Genes Genomes Genet 4:1307–1318

Anderson JE, Michno JM, Kono TJ, Stec AO, Campbell BW, Curtin SJ, Stupar RM (2016) Genomic variation and DNA repair associated with soybean transgenesis: a comparison to cultivars and mutagenized plants. BMC Biotechnol 16:41

Ashfield T, Bocian A, Held D, Henk AD, Marek LF et al (2007) Genetic and physical localization of the soybean Rpg1-b disease resistance gene reveals a complex locus containing several tightly linked families of NBS–LRR genes. Mol Plant Microbe Interact 16:817–826

Bernard RL, Cremeens CR (1988) Registration of "Williams 82" soybean. Crop Sci 28:1027–1028

Bikard D, Patel D, Le Metté C, Giorgi V, Camilleri C et al (2009) Divergent evolution of duplicate genes leads to genetic incompatibilities within *A. thaliana*. Science 323:623–626

Birchler JA, Veitia RA (2012) Gene balance hypothesis: connecting issues of dosage sensitivity across biological disciplines. Proc Natl Acad Sci USA 109:14746–14753

Bolon YT, Haun WJ, Xu WW, Grant D, Stacey MG et al (2011) Phenotypic and genomic analyses of a fast neutron mutant population resource in soybean. Plant Physiol 156:240–253

Bolon YT, Stec AO, Michno JM, Roessler J, Bhaskar PB et al (2014) Genome resilience and prevalence of segmental duplications following fast neutron irradiation of soybean. Genetics 198:967–981

Bomblies K, Weigel D (2007) Hybrid necrosis: autoimmunity as a potential gene-flow barrier in plant species. Nat Rev Genet 8:382–393

Boocock J, Chagné D, Merriman TR, Black MA (2015) The distribution and impact of common copy-number variation in the genome of the domesticated apple, *Malus* × *domestica* Borkh. BMC Genom 16:848

Cao J, Schneeberger K, Ossowski S, Günther T, Bender S et al (2011) Whole-genome sequencing of multiple *Arabidopsis thaliana* populations. Nat Genet 43:956–963

Chen K, Wallis JW, McLellan MD, Larson DE, Kalicki JM et al (2009) BreakDancer: an algorithm for high-resolution mapping of genomic structural variation. Nat Methods 6:677–681

Chia JM, Song C, Bradbury PJ, Costich D, de Leon N et al (2012) Maize HapMap2 identifies extant variation from a genome in flux. Nat Genet 44:803–807

Chung G, Singh RJ (2008) Broadening the genetic base of soybean: a multidisciplinary approach. Crit Rev Plant Sci 27:295–341

!Chung WH, Jeong N, Kim J, Lee WK, Lee YG et al (2014) Population structure and domestication revealed by high-depth resequencing of korean cultivated and wild soybean genomes. DNA Res 21:153–167

Cook DE, Lee TG, Guo X, Melito S, Wang K et al (2012) Copy number variation of multiple genes at Rhg1 mediates nematode resistance in soybean. Science 338:1206–1209

Cook DE, Bayless AM, Wang K, Guo X, Song Q et al (2014) Distinct copy number, coding sequence, and locus methylation patterns underlie Rhg1-mediated soybean resistance to soybean cyst nematode. Plant Physiol 165:630–647

Curtin SJ, Zhang F, Sander JD, Haun WJ, Starker C et al (2011) Targeted mutagenesis of duplicated genes in soybean with zinc-finger nucleases. Plant Physiol 156:466–473

Dangl JL, Jones JD (2001) Plant pathogens and integrated defence responses to infection. Nature 411:826–833

Díaz A, Zikhali M, Turner AS, Isaac P, Laurie DA (2012) Copy number variation affecting the Photoperiod-B1 and Vernalization-A1 genes is associated with altered flowering time in wheat (*Triticum aestivum*). PLoS ONE 7:e33234

Dunham MJ, Badrane H, Ferea T, Adams J, Brown PO et al (2002) Characteristic genome rearrangements in experimental evolution of *Saccharomyces cerevisiae*. Proc Natl Acad Sci USA 99:16144–16149

Epstein B, Sadowsky MJ, Tiffin P (2014) Selection on horizontally transferred and duplicated genes in sinorhizobium (ensifer), the root-nodule symbionts of medicago. Genome Biol Evol 6:1199–1209

Fang Z, Pyhajarvi T, Weber AL, Dawe RK, Glaubitz JC et al (2012) Megabase-scale inversion polymorphism in the wild ancestor of maize. Genetics 191:883–894

Findley SD, Cannon S, Varala K, Du J, Ma J, Hudson ME, Birchler JA, Stacey G (2010) A fluorescence in situ hybridization system for karyotyping soybean. Genetics 185:727–744

Findley SD, Pappas AL, Cui Y, Birchler JA, Palmer RG, Stacey G (2011) Fluorescence in situ hybridization-based karyotyping of soybean translocation lines. Genes Genomes Genet 1:117–129

Flor HH (1971) Current status of the gene-for-gene concept. Annu Rev Phytopathol 9:275–296

Gaines TA, Zhang W, Wang D, Bukun B, Chisholm ST et al (2010) Gene amplification confers glyphosate resistance in Amaranthus palmeri. Proc Natl Acad Sci USA 107:1029–1034

Gaines TA, Shaner DL, Ward SM, Leach JE, Preston C et al (2011) Mechanism of resistance of evolved glyphosate-resistant *Palmer amaranth* (*Amaranthus palmeri*). J Agric Food Chem 59:5886–5889

Garsmeur O, Schnable JC, Almeida A, Jourda C, D'Hont A et al (2014) Two evolutionarily distinct classes of paleopolyploidy. Mol Biol Evol 31:448–454

Gill N, Findley S, Walling JG, Hans C, Ma J et al (2009) Molecular and chromosomal evidence for allopolyploidy in soybean. Plant Physiol 151:1167–1174

Gordon SG, St. Martin SK, Dorrance AE (2006) Rps8 maps to a resistance gene rich region on soybean molecular linkage group F. Crop Sci 46:168–173

Gresham D, Dunham MJ, Botstein D (2008) Comparing whole genomes using DNA microarrays. Nat Rev Genet 9:291–302

Gu W, Zhang F, Lupski JR (2008) Mechanisms for human genomic rearrangements. Pathogenetics 1:4

Han JJ, Jackson D, Martienssen R (2012) Pod corn is caused by rearrangement at the *Tunicate1 locus*. Plant Cell 24:2733–2744

Hastings PJ, Lupski JR, Rosenberg SM, Ira G (2009) Mechanisms of change in gene copy number. Nat Rev Genet 10:551–564

Haun WJ, Hyten DL, Xu WW, Gerhardt DJ, Albert TJ et al (2011) The composition and origins of genomic variation among individuals of the soybean reference cultivar Williams 82. Plant Physiol 155:645–655

Haun W, Coffman A, Clasen BM, Demorest ZL, Lowy A et al (2014) Improved soybean oil quality by targeted mutagenesis of the fatty acid desaturase 2 gene family. Plant Biotechnol J 12:934–940

Hayes AJ, Jeong SC, Gore MA, Yu YG, Buss GR et al (2004) Recombination within a nucleotide-binding-site/leucine-rich-repeat gene cluster produces new variants conditioning resistance to soybean mosaic virus in soybeans. Genetics 166:493–503

Hirsch CN, Foerster JM, Johnson JM, Sekhon RS, Muttoni G et al (2014) Insights into the maize pan-genome and pan-transcriptome. Plant Cell 26:121–135

Huddleston J, Ranade S, Malig M, Antonacci F, Chaisson M et al (2014) Reconstructing complex regions of genomes using long-read sequencing technology. Genome Res 24:688–696

Hwang WJ, Kim MY, Kang YJ, Shim S, Stacey MG et al (2014) Genome-wide analysis of mutations in a dwarf soybean mutant induced by fast neutron bombardment. Euphytica 203:399–408

Hyten DL, Song Q, Zhu Y, Choi IY, Nelson RL et al (2006) Impacts of genetic bottlenecks on soybean genome diversity. Proc Natl Acad Sci USA 103:16666–16671

Hyten DL, Cannon SB, Song Q, Weeks N, Fickus EW et al (2010) High-throughput SNP discovery through deep resequencing of a reduced representation library to anchor and orient scaffolds in the soybean whole genome sequence. BMC Genom 11:38

Iafrate AJ, Feuk L, Rivera MN, Listewnik ML, Donahoe PK et al (2004) Detection of large-scale variation in the human genome. Nat Genet 36:949–951

Iovene M, Zhang T, Lou Q, Buell CR, Jiang J (2013) Copy number variation in potato—an asexually propagated autotetraploid species. Plant J 75:80–89

Jacobs TB, LaFayette PR, Schmitz RJ, Parrott WA (2015) Targeted genome modifications in soybean with CRISPR/Cas9. BMC Biotechnol 15:16

Jones JDG, Dangl JL (2006) The plant immune system. Nature 444:323–329

Katju V, Bergthorsson U (2013) Copy-number changes in evolution: rates, fitness effects and adaptive significance. Front Genet 4:273

Kim KS, Hill CB, Hartman GL, Hyten DL, Hudson ME et al (2010a) Fine mapping of the soybean aphid-resistance gene Rag2 in soybean PI 200538. Theor Appl Genet 121:599–610

Kim MY, Lee S, Van K, Kim TH, Jeong SC et al (2010b) Whole-genome sequencing and intensive analysis of the undomesticated soybean (*Glycine soja* Sieb. and Zucc.) genome. Proc Natl Acad Sci USA 107:22032–22037

Kirkpatrick M (2010) How and why chromosome inversions evolve. PLoS Biol 8:e1000501

Kirkpatrick M, Barton N (2006) Chromosome inversions, local adaptation and speciation. Genetics 173:419–434

Klein BA, Tenorio EL, Lazinski DW, Camilli A, Duncan MJ et al (2012) Identification of essential genes of the periodontal pathogen Porphyromonas gingivalis. BMC Genom 13:578

Knox AK, Dhillon T, Cheng H, Tondelli A, Pecchioni N et al (2010) CBF gene copy number variation at Frost Resistance-2 is associated with levels of freezing tolerance in temperate-climate cereals. Theor Appl Genet 121:21–35

Kyndt T, Quispe D, Zhai H, Jarret R, Ghislain M et al (2015) The genome of cultivated sweet potato contains Agrobacterium T-DNAs with expressed genes: an example of a naturally transgenic food crop. Proc Natl Acad Sci USA 112:5844–5849

Lam HM, Xu X, Liu X, Chen W, Yang G et al (2010) Resequencing of 31 wild and cultivated soybean genomes identifies patterns of genetic diversity and selection. Nat Genet 42:1053–1059

Leister D (2004) Tandem and segmental gene duplication and recombination in the evolution of plant disease resistance genes. Trends Genet 20:116–122

Li Y, Xiao J, Wu J, Duan J, Liu Y et al (2012) A tandem segmental duplication (TSD) in green revolution gene Rht-D1b region underlies plant height variation. New Phytol 196:282–291

Li Y, Zhao S, Ma J, Li D, Yan L et al (2013) Molecular footprints of domestication and improvement in soybean revealed by whole genome re-sequencing. BMC Genom 14:579

Li Y, Li G, Zhou J, Ma W, Jiang L et al (2014) De novo assembly of soybean wild relatives for pan-genome analysis of diversity and agronomic traits. Nat Biotechnol 32:1045–1052

Li Z, Liu ZB, Xing A, Moon BP, Koellhoffer JP et al (2015) Cas9-guide RNA directed genome editing in soybean. Plant Physiol 169:960–970

Lisch D (2013) How important are transposons for plant evolution? Nat Rev Genet 14:49–61

Liu B, Kanazawa A, Matsumura H, Takahashi R, Harada K et al (2008) Genetic redundancy in soybean photoresponses associated with duplication of the phytochrome A gene. Genetics 180:995–1007

Lynch M, Conery JS (2000) The evolutionary fate and consequences of duplicate genes. Science 290:1151–1155

Lynch M, Force A (2000) The probability of duplicate gene preservation by subfunctionalization. Genetics 154:459–473

Mahama AA, Deaderick LM, Sadanaga K, Newhouse KE, Palmer RG (1999) Cytogenetic analysis of translocations in soybean. J Hered 90:648–653

Makino T, McLysaght A, Kawata M (2013) Genome-wide deserts for copy number variation in vertebrates. Nat Commun 4:2283

Maldonado dos Santos JV, Valliyodan B, Joshi T, Khan SM, Liu Y et al (2016) Evaluation of genetic variation among Brazilian soybean cultivars through genome resequencing. BMC Genom 17:110

Maron LG, Guimaraes CT, Kirst M, Albert PS, Birchler JA et al (2013) Aluminum tolerance in maize is associated with higher MATE1 gene copy number. Proc Natl Acad Sci USA 110:5241–5246

Marroni F, Pinosio S, Morgante M (2014) Structural variation and genome complexity: is dispensable really dispensable? Curr Opin Plant Biol 18:31–36

McClintock B (1931) Cytological observations of deficiencies involving known genes, translocations and an inversion in *Zea mays*. Mo Agric Exp Stn Res Bull 163:1–30

McHale L, Tan X, Koehl P, Michelmore RW (2006) Plant NBS–LRR proteins: adaptable guards. Genome Biol 7:212

McHale LK, Haun WJ, Xu WW, Bhaskar PB, Anderson JE et al (2012) Structural variants in the soybean genome localize to clusters of biotic stress-response genes. Plant Physiol 159:1295–1308

Men AE, Laniya TS, Searle IR, Iturbe-Ormaetxe I, Gresshoff I et al (2002) Fast neutron mutagenesis of soybean (*Glycine soja* L.) produces a supernodulating mutant containing a large deletion in linkage group H. Genome Lett 1:147–155

Meyer JD, Silva DC, Yang C, Pedley KF, Zhang C et al (2009) Identification and analyses of candidate genes for rpp4-mediated resistance to Asian soybean rust in soybean. Plant Physiol 150:295–307

Michelmore RW, Meyers BC (1998) Clusters of resistance genes in plants evolve by divergent selection and a birth-and-death process. Genome Res 8:1113–1130

Michno JM, Wang X, Liu J, Curtin SJ, Kono TJ, Stupar RM (2015) CRISPR/Cas mutagenesis of soybean and *Medicago truncatula* using a new web-tool and a modified Cas9 enzyme. GM Crops Food 6:243–252

Morgante M, De Paoli E, Radovic S (2007) Transposable elements and the plant pan-genomes. Curr Opin Plant Biol 10:149–155

Munoz-Amatriain M, Eichten SR, Wicker T, Richmond TA, Mascher M et al (2013) Distribution, functional impact, and origin mechanisms of copy number variation in the barley genome. Genome Biol 14:R58

Ohno S (1970) Evolution by gene duplication. Springer, New York

Ossowski S, Schneeberger K, Lucas-Lledo JI, Warthmann N, Clark RM et al (2010) The rate and molecular spectrum of spontaneous mutations in *Arabidopsis thaliana*. Science 327:92–94

Palmer RG, Sun H, Zhao LM (2000) Genetics and cytology of chromosome inversions in soybean germplasm. Crop Sci 40:683–687

Pearce S, Saville R, Vaughan SP, Chandler PM, Wilhelm EP et al (2011) Molecular characterization of Rht-1 dwarfing genes in hexaploid wheat. Plant Physiol 157:1820–1831

Qi Y, Li X, Zhang Y, Starker CG, Baltes NJ et al (2013) Targeted deletion and inversion of tandemly arrayed genes in *Arabidopsis thaliana* using zinc finger nucleases. Genes Genomes Genet 3:1707–1715

Qiu J, Wang Y, Wu S, Wang YY, Ye CY et al (2014) Genome re-sequencing of semi-wild soybean reveals a complex Soja population structure and deep introgression. PLoS ONE 9:e108479

Qutob D, Tedman-Jones J, Dong S, Kuflu K, Pham H et al (2009) Copy number variation and transcriptional polymorphisms of *Phytophthora sojae* RXLR effector genes Avr1a and Avr3a. PLoS ONE 4:e5066

Ravensdale M, Nemri A, Thrall PH, Ellis JG, Dodds PN (2011) Co-evolutionary interactions between host resistance and pathogen effector genes in flax rust disease. Mol Plant Pathol 12:93–102

Rodin SN, Riggs AD (2003) Epigenetic silencing may aid evolution by gene duplication. J Mol Evol 56:718–729

Roulin A, Auer PL, Libault M, Schlueter J, Farmer A et al (2013) The fate of duplicated genes in a polyploid plant genome. Plant J 73:143–153

Santuari L, Pradervand S, Amiguet-Vercher AM, Thomas J, Dorcey E et al (2010) Substantial deletion overlap among divergent Arabidopsis genomes revealed by intersection of short reads and tiling arrays. Genome Biol 11:1–8

Schatz MC, Maron LG, Stein JC, Hernandez Wences A, Gurtowski J et al (2014) Whole genome de novo assemblies of three divergent strains of rice, *Oryza sativa*, document novel gene space of aus and indica. Genome Biol 15:506

Schmutz J, Cannon SB, Schlueter J, Ma J, Mitros T et al (2010) Genome sequence of the palaeopolyploid soybean. Nature 463:178–183

Sebat J, Lakshmi B, Troge J, Alexander J, Young J et al (2004) Large-scale copy number polymorphism in the human genome. Science 305:525–528

Severin AJ, Cannon SB, Graham MM, Grant D, Shoemaker RC (2011) Changes in twelve homoeologous genomic regions in soybean following three rounds of polyploidy. Plant Cell 23:3129–3136

Shao ZQ, Zhang YM, Hang YY, Xue JY, Zhou GC et al (2014) Long-term evolution of nucleotide-binding site-leucine-rich repeat (NBS–LRR) genes: understandings gained from and beyond the legume family. Plant Physiol 166:217–234

Shen X, Liu ZQ, Mocoeur A, Xia Y, Jing HC (2015) PAV markers in Sorghum bicolour: genome pattern, affected genes and pathways, and genetic linkage map construction. Theor Appl Genet 128:623–637

Shoemaker RC, Schlueter J, Doyle JJ (2006) Paleopolyploidy and gene duplication in soybean and other legumes. Curr Opin Plant Biol 9:104–109

Sims D, Sudbery I, Ilott NE, Heger A, Ponting CP (2014) Sequencing depth and coverage: key considerations in genomic analyses. Nat Rev Genet 15:121–132

Song Q, Hyten DL, Jia G, Quigley CV, Fickus EW et al (2013) Development and evaluation of SoySNP50 K, a high-density genotyping array for soybean. PLoS ONE 8:e54985

Stranger BE, Forrest MS, Dunning M, Ingle CE, Beazley C et al (2007) Relative impact of nucleotide and copy number variation on gene expression phenotypes. Science 315:848–853

Sutton T, Baumann U, Hayes J, Collins NC, Shi BJ et al (2007) Boron-toxicity tolerance in barley arising from efflux transporter amplification. Science 318:1446–1449

Swanson-Wagner RA, Eichten SR, Kumari S, Tiffin P, Stein JC et al (2010) Pervasive gene content variation and copy number variation in maize and its undomesticated progenitor. Genome Res 20:1689–1699

Takken F, Rep M (2010) The arms race between tomato and *Fusarium oxysporum*. Mol Plant Pathol 11:309–314

Tattini L, D'Aurizio R, Magi A (2015) Detection of genomic structural variants from next-generation sequencing data. Front Bioeng Biotechnol 3:92

Tettelin H, Masignani V, Cieslewicz MJ, Donati C, Medini D et al (2005) Genome analysis of multiple pathogenic isolates of *Streptococcus agalactiae*: implications for the microbial "pan-genome". Proc Natl Acad Sci USA 102:13950–13955

Tian D, Traw MB, Chen JQ, Kreitman M, Bergelson J (2003) Fitness costs of R-gene-mediated resistance in *Arabidopsis thaliana*. Nature 423:74–77

Todd JJ, Vodkin LO (1996) Duplications that suppress and deletions that restore expression from a chalcone synthase multigene family. Plant Cell 8:687–699

Tuteja JH, Vodkin LO (2008) Structural features of the endogenous CHS silencing and target loci in the soybean genome. Crop Sci 48:S49–S68

Tuteja JH, Zabala G, Varala K, Hudson M, Vodkin LO (2009) Endogenous, tissue-specific short interfering RNAs silence the chalcone synthase gene family in glycine max seed coats. Plant Cell 21:3063–3077

Valliyodan B, Qiu D, Patil G, Zeng P, Huang J et al (2016) Landscape of genomic diversity and trait discovery in soybean. Sci Rep 6:23598

Voytas DF (2013) Plant genome engineering with sequence-specific nucleases. Annu Rev Plant Biol 64:327–350

Wang J, Mullighan CG, Easton J, Roberts S, Heatley SL et al (2011) CREST maps somatic structural variation in cancer genomes with base-pair resolution. Nat Methods 8:652–654

Wang Y, Lu J, Chen S, Shu L, Palmer RG et al (2014) Exploration of presence/absence variation and corresponding polymorphic markers in soybean genome. J Integr Plant Biol 56:1009–1019

Wang M, Beck CR, English AC, Meng Q, Buhay C et al (2015a) PacBio-LITS: a large-insert targeted sequencing method for characterization of human disease-associated chromosomal structural variations. BMC Genom 16:214

Wang Y, Xiong G, Hu J, Jiang L, Yu H et al (2015b) Copy number variation at the GL7 locus contributes to grain size diversity in rice. Nat Genet 47:944–948

Wingen LU, Munster T, Faigl W, Deleu W, Sommer H et al (2012) Molecular genetic basis of pod corn (*Tunicate maize*). Proc Natl Acad Sci USA 109:7115–7120

Wu X, Ren C, Joshi T, Vuong T, Xu D et al (2010) SNP discovery by high-throughput sequencing in soybean. BMC Genom 11:469

Xu K, Xu X, Fukao T, Canlas P, Maghirang-Rodriguez R et al (2006) Sub1A is an ethylene-response-factor-like gene that confers submergence tolerance to rice. Nature 442:705–708

Yao H, Zhou Q, Li J, Smith H, Yandeau M et al (2002) Molecular characterization of meiotic recombination across the 140-kb multigenic a1-sh2 interval of maize. Proc Natl Acad Sci USA 99:6157–6162

Yu P, Wang CH, Xu Q, Feng Y, Yuan XP et al (2013) Genome-wide copy number variations in *Oryza sativa* L. BMC Genom 14:649

Zhang C, Grosic S, Whitham SA, Hill JH (2012) The requirement of multiple defense genes in soybean Rsv1-mediated extreme resistance to Soybean mosaic virus. Mol Plant Microbe Interact 25:1307–1313

Zhang Z, Mao L, Chen H, Bu F, Li G et al (2015) Genome-wide mapping of structural variations reveals a copy number variant that determines reproductive morphology in cucumber. Plant Cell 27:1595–1604

Zheng LY, Guo XS, He B, Sun LJ, Peng Y et al (2011) Genome-wide patterns of genetic variation in sweet and grain sorghum (*Sorghum bicolor*). Genome Biol 12:R114

Zhou H, Liu B, Weeks DP, Spalding MH, Yang B (2014) Large chromosomal deletions and heritable small genetic changes induced by CRISPR/Cas9 in rice. Nucl Acids Res 42:10903–10914

Zhou Z, Jiang Y, Wang Z, Gou Z, Lyu J et al (2015) Resequencing 302 wild and cultivated accessions identifies genes related to domestication and improvement in soybean. Nat Biotechnol 33:408–414

Żmieńko A, Samelak A, Kozłowski P, Figlerowicz M (2014) Copy number polymorphism in plant genomes. Theor Appl Genet 127:1–18

Sequencing, Assembly, and Annotation of the Soybean Genome

5

Babu Valliyodan, Suk-Ha Lee and Henry T. Nguyen

Abstract

Genome sequencing yields an exceptional resource of genetic information. Knowledge of whole genome sequence information helps characterize individual genomes, transcriptional states and genetic variation in populations and provide genetic architecture associated with each trait. After the release of the first human genome assembly, other model organism assemblies became available; including the model plant *Arabidopsis thaliana*. The soybean community published the first reference genome of the variety Williams 82 in 2010. Soybean has important syntenic relationships with the other legume species and is a model plant for the legumes. In this chapter, we discuss about the soybean genome assemblies and annotations, and its fine-tuning in view of the next-generation sequencing technologies and bioinformatics tools. In addition, comparison of the structural variations between the cultivated reference genome with the available wild soybean genome information will be discussed. This is followed by the discussion on the opportunities of next-generation sequencing technologies and challenges that we anticipate on the development of more pangenomes and reference genomes for soybean. This will significantly affect the discovery of rare alleles associated with key agronomic and quality traits and shape up the next-generation breeding technologies and crop improvement.

5.1 Introduction

Genetic code is the basis of organismal life. Deciphering the primary DNA sequence, i.e. the genome, and the associated gene, has become a fundamental resource in biology. The speed of genome sequencing is higher in the case of mammalian and microbial genomes when compared to

B. Valliyodan (✉) · H.T. Nguyen
Division of Plant Sciences, University of Missouri, Columbia, MO 65211, USA
e-mail: valliyodanb@missouri.edu

S.-H. Lee
Department of Plant Science, Research Institute for Agriculture and Life Sciences and Plant Genomics and Breeding Institute, Seoul National University, Seoul 151-921, Korea

H.T. Nguyen and M.K. Bhattacharyya (eds.), *The Soybean Genome*, Compendium of Plant Genomes, DOI 10.1007/978-3-319-64198-0_5

plants. Advances in the next-generation sequencing technologies made the whole genome sequencing technically feasible and cost-effective. Application of genomics and the genome data show significant impact on biological investigations irrespective of areas and disciplines. Whole genome sequencing combined with de novo assemblies generates newer reference genomes for genetic investigation and trait discovery in each organism. State-of-the-art sequencing techniques and the associated computational processes provide better resolution to transcriptome, epigenome, and organellar genomes (Varshney et al. 2009). In addition, targeted sequencing including the exomes and regulomes bring major shift to the high-throughput trait diagnostics (Hashmi et al. 2015). A total of 24,002-genome information is available in the NCBI, which includes eukaryotes, prokaryotes, viruses, plasmids, and organelles (https://www.ncbi.nlm.nih.gov/genome/browse/).

The release of the Arabidopsis genome in 2000 (Arabidopsis Genome Consortium 2000) and the technological advances in sequencing methods accelerated the number of sequenced plant genomes closely to 185. Out of these, more than 50% of them are crop genomes and the rest belongs to model plant genomes, orphan crops, and wild relative species. Availability of this genome information shifted the genetic investigation to the next level of research, precision genomics. Also, collection of these genomes representing various species, genera, and classes in the plant kingdom help reveal the evolutionary relationships between plants and also provide better clarity of synteny and gene family evolutions. Above all, the crop-specific genome sequence and transcript information enhances the power of gene and functional marker discovery that will greatly impact the next-generation breeding programs including the genomics-assisted breeding and genomic section.

5.2 Genome Sequencing Platform

The first human genome (Venter et al. 2001; HGSC 2004) and other genome sequencing projects including model plants as well as major crop species such as rice, soybean, sorghum, maize, and grape mainly depended upon Sanger sequencing methods (Sanger et al. 1977), which are expensive and time consuming for complex genomes. Introduction of the next-generation sequencing technologies improved the output/cost ratio of genome sequencing dramatically. In other words, the pyrosequencing approaches were replaced to certain extent by the "sequencing by synthesis" approaches (Margulies et al. 2005; Shendure et al. 2005). At the earlier phase, this technology provided multiplexed DNA fragment sequencing, generated short reads with low-quality data, and after improving the basic chemistry, now it is possible to acquire high-quality, short-read DNA fragment sequences (Fuller et al. 2009). However, the complex genome assemblies using these comparatively short reads show major drawbacks, including assembly and determination of complex genomic regions, identification of gene isoform and other structural rearrangements.

Single-molecule, real-time sequencing technology developed by Pacific BioSciences offers longer read lengths and higher consensus accuracies than the short-read technologies, making it well suited for better characterization in genome, transcriptome, and epigenetics research (Eid et al. 2009; Nakano et al. 2017). The major requirement for the de novo reference assembly of genomes is less percentages of gaps and the contigs generated using the longer reads significantly improve the assemblies by closing gaps. In addition, it helps to better characterize the structural variations in the genomes (Utturkar et al. 2014; Rhoads and Au 2015). More recently, development of optical map helped to detect mismatches in assemblies, providing genomic maps with high resolution and to allow assemblies with more accuracy and completeness (Schwartz et al. 1993; Tang et al. 2015; Jiao et al. 2017a). In general, hybrid-sequencing strategies including short reads, long reads, and optical maps are more affordable and scalable for genomic investigations. These high-quality sequences improve the quality of contigs and scaffolds yielding several fold better genome assembly.

5.3 Plant Genomes and Assemblies

The development of reference genome assemblies is essential to identify the DNA sequence variations. Plant genome assembly is a challenging task and even more challenging than the animal genomes. Larger genome size, polyploidy, highly repetitive genomic regions, and genome duplication are the major challenges with most of the plant genomes. For example, the repetitive fraction of the human genome varies between 35 and 45%, whereas in maize, it is 64–73% (Imelfort and Edwards 2009). Even though there are more than 180 plant reference genomes, assembly of only less number of them are at the chromosome level (Schnable et al. 2009; Jiao et al. 2017b). Recent advances in the next-generation sequencing technologies, especially the long-read sequencing, long-range scaffolding, and chromosome capturing (Burton et al. 2013) helped to overcome the challenges in the chromosome-level assemblies of genomes.

Genomes sequenced by Sanger sequencing technique were initially assembled using the TIGR Assembler (Sutton et al. 1995) and Celera Assembler (Myers 1995). Advances in the genome sequencing technology required newer assembly tools as well. In the case of genome assembly using the short-read, de Bruijn graph assemblers (Pevzner et al. 2001) such as Velvet (Zerbino and Birney 2008) and ABySS (Simpson et al. 2009) were used. There are de novo sequence assembly tools available to generate good quality and robust assembly of complex genomes using short reads. These assemblers include SSAKE (Warren et al. 2007), VCAKE (Jeck et al. 2007), Edena (Hernandez et al. 2008), EulerSR (Chaisson and Pevzner 2008) and All-Paths (Butler et al. 2008). Recently, a new assembler called Supernova (Weisenfeld et al. 2017) and new version of ABySS assembler explores the linked reads and optical mapping for improved scaffolding.

A combination of single-molecule sequencing with complementary technologies has also become a common strategy. Sequencing technologies such as PacBio, Nanopore, and optical mapping produce longer reads exceeding 10 kb, and this larger read length and increased error rate of these new technologies required updated assembly methods. New assembly tools such as Canu (Koren et al. 2017), HINGE (Kamath et al. 2017), and Racon (Vaser et al. 2017) designed specifically for long-read PacBio and Nanopore data assembly. In the case of plant genomes, Zinin et al. (2017) assembled the highly repetitive grass *Aegilops tauschii*, by combining PacBio long-read and Illumina short-read sequences. Also, combination of chromosome conformation capture and optical mapping generated a new version of the model plant *Arabidopsis thaliana* genome (Jiao et al. 2017a). This approach helped to improve assembly contiguity reaching chromosome-arm levels.

5.4 Soybean Genome Assembly and Annotation

5.4.1 Soybean Genome Before Whole Genome Sequencing

Grain legumes play significant role in the global food and nutritional security, and soybean is one of the leading oil seed crops in the world. The United States Department of Agriculture (USDA) estimates that the global soybean production in 2016/2017 will be 348.04 million metric tons, around 2.07 million tons more than the previous month's projection (USDA-FAS, May 2017). The development of several genomic tools including better genetic map and genome sequence information have immense contribution towards crop improvement. Soybean the cultivated species *Glycine max* and the close relatives including wild species *Glycine soja* and wild perennial *Glycine tomentella* are members of the tribe *Phaseoleae*, the most economically important of the legume tribes. The genus *Glycine* is paleopolyploid, with $2n = 40$ as its base chromosome number, as compared with other phaseoloid legumes, which are largely $2n = 20$ or 22 (Goldblatt 1981). The estimated average size of soybean genome was estimated at about 1.1×10^9 bp/C based on flow cytometry (Arumuganathan and Earle 1991), and similar values were predicted by DNA re-association kinetics (Cot analysis; Goldberg 1978; Gurley et al. 1979). Both these soybean Cot studies suggested that 40–60% of the genome is repetitive; re-association of different size fragments indicates that the majority of the repetitive sequences are physically linked while a smaller fraction is interspersed with single copy DNA. This latter conclusion is supported by 2700 sequence sampling of nearly 1000 loci across all 20 soybean linkage groups (Marek et al. 2001). The predictions from this research are that the gene rich and repetitive regions occasionally are interspersed, but gene rich "islands" are present as well. Cytological studies show that euchromatin represents ~65% of the genome with heterochromatin mainly localized to the pericentromeric regions and the short arms of four chromosomes (Singh and Hymowitz 1988). Collectively these studies suggest that gene space accounts for about one third of the soybean genome (i.e. $\sim 3.7 \times 10^8$ bp). Two large-scale genome-wide duplication events at 40–50 and 8–10 Mya occurred in the case of soybean and the duplicated regions were segmented and reshuffled after these events (Shultz et al. 2007).

The soybean community has developed substantial amount of marker information and generated physical and genetic maps of soybean, which contributed heavily to the whole genome sequencing and chromosome assembly. As a complimentary approach to the whole genome sequencing, a large number of expressed sequence tags were generated for soybean. Initially around 200,000 expressed sequence tags (ESTs) were generated from more than 50 cDNA libraries representing a wide range of organs, developmental stages, genotypes, and environmental conditions (Shoemaker et al. 2002). In the early 1990s, genome mapping of soybean based on the DNA markers was initiated and several genetic linkage maps of soybean have been published in the last decade. The early maps were developed based primarily on the restriction fragment length polymorphism (RFLP), amplified fragment length polymorphism (AFLP), and simple sequence repeat (SSR) markers, and the more recent maps include single nucleotide polymorphism (SNP) markers. In total, several thousand genetic markers (mostly SSR and SNP markers) were mapped in soybean before release of the first draft genome sequence (Keim et al. 1990; Cregan et al. 1999; Wu et al. 2004, 2010; Song et al. 2004; Kassem et al. 2006; Choi et al. 2007; Xia et al. 2007; Hisano et al. 2008; Yang et al. 2008). Initial physical map for the cultivar Williams 82 was developed from sequencing of the bacterial artificial chromosome (BAC) libraries (Luo et al. 2003; Warren 2006; Shoemaker et al. 2008). Soybean physical maps for "Forrest" and "Williams 82" representing the southern and northern US soybean germplasm base were constructed with different fingerprinting methods. These physical maps are complementary for coverage of gaps on the 20 soybean linkage groups. More than 5000 genetic markers have been anchored onto the Williams 82 physical map, but only a limited number of markers have been anchored to the Forrest physical map. A framework map with almost 1000 genetic markers was constructed using a core set of recombinant inbred lines (RILs) developed from a mapping population of Forrest × Williams 82 (Wu et al. 2011). High-resolution physical maps for both *G. max* and *G. soja* were developed later (Ha et al. 2012), and these maps served as a framework for ordering sequence fragments, comparative genomics, cloning genes, and evolutionary analyses of legume genomes.

5.4.2 Soybean Whole Genome Sequencing

5.4.2.1 Cultivated Soybean Genome Version 1

The first reference genome of the cultivated soybean was released in the year 2010 by the soybean research community in collaboration with the Department of Energy Joint Genome Institute (DOE-JGI) (Schmutz et al. 2010). The northern US cultivar Williams 82 represented the first soybean reference genome. Whole genome shot gun approach was adapted to sequence 1.1 gigabase (Gb) genome. Three sized insert

Table 5.1 Final assembly statistics of *G. max* cv. Williams 82 reference genome version 1 (Glyma1.01)

Total scaffold	1168
Total contig	16,311
Total scaffold sequence	973.3 Mb
Total contig sequence	955.1 Mb
Scaffold N/L50	10/47.8 Mb
Contig N/L50	1492/189.4 Kb

libraries and several BAC libraries were sequenced using Sanger sequencing protocols on ABI 3730XL capillary sequencing machines at the JGI. A total of 15,332,163 sequence reads were assembled into 3363 scaffolds using Arachne assembler (Batzoglou et al. 2002; Jaffe et al. 2003). This assembly covers 969.6 Mb of the soybean genome. To obtain 20 chromosome-level pseudomolecules, integrated the genome assembly with the physical map and high-density genetic map of soybean and the resulting assembly covered 937.3 Mb of the genome size. In addition, 1148 unmapped scaffolds that covered 17.7 Mb of the genome, added to make the total sequence coverage of 8.04×. After the genome assembly and chromosome assignments, this reference genome represented 85% (955 Mb) of the estimated genome size with 1.9% gap (Table 5.1).

Large fractions of plant genomes contain repeated DNA sequences of various types. A combination of genome structure analysis and homology studies identified that 59% of the soybean genome is repeat rich where majority of them are transposable elements (TEs). Analysis of the repetitive elements in the soybean reference genome revealed 57% of the repeat rich low recombination heterochromatic regions are around the centromeres. Percentage of repeat elements are more as compared to Arabidopsis (de la Chaux et al. 2012) and rice (Takata et al. 2007), and lesser compared to maize genome (Schnable et al. 2009). Long terminal repeat (LTR) retrotransposons are the majority of the repetitive elements in the soybean genome (42%) and consist of 510 families containing 14,106 intact elements, including 9733 gypsy-like and 4373 Copia-like transposable elements. Major role of transposable elements is to generate genomic novelty in organisms. This genomic novelty happens through a combination of chromosome rearrangements and associated gene regulation processes. TEs influence genome size, gene content, gene order, and several aspects of nuclear biology (Bennetzen and Wang 2014). Large-scale epigenetic changes due to polyploidization or stress responses affects TEs and all these influence the genomes to generate novel functions (Galindo-González et al. 2017).

The Williams 82 genome contains 46,430 high-confidence protein-coding loci and another ~20,000 predicted loci with low confidence. While comparing these loci with the angiosperm protein families, 283 legume specific gene families harbouring 448 soybean specific genes were identified. A 12.2% of the high-confidence protein-coding loci (5671) represents putative transcription factors in the genome (Schmutz et al. 2010). All these transcription factors from the 28 families were further annotated and a comprehensive soybean transcription factor database, SoyDB, was generated (Wang et al. 2010). The annotations in SoyDB include predicted tertiary structures, protein domains, multiple sequence alignments, DNA binding sites, and consensus sequences for each transcription factor family. Data providing experimental support to the gene content was available after the release of the soybean genome information (Libault et al. 2010; Severin et al. 2010). Transcriptome analysis of soybean tissues collected from 14 different biological conditions revealed the transcription of 55,616 annotated genes and demonstrated that 13,529 annotated soybean genes are putatively pseudogenes (Libault et al. 2010). Mining of the above gene expression atlas identified several tissue specific transcription factors and helped to understand the molecular

Table 5.2 Final assembly statistics of *G. max* cv. Williams 82 reference genome version 2 (Wm82.a2.v1)

Total scaffold	1190
Total contig	17,191
Total scaffold sequence	978.5 Mb
Total contig sequence	955.4 Mb
Scaffold N/L50	10/48.6 Mb
Contig N/L50	1548/189.4 Kb

basis of transcriptional regulation associated with various biological processes.

5.4.2.2 Cultivated Soybean Genome Version 2

Recently, a new version of the soybean genome assembly has been released (Wm82.a2.v1) after correcting several issues in the first version (Glyma1.01). The Glyma1.01 assembly of the whole genome sequence contains 236 unanchored scaffolds with lengths ranging from 10 to 100 kb and 51 unanchored scaffolds with lengths greater than 100 kb (Song et al. 2016). Even though the first assembly was generated using the integrated linkage maps and a genetic map with additional markers, the marker density was not enough to fully cover all regions of the soybean genome. Two high-density genetic linkage maps of soybean based on 21,478 SNP loci mapped in the *G. max* Williams 82 × *G. soja* PI479752 population with 1083 RILs and 11,922 SNP loci mapped in the *G. max* Essex × *G. max* Williams 82 population with 922 RILs were constructed. The high-density genetic linkage maps helped to identify false joins or misplaced scaffolds and unanchored scaffolds in the Glyma1.01 assembly, and the corresponding scaffolds were broken or reassembled to a new Wm82.a2.v1 assembly (Song et al. 2016).

The Wm82.a2.v1 was generated using the latest version of the ARCHNE assembler. A combination of high-density genetic linkage maps and *Phaseolus* synteny was used to identify the false joins in the Glyma1.01 assembly and the scaffold was broken based on these false joins. A total of 63 breaks were identified and broken, and the order and orientation of the broken scaffolds was achieved using the high-density markers and *Phaseolus* synteny. The total sequence including the 1170 unmapped scaffolds was 978.5 Mb and the new build of the 20 chromosomes captured 949.2 Mb (Table 5.2). In the new version of soybean genome assembly, 56,044 protein-coding loci and 88,647 transcripts were predicted. The Wm82.a2.v1 gene set integrates ~1.6 million ESTs, and 1.5 billion paired-end Illumina RNA sequence reads with homology-based gene predictions (https://phytozome.jgi.doe.gov).

5.4.2.3 Wild Soybean Whole Genome Sequencing

Domestication history of cultivated soybean traces back to around 5000 years ago in China and it is considered that wild soybean *Glycine soja* is the closest relative of the cultivated soybean, *Glycine max* (Hymowitz 1970). Both the species have 20 chromosomes ($2n = 40$) and belong to the primary gene pool, where the hybrids are vigorous, exhibit normal meiotic chromosome pairing and normal gene segregation. However, the wild and cultivated soybeans differ in several morphological characteristics (Hymowitz 2004). In the case of *G. soja* publically available DNA sequence resources are limited. It is considered that the wild species is the untapped genetic resource for crop improvement (Valliyodan et al. 2016). The whole genome of wild soybean, *G. soja* var. IT 812932 was sequenced using two platforms, Illumina-GA, and GS-FLX, after the release of the cultivated soybean genome (Kim et al. 2010). Around 48.8 Gb short DNA reads were aligned to the *G. max* reference genome and a consensus sequence was determined for *G. soja*. This consensus sequence spanned 915.4 Mb, with a coverage of 97.65% of the available *G. max* reference genome sequence and an average mapping depth of 43-fold. Also, 32.4 Mb of large deletions and 8.3 Mb of novel sequence contigs in the *G. soja* genome were detected. The *G. soja* unmapped and unpaired reads were assembled de novo by Velvet (Version 0.7.31) assembler (Zerbino and Birney 2008). The genome structural analysis revealed 5794 deletions and 194 inversions and predicted 8554 insertions in the *G. soja* genome. More than 48% of the deleted regions in the wild soybean genome contain repetitive elements including the major classes of retrotransposons (20.74%). In another study, the 5794 deletions were compared against *G. max* gene positions for regions of overlap. This comparative genomic analysis identified 425 unique genes from the list of higher confidence 46,430 gene model predictions of the Glyma1.01, that are absent in *G. soja* and unique in *G. max* (Joshi et al. 2013). In addition, this study showed that there are significant genomic level differences between *G. max* and *G. soja* that are associated with some functionally important genes for seed, oil, and protein traits. Further investigation of these genes can explain some of the phenotypic differences observed between *G. soja* and *G. max* especially in terms of the major seed composition and agronomic traits.

Recently, de novo assembly of a salt tolerant wild soybean (*G. soja*) accession, W05 genome was reported (Qi et al. 2014). The genome size of this wild soybean was estimated at 1.17 Gb using k-mer statistics (Lander and Waterman 1988), and it is close to the estimated size (1.12 Gb) of the cultivated soybean reference genome, Williams 82. SOAPdenovo software (Li et al. 2010) assembled the 868 Mb of the W05 genome (Table 5.3). About 43.41% of the genome contains repeat elements and majority of them are LTRs (30.89%). The W05 genome contains 52,395 protein-coding genes, and out of these, 49,560 are functionally annotated and the average coding sequence length is 1083.9 bp. Development of wild soybean de novo genomes can facilitate the identification of genetic materials from the untapped genomic resource for crop improvement since these genomes can identify small and large genomic variations.

In order to capture the entire genomic sequence present within the species, including the complete gene set, the pangenome needs to be sequenced (Golicz et al. 2016). Tettelin et al. (2005) introduced the concept of pangenome defining it as a full genomic (genic) makeup of a species. A single genome is insufficient to represent the genomic content in the case of both cultivated and wild soybean where individuals are distinct from one another due to low levels of genetic exchange and recombination. Pangenomes of seven *G. soja* accessions were generated (Li et al. 2014) using the Illumina HiSeq2000 sequencing reads and SOAPdenovo assembler (Li et al. 2010). The average genome coverage was 119.9× and the estimated genome size ranged from 889.33 Mbp for the accession GsojaG (93.6% of the GmaxW82) to 1118.34 Mbp for the accession GsojaD (117.7% of GmaxW82) (Li et al. 2014). In this study, the estimated average number of genes per *G. soja* genome is 55,570, slightly more than the *G. max* Williams 82 genome. These *G. soja* assemblies enabled accurate detection of variation and comparative analyses within genic regions and found significant structural variations when compared to the *G. max* assembly.

Table 5.3 Final assembly statistics of *G. soja*, accession W05 reference genome

Total scaffold	51,178
Total contig	9008
Total scaffold sequence	868 Mb
Total contig sequence	808.67 Mb
Scaffold N/L50	442/401.3 Kb
Contig N/L50	8897/24.17 Kb

5.5 Conclusion and Future Perspectives

To meet the increasing global demand of grain legumes, including soybean, we have to maximize the development of high-yielding cultivars with better plant performances under the biotic and abiotic stress conditions. In order to achieve this, goal precise dissection of genomic regions associated with major traits to the haplotype/allelic level is needed leading to the next-generation breeding strategies. Advances in the next-generation sequencing platforms, cost-effectiveness per bp of genome sequences, and the development of various computational tools including better genome assemblers will expedite the generation of individual crop genomes. Generation of long-sequence reads and the associated genome assemblers will substantially reduce the major challenges in genome assembly including the one posed by genomic repeats and help improve the haplotype resolution and analysis of genomic variation. The accurate detection of genomic variation and comparative analyses within genic region is essential for the discovery of novel genes or alleles associated with the traits of interest and development of new varieties.

A single reference genome is not enough to capture the genomic diversity due to a large amount of structural variation including the copy number variations and presence/absence variation, which significantly alter the individual genomic sequences. Pangenomes representing landrace, elite, and wild soybean from the global collection are needed to address the above issue of genome structure variation and mining for rare alleles. Large-scale sequencing of soybean germplasm and additional reference genome build for cultivated and wild soybean is in pipeline and will be available soon to the soybean community. From these efforts, it is found that long-read sequencing of soybean genome using the PacBio platform helped to improve the whole genome assembly far better. Wild soybean is important resource for novel alleles due to its higher genomic diversity. More reference genome or pangenomes for wild soybean will significantly influence the tapping of rare genomic resources towards crop improvement. Soybean genomes exhibit higher linkage disequilibrium (LD) and more germplasm lines with deep sequencing will help enable precise genotype–phenotype association studies towards pinpointing the candidate gene associate with the specific trait and for the genomic predictions. In addition, development of a robust haplotype map for soybean (HapMap2) is inevitable to find specific genetic variants that affect plant developmental processes and response to disease and climate changes. Generation of large-scale sequencing of germplasm and various genetic populations need a better data visualization tool to compare and mine the genomic information at the allelic level. This will enable breeders and soybean community to easily associate genome sequence information with traits of interest and apply to genome enabled crop improvement programs including genome editing and next-generation breeding strategies.

Acknowledgements The authors are grateful to the United Soybean Board and the United States Department of Agriculture for project support.

References

Arumuganathan K, Earle ED (1991) Nuclear DNA content of some important plant species. Plant Mol Biol Rep 9:208–219

Batzoglou S, Jaffe DB, Stanley K, Butler J, Gnerre S et al (2002) ARACHNE: a whole-genome shotgun assembler. Genome Res 12(1):177–189

Bennetzen JL, Wang H (2014) The contributions of transposable elements to the structure, function, and evolution of plant genomes. Annu Rev Plant Biol 65:505–530

Butler J, MacCallum I, Kleber M, Shlyakhter IA, Belmonte MK et al (2008) ALLPATHS: de novo assembly of whole-genome shotgun microreads. Genome Res 18:810–820

Chaisson MJ, Pevzner PA (2008) Short read fragment assembly of bacterial genomes. Genome Res 18:324–330

de la Chaux N, Tsuchimatsu T, Shimizu KK, Wagner A (2012) The predominantly selfing plant *Arabidopsis thaliana* experienced a recent reduction in transposable element abundance compared to its outcrossing relative *Arabidopsis lyrata*. Mob DNA 3:2

Eid J, Fehr A, Gray J, Luong K, Luong K et al (2009) Real-time DNA sequencing from single polymerase molecules. Science 323:133–138

Fuller CW, Middendorf LR, Benner SA, Church GM, Harris T et al (2009) The challenges of sequencing by synthesis. Nat Biotechnol 27:1013–1023

Galindo-González L, Mhiri C, Deyholos MK, Grandbastien MA (2017) LTR-retrotransposons in plants: engines of evolution. Gene S0378–1119(17):30322

Goldberg RB (1978) DNA sequence organization in the soybean plant. Biochem Genet 16:45–51

Goldblatt P (1981) Cytology and phylogeny of Leguminosae. In: Polhill RM, Raven PH (eds) Advances in legume systematics, part 2. Royal Botanic Gardens, Kew, pp 427–463

Golicz AA, Batley J, Edwards D (2016) Towards plant pangenomics. Plant Biotechnol J 14:1099–1105

Ha J, Abernathy B, Nelson W, Grant D, Wu X et al (2012) Integration of the draft sequence and physical map as a framework for genomic research in soybean (*Glycine max* (L.) Merr.) and wild soybean (*Glycine soja* Sieb. and Zucc.). Genes Genomes Genet (Bethesda) 2:321–329

Hashmi U, Shafqat S, Khan F, Majid M, Hussain H et al (2015) Plant exomics: concepts, applications and methodologies in crop improvement. Plant Signal Behav 10(1):e976152

Hernandez D, Francois P, Farinelli L, Osterås M, Schrenzel J (2008) De novo bacterial genome sequencing: millions of very short reads assembled on a desktop computer. Genome Res 18:802–809

Hymowitz T (1970) On the domestication of soybean. Econ Bot 24:408–421

Hymowitz T (2004) Speciation and cytogentics. In Boerma HR, Specht JE (eds) Soybeans: improvement, production and uses. American Society of Agronomy, Madison, pp 97–136

Imelfort M, Edwards D (2009) De novo sequencing of plant genomes using second-generation technologies. Brief Bioinform 10:609–618

International Rice Genome Sequencing Project (2005) The map-based sequence of the rice genome. Nature 436:793–800

Jackman SD, Vandervalk BP, Mohamadi H, Chu J, Yeo S et al (2017) ABySS 2.0: resource-efficient assembly of large genomes using a Bloom filter. Genome Res. doi:10.1101/gr.214346.116

Jaffe DB, Butler J, Gnerre S, Mauceli E, Lindblad-Toh K et al (2003) Whole-genome sequence assembly for mammalian genomes: Arachne 2. Genome Res 13:91–96

Jeck WR, Reinhardt JA, Baltrus DA, Hickenbotham MT, Magrini V et al (2007) Extending assembly of short DNA sequences to handle error. Bioinformatics 23:2942–2944

Jiao WB, Accinelli G, Hartwig B, Kiefer C, Baker D et al (2017a) Improving and correcting the contiguity of long-read genome assemblies of three plant species using optical mapping and chromosome conformation capture data. Genome Res 27:778–786

Jiao WB, Accinelli G, Hartwig B, Kiefer C, Baker D et al (2017b) Improving and correcting the contiguity of long-read genome assemblies of three plant species using optical mapping and chromosome conformation capture data. Genome Res. doi:10.1101/gr.213652.116

Joshi T, Valliyodan B, Wu JH, Lee SH, Xu D, Nguyen HT (2013) Genomic differences between cultivated soybean, *G. max* and its wild relative *G. soja*. BMC Genomics 14(Suppl 1):S5

Keim P, Diers BW, Olson TC, Shoemaker RC (1990) RFLP mapping in soybean: association between marker loci and variation in quantitative traits. Genetics 126:735–742

Keim P, Schupp JM, Travis SE, Clayton K, Zhu T et al (1997) A high density soybean genetic map based on AFLP markers. Crop Sci 37:537–543

Kim MY, Lee S, Van K, Kim TH, Jeong SC et al (2010) Whole-genome sequencing and intensive analysis of the undomesticated soybean (*Glycine soja* Sieb. and Zucc.) genome. Proc Natl Acad Sci USA 107:22032–22037

Lander ES, Waterman MS (1988) Genomic mapping by fingerprinting random clones: a mathematical analysis. Genomics 2:231–239

Li R, Zhu H, Ruan J, Qian W, Fang X et al (2010) De novo assembly of human genomes with massively parallel short read sequencing. Genome Res 20:265–272

Li YH, Zhou G, Ma J, Jiang W, Jin LG et al (2014) De novo assembly of soybean wild relatives for pan-genome analysis of diversity and agronomic traits. Nat Biotechnol 10:1045–1052

Libault M, Farmer A, Joshi T, Takahashi K, Langley RJ et al (2010) An integrated transcriptome atlas of the crop model *Glycine max*, and its use in comparative analyses in plants. Plant J 63:86–99

Luo MC, Thomas C, You FM, Hsiao J, Ouyang S et al (2003) High-throughput fingerprinting of bacterial artificial chromosomes using the SNaPshot labeling kit and sizing of restriction fragments by capillary electrophoresis. Genomics 82:378–389

Marek LF, Mudge J, Darnielle L, Grant D, Hanson N et al (2001) Soybean genomic survey: BAC-end sequences near RFLP and SSR markers. Genome 44:572–581

Margulies M, Margulies M, Egholm M, Altman WE et al (2005) Genome sequencing in microfabricated high-density picolitre reactors. Nature 437:376–380

Myers EW (1995) Toward simplifying and accurately formulating fragment assembly. J Comput Biol 2:275–290

Nakano K, Shiroma A, Shimoji M, Tamotsu H, Ashimine N et al (2017) Advantages of genome sequencing by long-read sequencer using SMRT technology in medical area. Hum Cell. doi:10.1007/s13577-017-0168-8

Pevzner PA, Tang H, Waterman MS (2001) An Eulerian path approach to DNA fragment assembly. Proc Natl Acad Sci USA 98:9748–9753

Qi X, Li MW, Xie M, Liu X, Ni M et al (2014) Identification of a novel salt tolerance gene in wild soybean by whole-genome sequencing. Nat Commun 5:4340

Rhoads A, Au KF (2015) PacBio sequencing and its applications. Genom Proteom Bioinform 13:278–289

Sanger F, Air GM, Barrell BG, Brown NL, Coulson AR et al (1977) Nucleotide sequence of bacteriophage φX174 DNA. Nature 24:687–695

Schmutz J, Cannon SB, Schlueter J, Ma J, Mitros T et al (2010) Genome sequence of the palaeopolyploid soybean. Nature 463:178–183

Schnable PS, Ware D, Fulton RS, Stein JC, Wei F et al (2009) The B73 maize genome: complexity, diversity, and dynamics. Science 326:1112–1115

Schwartz DC, Li X, Hernandez LI, Ramnarain SP, Huff EJ, Wang YK (1993) Ordered restriction maps of *Saccharomyces cerevisiae* chromosomes constructed by optical mapping. Science 262:110–114

Severin AJ, Woody JL, Bolon YT, Joseph B, Diers BW et al (2010) RNA-Seq atlas of *Glycine max*: a guide to the soybean transcriptome. BMC Plant Biol 10:160

Shendure J, Porreca GJ, Reppas NB, Lin X, McCutcheon JP et al (2005) Accurate multiplex polony sequencing of an evolved bacterial genome. Science 309:1728–1732

Shoemaker R, Keim P, Vodkin L, Retzel E, Clifton SW et al (2002) A compilation of soybean ESTs: generation and analysis. Genome 45:329–338

Shoemaker RC, Grant D, Olson T, Warren WC, Wing R et al (2008) Microsatellite discovery from BAC end sequences and genetic mapping to anchor the soybean physical and genetic maps. Genome 51:294–302

Shultz JL, Ray JD, Lightfoot DA (2007) A sequence based synteny map between soybean and *Arabidopsis thaliana*. BMC Genom 8:8

Simpson JT, Wong K, Jackman SD, Schein JE, Jones SJ, Birol I (2009) ABySS: a parallel assembler for short read sequence data. Genome Res 19:1117–1123

Singh RJ, Hymowitz T (1988) The genomic relationship between *Glycine max* (L.) Merr. and *G. soja* Sieb. and Zucc. as revealed by pachytene chromosome analysis. Theor Appl Genet 76:705–711

Song Q, Jenkins J, Jia G, Hyten DL, Pantalone V et al (2016) Construction of high resolution genetic linkage maps to improve the soybean genome sequence assembly Glyma1.01. BMC Genom 17:33

Sutton GG, White O, Adams MD, Kerlavage AR (1995) TIGR assembler: a new tool for assembling large shotgun sequencing projects. Genome Sci Technol 1:9–19

Takata M, Kiyohara A, Takasu A, Kishima Y, Ohtsubo H, Sano Y (2007) Rice transposable elements are characterized by various methylation environments in the genome. BMC Genom 8:469

Tang H, Lyons E, Town CD (2015) Optical mapping in plant comparative genomics. Gigascience 4:3

The Arabidopsis Genome Initiative (2000) Analysis of the genome sequence of the flowering plant *Arabidopsis thaliana*. Nature 408:796–815

The International Human Genome Sequencing Consortium (2004) Finishing the euchromatic sequence of the human genome. Nature 431:931–945

United States Department of Agriculture-Foreign Agricultural Service (2017) World agricultural production, Circular Series WAP 05-17

Utturkar SM, Klingeman DM, Land ML, Schadt CW, Doktycz MJ, Pelletier DA, Brown SD (2014) Evaluation and validation of de novo and hybrid assembly techniques to derive high-quality genome sequences. Bioinformatics 30:2709–2716

Valliyodan B, Qiu D, Patil G, Zeng P, Huang J et al (2016) Landscape of genomic diversity and trait discovery in soybean. Sci Rep 6:23598

Varshney RK, Nayak SN, May GD, Jackson SA (2009) Next-generation sequencing technologies and their implications for crop genetics and breeding. Trends Biotechnol 27(9):522–530

Venter JC, Adams MD, Myers EW, Li PW, Mural RJ et al (2001) The sequence of the human genome. Science 291:1304–1351

Wang Z, Libault M, Joshi T, Valliyodan B, Nguyen HT et al (2010) SoyDB: a knowledge database of soybean transcription factors. BMC Plant Biol 10:14

Warren WC, The Soybean Mapping Consortium (2006) A physical map of the "Williams 82" soybean (*Glycine max*) genome. Abstract W151. In: Plant and animal genomes XIV conference, San Diego, CA, 14–18 Jan 2006

Warren RL, Sutton GG, Jones SJM, Holt RA (2007) Assembling millions of short DNA sequences using SSAKE. Bioinformatics 23:500–501

Weissensteiner MH, Pang AWC, Bunikis I, Höijer I, Vinnere-Petterson O, Suh A, Wolf JBW (2017) Combination of short-read, long-read, and optical mapping assemblies reveals large-scale tandem repeat arrays with population genetic implications. Genome Res. doi:10.1101/gr.215095.116

Wu X, Ren C, Joshi T, Vuong T, Xu D, Nguyen HT (2010) SNP discovery by high-throughput sequencing in soybean. BMC Genom 11:469

Wu X, Vuong TD, Leroy JA, Shannon GJ, Sleper DA, Nguyen HT (2011) Selection of a core set of RILs from Forrest × Williams 82 to develop a framework map in soybean. Theor Appl Genet 122:1179–1187

Zerbino DR, Birney E (2008) Velvet: algorithms for de novo short read assembly using de Bruijn graphs. Genome Res 18:821–829

Comparative Genomics of Soybean and Other Legumes

6

Rick E. Masonbrink, Andrew J. Severin and Arun S. Seetharam

Abstract

Comparative genomics is the leveraging of genomic data between species to understand the evolution of genomes and species. With the increasing availability of genomics resources (genomes, transcriptomes, epigenomes, proteomes, etc.), opportunities exist to explore species relationships using comparative genomics. Comparative genomics is most commonly used to determine structural and functional variation between genomes. Traditional approaches that study genomes in isolation are limiting in both the kind of questions that can be answered, as well as the transferability of knowledge between species. Herein, we will address the recent advances in comparative genomics research, specifically in legumes, and how this wealth of knowledge can further expand our understanding of biological diversity. Comparative genomics can be performed at the genic or at genomic level, for which there are numerous workflows to exploit, including gene prediction and annotation, orthologous gene relationships, building gene and species phylogenetic trees, synteny, finding lineage specific genes, and pan-genomic analyses.

6.1 Introduction

Comparative genomics is essentially the study of variation in the genomic features between related species and can be used to better understand the connections between genotype and phenotype, leading to the improvement of agronomic traits in legumes. Comparative genomics can be used to examine the interplay between genome structure, mutations, and phenotype. By comparing multiple sources of data (transcriptomes, proteomes, metabolomes, physiomes, phenomes,

R.E. Masonbrink (✉) · A.J. Severin · A.S. Seetharam
Genome Informatics Facility, Iowa State University, Ames, IA 50011, USA
e-mail: remkv6@iastate.edu

A.J. Severin
e-mail: severin@iastate.edu

A.S. Seetharam
e-mail: arnstrm@iastate.edu

H.T. Nguyen and M.K. Bhattacharyya (eds.), *The Soybean Genome*, Compendium of Plant Genomes, DOI 10.1007/978-3-319-64198-0_6

etc.) from many phylogenetically related species, connections between phenotype and genotype are more apparent than studying a single species in isolation.

For example, the loci associated with a trait (quantitative trait loci, QTL) in one species, can be used to identify conserved gene order between two genomes, synteny (Khazaei et al. 2014). This reciprocal transfer of knowledge is not usually complete, but valuable insight can be had and used for the improvement of crop species traits (Gondo et al. 2007; Isemura et al. 2007; Maughan et al. 1996; Schmutz et al. 2010). Thus, comparative genomics can significantly increase our ability to translate the unique findings from one species to another. The massive and growing collection of genomic data enhances our capacity to manipulate and improve traits of interest like plant physiological responses to abiotic and biotic stress to increase yield, tolerance, and resistance.

By incorporating a combined knowledge on all related species, analyses on nonmodel organisms suddenly become tractable, as they can be analyzed with less funding and time. For example, a commonly employed comparative genomics technique involves the identification of gene structure and function. By first identifying a model organism within a clade and making comparisons to multiple genera, gene structure information can then be extrapolated to the whole clade. With a preexisting model genome in a clade, the conserved protein/DNA homology will allow researchers to further expand the genotype and phenotype connection.

6.2 Big Data and Comparative Genomics

Prior to the genomics era, most plant genome studies were accomplished through the use of cytogenetics, molecular biology, and breeding. With the advancement of high-throughput sequencing technology, genomic resources have become cheaper, faster, and easier to attain, thus exponentially increasing the number of species available for comparative analyses. Since the release of the first prokaryotic genome, 92,697 species' genomes have been deposited in NCBI (last accessed Feb 6, 2017). With this explosion of genomic data, it has become apparent that many mechanisms result in genomic variability between species, some of which lead to survival or a selective advantage. As we understand the mechanisms of genomic variation, we can better attribute genomic structure with traits of interest. Genome variation can include multiple ancestral whole-genome duplications, recent polyploid events, and genomic fractionation. For example, intra-strand and illegitimate recombination contribute to the evolutionary divergence between species (Hawkins et al. 2008; Vitte and Bennetzen 2006) which is positively correlated with increasing phylogenetic distance. In addition, genomic variation can arise from differences in recombination rates and have also been attributed to variation in genome size (Ross-Ibarra 2007; Tiley and Burleigh 2015; Zenil-Ferguson et al. 2016). However, transposable elements claim the primary mechanism in genome size variation, as their expansions and contractions can cause significant genome size fluctuations among species.

6.2.1 Availability of Sequenced Legumes and Current Polyploid Knowledge

Fabaceae is one of the most studied plant families due to its agricultural importance and ability to fix nitrogen through symbiosis with rhizobial bacteria. It is also one of the largest plant families, comprising ~19,000 species and 727 genera, divided into 3 subfamilies: Papilionoideae, Caesalpinioideae, and Mimosoideae (Doyle and Luckow 2003). The first step in any comparative genomics study is to sequence the genome, the quality of which is highly dependent on the complexity and size of the genome. In legumes, several whole-genome duplication events and a triplication event that occurred early after the split between monocots and eudicots (Severin et al. 2011). Several additional whole-genome

Table 6.1 All legume genomes available on NCBI

Organism	Size (Mb)	Chrs	Assemblies
Vicia faba	~13,000	6	1
Trifolium pratense	345.9	7	2
Lotus japonicus	394.4	6	1
Medicago truncatula	412.9	8	1
Vigna radiata	463.6 (NCBI)—579	11	3
Vigna angularis	467.3	11	3
Trifolium subterraneum	471.8	8	1
Phaseolus vulgaris	521.1	11	2
Cicer arietinum	530.8	8	2
Cajanus cajan	592.8	11	2
Lupinus angustifolius	609.2	20	2
Vigna unguiculata	695.1	10	1
Glycine soja	863.6	20	1
Glycine max	979.0	20	2
Arachis duranensis	1068.4	10	2
Arachis ipaensis	1349.1	10	1

Genome size, chromosome counts, and number of assemblies are shown as columns. Genome sizes were rounded up to the nearest 100 kb

duplication events have been found in the Papilionoideae clade (~48 mya) (Cannon et al. 2010), with soybean having a whole-genome duplication event as recent as ~13 mya (Doyle et al. 2000).

Despite the challenges of paleopolyploidy, by 2011 three legume species (*G. max, Medicago, Lotus japonicus*) were sequenced and assembled. Since then, an additional 13 genome assemblies have been deposited in NCBI. There are currently 16 legume species with a chromosomal content varying from $n = 6$ to $n = 20$, and genome sizes that span from 345 Mb to ~13 Gb, a sevenfold difference (NCBI 2013) (Table 6.1).

6.3 Gene-by-Gene Analyses

6.3.1 Gene Annotation and Function

After genome assembly, the second foundational step necessary for comparative genomics lies with gene annotation, which is comprised of gene model prediction and gene function prediction. Gene model prediction specifies the start and stop positions that correspond to exons, introns, and UTRs along the chromosomes. The ease of gene model prediction has increased due to the vast quantities of transcriptomic data available and the low cost to generate sequencing data. Gene model prediction includes the identification of novel genes, noncoding RNAs, small RNAs, and complex varieties of repeats.

A number of open-source programs can predict genes in newly assembled genomes, and they are broadly classified as ab initio gene predictors and extrinsic gene predictors. The methods used to identify genes rely on either the intrinsic properties of DNA sequences or external genetic factors. In an attempt to get "the best of both worlds" there are programs that incorporate both methods. MAKER-P (Holt and Yandell 2011) is a popular and robust tool that has been for gene prediction in many genome assemblies. Predictions are carried out in several rounds, with each round training and retraining ab initio predictors, until consensus gene sets are specified. Distinguishing genes from transposons can increase the resolution of consensus gene predictions that are more highly conserved, and therefore, the masking of highly repetitive sequences (repeats) prior to prediction is recommended in comparative studies. The repeat masked genome is BLAST (Boratyn et al. 2013) searched with the transcriptome and protein alignments are calculated. This information is then used to train a neural network of gene prediction programs including GeneMark-ES (Borodovsky and Lomsadze 2011), SNAP (Korf 2004), Augustus (Stanke et al. 2006), and FGENESH (SoftBerry Inc.).

The prediction of gene function is based on homology to proteins of known function. Homology-based searches are performed to databases, (Gramene (Tello-Ruiz et al. 2016), Phytozome (Goodstein et al. 2012), and SwissProt/UniProt) (Suzek et al. 2007) to identify proteins with known function that can be associated with the newly predicted gene models. InterProScan (Jones et al. 2014) can also be used to classify genes into families and associate proteins with conserved domains from multiple databases. Genome structure is critical to identifying candidate genes that are related to traits and can be useful when a predicted function can narrow down the genes responsible for a trait using marker-assisted selection (MAS). An interesting side effect of multiple-genome assemblies in multiple species is that due to the multitude of methods available, no two genomes are alike. Thus, it is important to standardize the gene prediction methods, so that the results of downstream analyses will not be skewed by bioinformatic error. By standardizing at least this one step, the error in prediction is minimized and the appearance of false-positive gene losses or gains can be avoided.

6.3.2 Orthologous Gene Family Prediction

After gene models are predicted and standardized, the next comparative analysis involves the identification of orthologous gene families. Orthologous gene families are comprised of a group of genes predicted to have origins with a common ancestor. This is based on gene model similarity within and between multiple-related genomes using programs like BLAST (Boratyn et al. 2013). This information can then be used to study gene loss and gain, gene family expansion and contraction, and syntenic analyses.

Previous attempts at resolving genetic relationships between sequences were limited to DNA markers including restriction fragment length polymorphism (RFLP), amplified fragment length polymorphism (AFLP), random amplified polymorphic DNA (RAPD), simple sequence repeat (SSR) that would sometimes land in transposable elements, which are not necessarily conserved among even closely related species. More recently, through technological and scientific advancement and the rapid publication of genome assemblies, comparisons of more conserved elements in the genome have become possible. Because exons are more likely to be conserved, many laboratories have employed the "exomes" of the genome to predict orthologs. In the recent past, many of these approaches have used 1:1 reciprocal best blast hits between species to predict orthologs, yet this method can also be problematic. With the natural variation among and even within species including tandem duplications, gene conversions, genomic rearrangements, and homeologous/paralogous genes resulting from multiple polyploid events, the chance is high for complications/errors in prediction.

Finer resolution can be generated by including all possible homologous genes in both species, rather than forcing 1:1 orthologous gene pairs. Familial analyses of gene relationships can be calculated using a phylogenetic approach, a homology/graph approach, or a hybrid of the two (e.g., Ensembl Compara), as described by Kuzniar et al. (2008). Phylogeny-based methods tend to provide the best predictions for orthology, although they can be computationally complex and time-consuming. For a newly sequenced species with a rapid evolution or highly divergent genome, homology or graph-based approaches are more feasible. OrthoMCL (Fischer et al. 2011; Li et al. 2003) can reliably predict homologues using BLAST, an MCL [Markov Cluster

Algorithm (Borodovsky and Peresetsky 1994)], and a relational database (Oracle or MySQL). After 1:1 reciprocal best hit (RBH) orthologs are established, the relationships can be converted to a graph format where the protein sequences are nodes and their relationships are denoted by weighted edges. After clustering, OrthoMCL can consider the global relationships in the graph simultaneously to enable the differentiation of paralogs, orthologs, and tandem duplications. The users can control the cluster tightness, by increasing the inflation value while running MCL clustering step.

After orthologous gene family prediction, analyses can be performed to differentiate young from ancient tandem duplications, and ancient paleologs from more recent homeologs. With this knowledge, a likelihood score can be calculated and provide greater insight into the evolutionary dynamics of chromosomal evolution and the translatability of gene function between species.

6.3.3 Phylogenetic Trees and Genome Relatedness

Single-copy orthologous genes can be used to infer phylogenetic relatedness between species. Alternatively, since identifying the presence of highly conserved BUSCO (Benchmarking Universal Single-Copy Orthologs) genes (Simao et al. 2015) is a common measure of genome completeness, these genes are readily available for quickly estimating species trees. When inferring evolutionary relationships in a newly sequenced species, phylogenetic approaches can either be simple neighbor joining/UPGMA-based or statistically rigorous maximum likelihood (ML), and Bayesian approaches. When tackling a complex of closely related species, like orchids (Li et al. 2016), generally a Bayesian or ML approach is recommended. Many programs are available that can construct phylogenetic trees and offer these approaches (RaxML (Stamatakis 2014) and FastTree (Price et al. 2010)). However, the validity of any phylogenetic reconstruction is dependent on the quality and accuracy of protein/DNA alignments, for which MAFFT (Katoh and Standley 2013) and PRANK (Loytynoja 2014) can be used, respectively. To assess alignment accuracy, programs such as GUIDANCE (Penn et al. 2010) can assign confidence scores to alignments, thereby providing a threshold for generating consensus gene trees. Below is a phylogenetic tree of all sequenced legume species using BUSCO genes.

BUSCO was used to identify single-copy ortholog genes from all the sequenced legume genomes. The identified genes' protein translation was aligned using MAFFT, and GUIDENCE was used for selecting optimal alignment. Phylogenetic trees were then reconstructed with a maximum likelihood (ML) approach using RAxML (version 8.1.16). A rapid bootstrap analysis and search for the best-scoring ML tree was conducted with 1000 bootstrap replicates. Best protein model with respect to the likelihood of a fixed, reasonable tree was automatically determined using the PROTGAMMAAUTO option of RAxML. *Arabidopsis thaliana* was set as the outgroup, and individual gene trees were then processed with ASTRAL to estimate a coalescent-based species tree.

6.4 Genomic Analyses

6.4.1 Synteny

Synteny has multiple definitions in the literature, which are mainly based on the methodology used to compute blocks of synteny. The first definition relies on the conservation of genomic sequence and is therefore defined as a set of conserved genomic blocks found through whole-genome alignments. The second definition relies on the conserved order of genes between species and can be more accurate, considering that the expansion, contraction, birth, and death of transposable elements are thought to be frequent between two species (Bennetzen 2002; Li et al. 2014). This second approach requires the calculation of orthologous gene families, the understanding of which is required to compute synteny.

Many studies have been performed to address relatedness among legume species and thus assess the efficacy of exchanging functional gene annotations. Prior to the availability of genome sequences, Choi et al. used bacterial artificial chromosomes that included 533 protein pairs to identify conserved synteny across 7 legume species (*M. truncatula, M. sativa, P. sativum, G. max, V. radiata, P. vulgaris, and L. japonicus*). With this data, they found multiple conserved syntenic microstructures between four species: *M. trunculata, L japonicus, M. truncatula* and *G. max* using 24 gene-specific markers to construct a phylogeny (Choi et al. 2004).

Synteny can also be used to identify highly conserved regions within a genome to explore paleopolyploidy and ancient genome duplication/triplication events. A region in soybean was conserved after a triplication and two whole-genome duplication events, resulting in 12 syntenic regions (Severin et al. 2011). Understanding why some genes are retained while others are lost can be studied using comparative genomics to provide insight into gene regulation. For instance, the genes in these 12 soybean regions agree with the gene balance hypothesis, which states that genes involved in many interactions biochemically or stoichiometrically are more likely to be retained (Birchler and Veitia 2007, 2010, 2014; Freeling and Thomas 2006; Rosado and Raikhel 2010; Severin et al. 2011).

6.4.2 A Composite Map of Legumes

Comparative genomics has been used to better understand the variation in genome size and chromosome count across legumes. The latest comparative genomics study in legumes used the genomes of 10 sequenced species to encompass a diversity of genome sizes ($\sim$13 Gb to 333 Mb and haploid chromosomal complements of 6–20) in the Fabaceae. The authors increased the resolution of a previous comparative genomics study by filtering previously developed markers and using only those found in genes to predict: orthologous relationships, dissect synteny, and gain insight into chromosomal evolution. Using 209 cross-species markers, 15,441 orthologous groups, and four of the most contiguous genome sequences, the authors formed 93 linkage groups and 110 syntenic blocks among the species. To construct syntenic relationships, they used the simplest and smallest genome (*M. truncatula*) to redraw syntenic relationships in the four species by locating genomic positions of cross-species markers and verifying orthology with BLAST.

Despite the large differences in genome size, syntenic relationships were not dramatically affected (Fig. 6.1). Interestingly, the Viceae tribe displayed the greatest expansion of transposable elements resulting in large genome sizes (Choi et al. 2004; Lee et al. 2017). This result indicates that particularly the Viceae tribe seems to have experienced genome expansion through the accumulation of repetitive elements, predominantly in noncoding regions. The authors inferred from the synteny, phylogenetic relationships, karyotype size, and chromosome counts that the ancestral legume genome likely had fewer and smaller chromosomes than extant species. Chromosome rearrangements, fissions, and fusions were implicated in the reduction of chromosome numbers in *L. japonicus*, and chromosomal fissions giving rise to increases in chromosome number in the Phaseoloid linage (Lee et al. 2017).

6.4.3 Single Genomes Are Unique

An overall outlook on the legume lineage offers many benefits over a single genome analysis in isolation. Each species is distinct and harbors adaptations to a unique niche, which is important to evaluate, considering that this variation gives rise to convergent and divergent phenotypes. The first legume genome assembly, *Lotus japonicus,* highlighted the expansion and contraction of gene families and revealed genes essential to symbiotic nitrogen fixation in legumes. The assembled genomes of *Glycine max* and a wild relative, *Glycine soja*, gave a unique insight into domestication and identified candidates for the

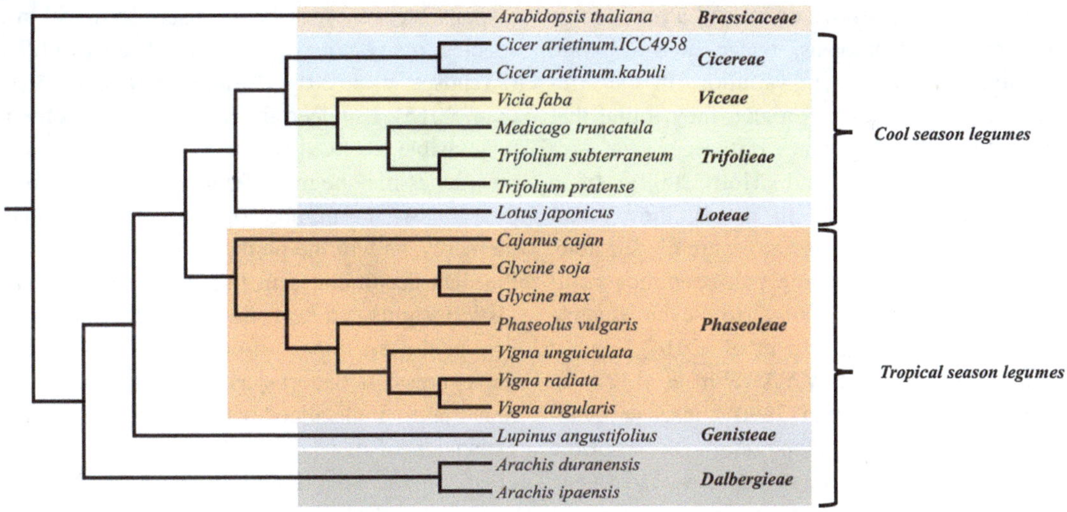

Fig. 6.1 Phylogenetic tree for sequenced legume species

improvement of a major crop species (Kim et al. 2010; Schmutz et al. 2010; Stupar 2010). While many of the Nod factors necessary for initiating root nodulation were attributed to a 58 MYA WGD through the sequencing of the *M. truncatula* genome, 593 cysteine-rich peptides were identified that play essential roles in the regulation of bacterial endosymbionts (Young et al. 2011). Comparative genomics using *C. cajanus*, a milletoid within papilionoideae (soybean, cowpea, common bean, mung bean), led to the identification of 28 duplicated syntenic blocks in the soybean genome, yet evidence was lacking for a whole-genome duplication in the *C. cajanus* genome. Interestingly, they found evidence for a massive rearrangement in soybean chromosome 11, which is syntenic with chromosome 5 in pigeon pea (Varshney et al. 2012). Thus, without these fine resolution studies of a single genome in a comparative context, much of the diversity and uniqueness in a species would be overlooked.

6.4.4 Pan-Genomes

The term pan-genome was first used by Tettelin et al. to describe the genome of *Streptococcus agalactiae* and denoted the repertoire of genes or ORFs that encompass the diversity of multiple genomes of a species. While previous studies have documented the extreme variation in the repetitive portion of the eukaryotic genome (Bennetzen and Wang 2014; Birchler et al. 2008; Leitch and Leitch 2013; Wendel et al. 2016), the genic portion also appears to be extremely variable (Vernikos et al. 2015). To take this variation into account, researchers must consider multiple individuals encompassing the diversity of a species' habitat. This has resulted in the birth of pan-genomics, defined as a genome comprised of a conserved "core" of genes found in all individuals of a species, and a group of unique "dispensable" genes found in one or few individuals (Vernikos et al. 2015). These unique and dispensable genes have arisen through a plethora of mechanisms, including the genomic integration of messenger RNAs, transposable element to gene conversion, genomic rearrangements (Biémont 2010). With the differences between transposable elements and genes becoming less distinct, random genetic mutations in transposable elements, paralogous genes, and even previously nonfunctional sequences can give rise to novel variation. We can gain further insight into the function, adaptive role, and agronomic potential of dispensable genes by analyzing multiple individuals from multiple habitats.

Many plant studies have addressed a part of this variation in Arabidopsis, maize, and *Glycine*, reporting wide ranges in variation. In the Arabidopsis 1001 genomes project, they found that up to 10% of coding genes in the genome were dispensable (Weigel and Mott 2009). In a pan-transcriptome study in maize, they found that up to ~83% of transcripts were dispensable.

A key study in legume pan-genomics compared the *G. max* genome to seven diverse accessions of *G. soja* (Li et al. 2014), a wild relative diverging ~8 MYA (Kim et al. 2010). Interestingly, they found 3990 nonsense mutations, genes with early stop codons in one of the two species. Between the two species, 1978 gene copy number variants were found, many of which were associated with genes involved in biotic and abiotic stress tolerance. In total, the authors found 2.3–3.9 Mb of *G. soja* presence–absence variation including 338 genes, with 34, 33, and 15 involved in defense response, cell growth, and photosynthesis, respectively. Comparisons among the core, the dispensable, and the pan-genome showed that the average pan-genome size between any two accessions had ~78% of the gene content of that found in all *G. soja* accessions. Eighty percent of the sequences conserved across all seven *G. soja* accessions comprised the core genome, and the dispensable genome was composed of 30,364 genes, present in at least two *G. soja* accessions. This dispensable genome had significantly higher mutation (d_n, d_s, and d_n/d_s) rates and was more variable at 4.12 SNPs/kb, as opposed to 2.67 SNPs/kb in the core. The enrichment of these types of genes suggests that the dispensable genome may play an evolutionary role in local adaptation (Li et al. 2014).

6.4.5 Novel Genes

The abundance of high-throughput RNA-Seq has opened another scientific frontier, the study of novel genes. With the unclear boundaries between transposable elements and genes, and dispensable genomes being discovered within a species, many novel genes have been identified (Arendsee et al. 2014; Li and Wurtele 2015; Li et al. 2015). If indeed the dispensable genome is responsible for local adaptation, the evolution of new genes may be more frequent than previously thought. As the number of pan-genomic studies increases, so will the plethora of novel genes. As the data accumulates in this area, more mechanistic insights can be had about the evolution of new genes, the impact of noncoding regions, and the expression patterns associated with adaptation and molecular-level innovations. The expansion of this field will greatly improve our grasp on the sought-after relationship between genotype and phenotype.

6.5 Comparative Genomics Resources

The call to integrate genetic and genomic resources for all legume species has been met by legumeinfo.org (LIS) (Dash et al. 2016; Gonzales et al. 2005; Waugh et al. 2001). Twelve sequenced legume species are available and have been annotated and compared with a multitude of analyses. Some of these resources include synteny mapping, phylogenetic trees, a genome context viewer, genetic maps, mapped traits. A dramatically useful aspect of LIS is that each of these tools is highly integrated. For example, synteny calculations using orthologous genes between multiple species are available, and complemented with functional annotations and calculations of synonymous nucleotide changes to adjust for polyploidy (Dash et al. 2016). This online interface makes comparative genomics accessible to all researchers working on a sequenced legume, regardless of programming ability. Thus, LIS provides a series of precomputed and standardized comparative genomics analyses to be explored across the scientific spectrum, thereby increasing collaborations and the general understanding of genomic phenomena.

References

Arendsee ZW, Li L, Wurtele ES (2014) Coming of age: orphan genes in plants. Trends Plant Sci 19:698–708

Bennetzen JL (2002) Mechanisms and rates of genome expansion and contraction in flowering plants. Genetica 115:29–36

Bennetzen JL, Wang H (2014) The contributions of transposable elements to the structure, function, and evolution of plant genomes. Annu Rev Plant Biol 65:505–530

Biémont C (2010) A brief history of the status of transposable elements: from junk DNA to major players in evolution. Genetics 186:1085–1093

Birchler JA, Veitia RA (2007) The gene balance hypothesis: from classical genetics to modern genomics. Plant Cell 19:395–402

Birchler JA, Veitia RA (2010) The gene balance hypothesis: implications for gene regulation, quantitative traits and evolution. New Phytol 186:54–62

Birchler JA, Veitia RA (2014) The gene balance hypothesis: dosage effects in plants. Methods Mol Biol 1112:25–32

Birchler JA, Albert PS, Gao Z (2008) Stability of repeated sequence clusters in hybrids of maize as revealed by FISH. Trop Plant Biol 1:34

Boratyn GM, Camacho C, Cooper PS, Coulouris G, Fong A, Ma N, Madden TL, Matten WT, McGinnis SD, Merezhuk Y, Raytselis Y, Sayers EW, Tao T, Ye J, Zaretskaya I (2013) BLAST: a more efficient report with usability improvements. Nucl Acids Res 41:W29–W33

Borodovsky M, Lomsadze A (2011) Eukaryotic gene prediction using GeneMark.hmm-E and GeneMark-ES. Curr Protoc Bioinformatics Chapter 4 (Unit 4.6):1–10

Borodovsky M, Peresetsky A (1994) Deriving non-homogeneous DNA Markov chain models by cluster analysis algorithm minimizing multiple alignment entropy. Comput Chem 18:259–267

Cannon SB, Ilut D, Farmer AD, Maki SL, May GD, Singer SR, Doyle JJ (2010) Polyploidy did not predate the evolution of nodulation in all legumes. PLoS ONE 5:e11630

Choi H-K, Mun J-H, Kim D-J, Zhu H, Baek J-M, Mudge J, Roe B, Ellis N, Doyle J, Kiss GB (2004) Estimating genome conservation between crop and model legume species. Proc Natl Acad Sci USA 101:15289–15294

Dash S, Campbell JD, Cannon EK, Cleary AM, Huang W, Kalberer SR, Karingula V, Rice AG, Singh J, Umale PE (2016) Legume information system (Legumelnfo. org): a key component of a set of federated data resources for the legume family. Nucl Acids Res 44:D1181–D1188

Doyle JJ, Luckow MA (2003) The rest of the iceberg. Legume diversity and evolution in a phylogenetic context. Plant Physiol 131:900–910

Doyle JJ, Doyle JL, Brown AHD, Pfeil BE (2000) Confirmation of shared and divergent genomes in the *Glycine tabacina* polyploid complex (Leguminosae) using histone H3-D sequences. Syst Bot 25:437–448

Fischer S, Brunk BP, Chen F, Gao X, Harb OS, Iodice JB, Shanmugam D, Roos DS, Stoeckert CJ Jr (2011) Using OrthoMCL to assign proteins to OrthoMCL-DB groups or to cluster proteomes into new ortholog groups. Curr Protoc Bioinformatics Chapter 6(Unit 6.12):11–19

Freeling M, Thomas BC (2006) Gene-balanced duplications, like tetraploidy, provide predictable drive to increase morphological complexity. Genome Res 16:805–814

Gondo T, Sato S, Okumura K, Tabata S, Akashi R, Isobe S (2007) Quantitative trait locus analysis of multiple agronomic traits in the model legume *Lotus japonicus*. Genome 50:627–637

Gonzales MD, Archuleta E, Farmer A, Gajendran K, Grant D, Shoemaker R, Beavis WD, Waugh ME (2005) The Legume Information System (LIS): an integrated information resource for comparative legume biology. Nucl Acids Res 33:D660–D665

Goodstein DM, Shu S, Howson R, Neupane R, Hayes RD, Fazo J, Mitros T, Dirks W, Hellsten U, Putnam N, Rokhsar DS (2012) Phytozome: a comparative platform for green plant genomics. Nucl Acids Res 40:D1178–D1186

Hawkins JS, Grover CE, Wendel JF (2008) Repeated big bangs and the expanding universe: directionality in plant genome size evolution. Plant Sci 174:557–562

Holt C, Yandell M (2011) MAKER2: an annotation pipeline and genome-database management tool for second-generation genome projects. BMC Bioinformatics 12:491

Isemura T, Kaga A, Konishi S, Ando T, Tomooka N, Han OK, Vaughan DA (2007) Genome dissection of traits related to domestication in azuki bean (*Vigna angularis*) and comparison with other warm-season legumes. Ann Bot 100:1053–1071

Jones P, Binns D, Chang HY, Fraser M, Li W, McAnulla C, McWilliam H, Maslen J, Mitchell A, Nuka G, Pesseat S, Quinn AF, Sangrador-Vegas A, Scheremetjew M, Yong SY, Lopez R, Hunter S (2014) InterProScan 5: genome-scale protein function classification. Bioinformatics 30:1236–1240

Katoh K, Standley DM (2013) MAFFT multiple sequence alignment software version 7: improvements in performance and usability. Mol Biol Evol 30:772–780

Khazaei H, O'Sullivan DM, Sillanpää MJ, Stoddard FL (2014) Use of synteny to identify candidate genes underlying QTL controlling stomatal traits in faba bean (*Vicia faba* L.). Theor Appl Genet 127:2371–2385

Kim MY, Lee S, Van K, Kim T-H, Jeong S-C, Choi I-Y, Kim D-S, Lee Y-S, Park D, Ma J (2010) Whole-genome sequencing and intensive analysis of the undomesticated soybean (*Glycine soja* Sieb. and Zucc.) genome. Proc Natl Acad Sci USA 107:22032–22037

Korf I (2004) Gene finding in novel genomes. BMC Bioinformatics 5:59

Kuzniar A, van Ham RC, Pongor S, Leunissen JA (2008) The quest for orthologs: finding the corresponding gene across genomes. Trends Genet 24:539–551

Lee C, Yu D, Choi H-K, Kim RW (2017) Reconstruction of a composite comparative map composed of ten legume genomes. Genes Genomics 39:111–119

Leitch IJ, Leitch AR (2013) Genome size diversity and evolution in land plants. Plant genome diversity, vol 2. Springer, Berlin, pp 307–322

Li L, Wurtele ES (2015) The QQS orphan gene of Arabidopsis modulates carbon and nitrogen allocation in soybean. Plant Biotechnol J 13:177–187

Li L, Stoeckert CJ Jr, Roos DS (2003) OrthoMCL: identification of ortholog groups for eukaryotic genomes. Genome Res 13:2178–2189

Li Y-h, Zhou G, Ma J, Jiang W, Jin L-g, Zhang Z, Guo Y, Zhang J, Sui Y, Zheng L (2014) De novo assembly of soybean wild relatives for pan-genome analysis of diversity and agronomic traits. Nat Biotechnol 32:1045–1052

Li L, Zheng W, Zhu Y, Ye H, Tang B, Arendsee ZW, Jones D, Li R, Ortiz D, Zhao X, Du C, Nettleton D, Scott MP, Salas-Fernandez MG, Yin Y, Wurtele ES (2015) QQS orphan gene regulates carbon and nitrogen partitioning across species via NF-YC interactions. Proc Natl Acad Sci USA 112:14734–14739

Li Y, Tong Y, Xing F (2016) DNA barcoding evaluation and its taxonomic implications in the recently evolved genus Oberonia Lindl. (Orchidaceae) in China. Front Plant Sci 7:1791

Loytynoja A (2014) Phylogeny-aware alignment with PRANK. Methods Mol Biol 1079:155–170

Maughan P, Maroof MS, Buss G (1996) Molecular-marker analysis of seed-weight: genomic locations, gene action, and evidence for orthologous evolution among three legume species. Theor Appl Genet 93:574–579

Ncbi RC (2013) Database resources of the National Center for Biotechnology Information. Nucl Acids Res 41:D8

Penn O, Privman E, Landan G, Graur D, Pupko T (2010) An alignment confidence score capturing robustness to guide tree uncertainty. Mol Biol Evol 27:1759–1767

Price MN, Dehal PS, Arkin AP (2010) FastTree 2—approximately maximum-likelihood trees for large alignments. PLoS ONE 5:e9490

Rosado A, Raikhel NV (2010) Application of the gene dosage balance hypothesis to auxin-related ribosomal mutants in Arabidopsis. Plant Signaling Behav 5: 450–452

Ross-Ibarra J (2007) Genome size and recombination in angiosperms: a second look. J Evol Biol 20:800–806

Schmutz J, Cannon SB, Schlueter J, Ma J, Mitros T, Nelson W, Hyten DL, Song Q, Thelen JJ, Cheng J (2010) Genome sequence of the palaeopolyploid soybean. Nature 463:178–183

Severin AJ, Cannon SB, Graham MM, Grant D, Shoemaker RC (2011) Changes in twelve homoeologous genomic regions in soybean following three rounds of polyploidy. Plant cell 23:3129–3136

Simao FA, Waterhouse RM, Ioannidis P, Kriventseva EV, Zdobnov EM (2015) BUSCO: assessing genome assembly and annotation completeness with single-copy orthologs. Bioinformatics 31:3210–3212

Stamatakis A (2014) RAxML version 8: a tool for phylogenetic analysis and post-analysis of large phylogenies. Bioinformatics 30:1312–1313

Stanke M, Keller O, Gunduz I, Hayes A, Waack S, Morgenstern B (2006) AUGUSTUS: ab initio prediction of alternative transcripts. Nucl Acids Res 34: W435–W439

Stupar RM (2010) Into the wild: the soybean genome meets its undomesticated relative. Proc Natl Acad Sci USA 107:21947–21948

Suzek BE, Huang H, McGarvey P, Mazumder R, Wu CH (2007) UniRef: comprehensive and non-redundant UniProt reference clusters. Bioinformatics 23:1282–1288

Tello-Ruiz MK, Stein J, Wei S, Preece J, Olson A, Naithani S, Amarasinghe V, Dharmawardhana P, Jiao Y, Mulvaney J, Kumari S, Chougule K, Elser J, Wang B, Thomason J, Bolser DM, Kerhornou A, Walts B, Fonseca NA, Huerta L, Keays M, Tang YA, Parkinson H, Fabregat A, McKay S, Weiser J, D'Eustachio P, Stein L, Petryszak R, Kersey PJ, Jaiswal P, Ware D (2016) Gramene 2016: comparative plant genomics and pathway resources. Nucl Acids Res 44:D1133–D1140

Tiley GP, Burleigh JG (2015) The relationship of recombination rate, genome structure, and patterns of molecular evolution across angiosperms. BMC Evol Biol 15:194

Varshney RK, Chen W, Li Y, Bharti AK, Saxena RK, Schlueter JA, Donoghue MT, Azam S, Fan G, Whaley AM (2012) Draft genome sequence of pigeonpea (*Cajanus cajan*), an orphan legume crop of resource-poor farmers. Nat Biotechnol 30:83–89

Vernikos G, Medini D, Riley DR, Tettelin H (2015) Ten years of pan-genome analyses. Curr Opin Microbiol 23:148–154

Vitte C, Bennetzen JL (2006) Analysis of retrotransposon structural diversity uncovers properties and propensities in angiosperm genome evolution. Proc Natl Acad Sci USA 103:17638–17643

Weigel D, Mott R (2009) The 1001 genomes project for *Arabidopsis thaliana*. Genome Biol 10:107

Wendel JF, Jackson SA, Meyers BC, Wing RA (2016) Evolution of plant genome architecture. Genome Biol 17:37

Waugh M, Anderson W, Bell C, Inman J, Schilkey F, Sullivan J, May G (2001) Legume information system. NAR molecular biology database collection 80

Young ND, Debellé F, Oldroyd GE, Geurts R, Cannon SB, Udvardi MK, Benedito VA, Mayer KF, Gouzy J, Schoof H (2011) The Medicago genome provides insight into the evolution of rhizobial symbioses. Nature 480:520–524

Zenil-Ferguson R, Ponciano JM, Burleigh JG (2016) Evaluating the role of genome downsizing and size thresholds from genome size distributions in angiosperms. Amer J Bot 103:1175–1186

From Hype to Hope: Genome-Wide Association Studies in Soybean

7

Chengsong Zhu, Babu Valliyodan, Yan Li, Junyi Gai and Henry T. Nguyen

Abstract

Association mapping studies in plants including soybean contribute to not only detecting the genetic basis of variation in yield, physiological, developmental, and morphological traits but also bringing together researchers to assemble core collections and develop genetic platforms for genotyping, phenotyping, analysis, and interpretation. The establishment of the unified mixed model greatly facilitated association mapping studies in plants and further methodology work in general. Association mapping is well positioned to exploit the advances in next-generation genomic technologies and high-through-put phenotyping. Genome-wide association studies are expected to increase dramatically once genome sequences are obtained. Moving forward, researchers in soybean and all other major plant genetics need to develop improved genetic designs and computational tools to address several challenges such as missing heritability, new gene identification, genotyping-by-sequencing, rare variants, imputation, high-throughput phenotyping, and integration of collective biological information and analytical tools into GWAS. In this chapter, we describe major progress in understanding population structure, advancements in design, and implementation of association mapping and summarize examples of association mapping in soybean. Finally, major opportunities with potential implications in soybean genetics are discussed as well.

7.1 Introduction

The priority of the plant genetics and breeding research community is to increase yield and stability of major crops to ensure food security for the fast-growing population (Varshney et al. 2014). To meet the challenge, plant breeding

C. Zhu · B. Valliyodan · H.T. Nguyen (✉)
Division of Plant Sciences, University of Missouri, Columbia, MO 65203, USA
e-mail: nguyenhenry@missouri.edu

Y. Li · J. Gai
College of Agriculture, Nanjing Agricultural University, Nanjing 210095, China

H.T. Nguyen and M.K. Bhattacharyya (eds.), *The Soybean Genome*, Compendium of Plant Genomes, DOI 10.1007/978-3-319-64198-0_7

methods, genetic designs, genomics, and biotechnologies need to be integrated to modify the adaptive, agronomic, and economic characteristics of crops. Two connected components of this endeavor are gene identification and complex trait dissection. While the former focuses on individual genes, the latter emphasizes genetic contribution and modes of action from many loci that result in phenotypic variation. Linkage mapping and association mapping are the most commonly used approaches in dissecting complex traits and identifying loci/genes underlying trait variation in plants, animals, and human (Risch and Merikangas 1996).

Association mapping provides a great platform to exploit genomic technologies and plant germplasm resources simultaneously (Zhu et al. 2008). Association mapping populations are typically assembled with diverse lines from breeding programs or sampled accessions from germplasm banks. Researchers can initiate genotyping and phenotyping activities with this approach while developing complementary linkage mapping populations. Compared with the traditional biparental linkage analysis, association mapping offers several advantages, (1) generally higher mapping resolution; (2) less research time; (3) greater allele number (Myles et al. 2009; Yu and Buckler 2006). Since its introduction to plants (Thornsberry et al. 2001), association mapping has continued to gain favorability in genetic research because of advances in high-throughput genomic technologies, interests in identifying novel and superior alleles, and improvements in statistical methods. In this chapter, we will focus on major achievements: understanding of the population structure, research strategies, and examples and progresses in soybeans. We then outline the main challenges that have broad implications.

7.2 Principles of Association Analysis

7.2.1 Population Structure and Association Methods

Because an association mapping panel is often an assembled population, rather than a random mated or a designed population, the presence of population structure can lead to false positive discoveries if this structure is not adequately accounted for during marker-trait association analysis (Fig. 7.1). In general, population structure can arise due to differences in geographical origins, local adaptations, or breeding history of the lines in the panel (Yu and Buckler 2006). Although the complex genealogical history of a species or a population prevents us from drawing a clear delineation among various scenarios, association mapping samples are generally pragmatically classified into five categories based on population structure and familial relatedness. (1) ideal samples with subtle population structure and familial relatedness, (2) samples with familial relationship, (3) samples with population structure, (4) samples with both population structure and familial relationships, and (5) samples with severe population structure and familial relationships (Zhu and Yu 2009). It is possible to quantify population structure using neutral markers and then account for the structure statistically in identifying marker-trait associations. Several methods have been used to control for population structure in association mapping. These include genomic control (Devlin and Roeder 1999; Devlin et al. 2004), structured association (Pritchard et al. 2000), principal component analysis (PCA) (Price et al. 2006), unified mixed model (Yu et al. 2006), non-metric multidimensional scaling (nMDS) (Zhu and Yu

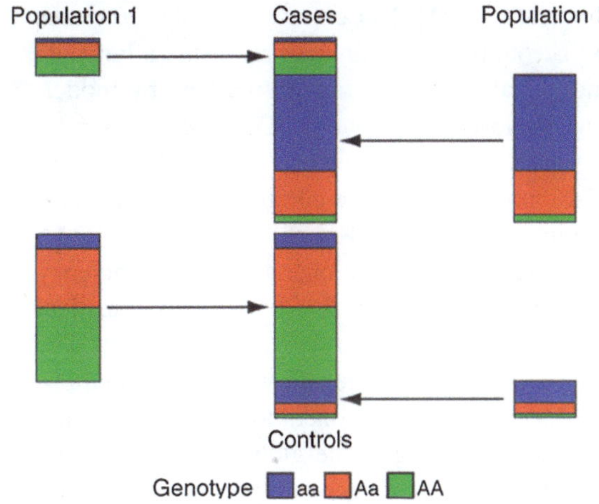

Fig. 7.1 Effect of population structure at a SNP locus which mimics the false signal of association in that there is a significant difference in allele and genotype frequencies between cases and controls, but there is truly no association in each subpopulation

2009). And the computational speed and power of the mixed model have also been increased (Zhang et al. 2010; Zhou and Stephens 2012; Loh et al. 2015).

Recent methods mainly use the mixed linear model (MLM) to account for population structure and familial relatedness (Price et al. 2010a, b; 2013). The unified mixed-model (Yu et al. 2006) approach for association mapping considers both population structure and pair-wise relatedness to account for genetic relationship. With this general framework, population structure (*Q* matrix) estimated using STRUCTURE software (Pritchard et al. 2000) is fitted as a fixed effect, and kinship (*K* matrix) estimated using SPAGeDi (Hardy and Vekemans 2002) is used to define the variance–covariance structure of the random effects among individuals for association mapping.

Principal component analysis (PCA) of the genotypic data transforms the variation into a series of orthogonal continuous axes, and these components, typically the first few, can then be used to replace *Q* to adjust for population structure. EIGENSTRAT is one software package that can be used to infer PCA (Patterson et al. 2006; Price et al. 2006). Unlike *Q*, computing of PCA needs no assumptions about the number of groups in a population. With the unified mixed-model framework, top eigenvectors can be fit as fixed effects to complement the random effect (Zhu and Yu 2009). Alternatively, nMDS can be used. A comparison of nMDS and PCA using the simulated and empirical data from cross- and self-pollinated species showed that models within MDS resulted in slightly higher power and fewer false positives (Zhu and Yu 2009). One question emerges as PCA or nMDS is used for population structure control: How many components one must fit into the model to account for the population structure, and should this number be the same for all traits, or vary for different traits? A two-stage dimension determination approach was proposed for PCA and nMDS (Zhu and Yu 2009). In general, a model testing should be conducted to select the most appropriate models based on the Bayesian Information Content (BIC), the lower the better. All relevant models are compared under maximum likelihood (ML). With the selected model, individual markers can then be tested for the marker-trait association analysis. With computer simulations, the corresponding quantile–quantile (Q–Q) plot of the selected model can be compared with the plots of other models to demonstrate that different numbers of PCAs are needed to have adequate, but not excessive, control for population structure.

With the unified mixed-model framework, newer and faster algorithms have been developed to speed up genome-wide association studies (GWAS) analysis, particularly when hundreds of thousands of single-nucleotide polymorphisms (SNPs) are tested. Efficient mixed-model association (EMMA) corrects sample structures by accounting for pair-wise relatedness between individuals and uses enough markers by modeling phenotype distribution. This method is related to a method developed to simulate a null distribution of variance-component test statistics (Crainiceanu and Ruppert 2004). EMMA increases the computational speed and efficiency of mixed-model analysis by enabling statistical tests with single-dimensional optimization. The method also avoids the redundant, computationally expensive matrix operations at iteration and allows converging to the global optimum of the likelihood in variance-component estimation with high confidence. This capability was demonstrated in in silico whole-genome association mapping of mouse, Arabidopsis, and maize datasets. Results from the EMMA method are consistent with published results in reducing false positives and are faster than the previous methods while performing near global optimization (Kang et al. 2008, 2010). Compressed MLM is another approach that clusters individuals into groups based on kinship estimates, thereby reducing the effective sample size for computation to improve speed in subpopulation determination. This method is an extension of the pedigree-based sire model with modifications (Henderson 1984; Zhang et al. 2010).

7.2.2 Software for Association Mapping

In this section, we will mention new algorithms implemented in a set of software packages. Some detailed information about these algorithms is further explained in the next section. The most commonly used and frequently updated software for association mapping is trait analysis by association, evolution and linkage (TASSEL), which is written in Java and can be used in virtually any operating system (Bradbury et al. 2007). TASSEL implements general linear model (GLM), MLM, compressed MLM, and P3D approaches for marker-trait association analysis. Other notable functions include evolutionary analysis, computation of LD, imputation of missing data, and data visualization. The program allows calculating and visualizing LD graphically. Structured association (Pritchard et al. 2000) as well as the unified mixed model (Yu et al. 2006) was first implemented in TASSEL to reduce the risk of false positives. The $Q + K$ method was implemented in TASSEL as a MLM function. TASSEL earlier employed an EM (expectation–maximization) algorithm for MLM analysis. To increase computing speed and analyze larger datasets, the EMMA algorithm was incorporated into TASSEL. As indicated earlier, compressed MLM was added recently to increase computational speed and this procedure can also be optimized to increase the power. For even larger datasets, a newer method estimates the population parameters prior to estimating the parameters for test markers. Termed as population parameters previously determined (P3D) (Zhang et al. 2010), this method is available in the newer version of TASSEL, and compressed estimation of variance components are available in software EMMA expedited (EMMAX) (Kang et al. 2010). EMMAX, publicly available software, implements a variance-component approach for GWAS. EMMAX is built on EMMA's previous approach.

ASREML is a complete package with different modules for mixed-model analysis (Gilmour 2007). SAS and R software are generic tools that can be used for association mapping. Genome Association and Prediction Integrated Tool (GAPIT) is a new tool in the R package that can perform genome-wide association study (Lipka et al. 2012). It integrates the unified mixed model, EMMA, P3D, and compressed MLM with genomic prediction. This software handles large genotypic datasets by subdividing them into multiple files, but the memory requirement remains the same. GAPIT can conduct hierarchical clustering and kinship matrices based on user input and linkage information. The results are produced in the form of Q–Q plot, Manhattan plot, PCA, and association tables. New state-of-the-art statistical methods have now been implemented in a new, enhanced version of GAPIT (Tang et al. 2016). These methods include factored spectrally transformed linear mixed models (FaST-LMM), enriched CMLM (ECMLM), FaST-LMM-Select, and settlement of mixed linear models under progressively exclusive relationship (SUPER). The genomic prediction methods implemented in this new release of the GAPIT include gBLUP based on CMLM, ECMLM, and SUPER. These enhancements make the GAPIT a valuable resource for determining appropriate experimental designs and performing GWAS and genomic prediction.

Genome-wide efficient mixed-model association (GEMMA) is a method *n* (sample size) times faster than the EMMA method. It is not an approximation method similar to genome-wide rapid association using mixed model and regression (GRAMMER) but an exact test method. This requires complete or imputed SNP data, and for each SNP tested it replaces the additional eigen-decomposition step in EMMA with one-matrix vector multiplication. After that, each iteration of the optimization requires inexpensive operations. GEMMA was built on EMMA framework to facilitate genome-wide application (Zhou and Stephens 2012).

Multi-locus mixed model (MLMM) is a method which can outperform the existing methods in analyzing GWAS data in power and false discovery rate. The existing methods to account for population structure are good when small number of loci control the traits. But for complex traits controlled by several large-effect loci, MLMM is efficient. The principle of this approach is similar to the use of cofactors in the statistical models of quantitative trait loci (QTL) mapping in composite interval mapping. Likewise including multiple cofactors on a genome-wide scale can increase the power and reduce false discovery rates in GWAS (Segura et al. 2012).

7.2.3 Computational Speed

The unified mixed-model method[1] is widely used to correct for the effects of genetic relatedness in association mapping studies; however, in analyzing genome-wide datasets, solving the mixed model requires a huge amount of computing power. The computing time for solving an MLM increases with the cube of the number of individuals. One approach to reducing computing time is compressed MLM (Zhang et al. 2010), which decreases the effective sample size of such datasets by clustering individuals into groups. The rationale behind this method has its roots in the sire-model approach (Quaas and Pollak 1981). As a complementary approach to compressed MLM, the population parameters previously determined (P3D) approach reduces computing time by skipping the iteration process in each individual marker test. In the first step, a base MLM without fitting any marker effect is solved for the variance components. In the second step, an individual marker test with MLM simply uses variance components from the first step without solving the specific mixed model again (Zhang et al. 2010). This practice has been used in previous mixed-model analyses to save computing time (Yu et al. 2006), but the need to reduce the computational burden of MLM is much higher in GWAS. Compressed MLM and P3D, when implemented jointly, significantly reduce computing time and maintain statistical power (Zhang et al. 2010). These methods are implemented in the software program TASSEL (Bradbury et al. 2007). A different residual analysis approach was also proposed to conduct fast genome-wide pedigree-based association analysis (Aulchenko et al. 2007).

A variance-component approach implemented in a publicly available software package, EMMAX (Kang et al. 2010), reduces computing time for analyzing large GWAS datasets. First, a pair-wise relatedness matrix is computed from high-density markers and used to represent the sample structure. Secondly, the contribution of the sample structure to the phenotype using a variance-component model is estimated, resulting in an estimated covariance matrix of phenotypes that models the effect (Kariya and Kurata 2004) of genetic relatedness on the phenotypes. Thirdly, a generalized least square (GLS) *F*-test (Kariya and Kurata 2004) is applied to each marker to detect associations accounting for the

[1]It seems that a list of abbreviations (or adding the complete terms in parentheses after some of the abbreviations) is needed.

sample structure using the covariance matrix. A study on the welcome trust consortium data (Browning and Browning 2008) found that EMMAX outperforms both PCA (Price et al. 2006) and genomic control (Devlin and Roeder 1999). FaST-LMM, a factored spectrally transformed linear mixed model, was recently proposed to further address the computational issues of MLM (Lippert et al. 2011).

FaST-LMM is an algorithm for genome-wide association studies that scales linearly with sample size in both runtime and memory use. With data from 15,000 individuals, FaST-LMM ran an order of magnitude faster than current algorithms; whereas data for 120,000 individuals were analyzed with FaST-LMM in few hours, current algorithms failed. The LMM corrects for confounding by measuring genetic similarity using methods of identity by descent and a realized genetic relationship matrix (GRM), estimated by using a small sample of markers. FaST-LMM can produce results similar to the LMM by reformulating the LMMs with two conditions: (1) the number of SNPs used to estimate genetic similarity is less than the number of individuals in the dataset, and (2) the GRM is used to determine these similarities. This method requires a single spectral decomposition but does not assume variance parameters are same across the SNPs.

7.2.4 Threshold for Declaring the Signals

GWAS have long relied on proposed statistical significance thresholds to be able to differentiate true positives from false positives. For work focused on gene discovery, minimizing the false positive rate is a more important consideration than controlling the false negative rate. A variety of statistical approaches accounting for multiple testing in the genome-wide setting have been developed, including statistical independence methods; methods for adjusting the error rate, such as Bonferroni correction, Sidak correction, false discovery rate, weighted multiple hypothesis testing; and data reduction methods (Sham and Purcell 2014). At a practical level, some early GWAS used a threshold of $P \leq 10^{-7}$ or Bonferroni correction for small scale of SNP sets like SoySNP50K Bead Chip in soybean (Ray et al. 2015; Song et al. 2015; Rincker et al. 2016), but the current practice seems to prefer routinely a threshold of $P \leq 5 \times 10^{-8}$ (Barsh et al. 2012), which is equivalent to the Bonferroni corrected threshold ($\alpha = 0.05$) for 1 million independent variants. The experimental systems in plants are diverse and include structured populations, advanced intercrosses, recombinant inbred collections. Overall, when genome-wide information is used to analyze these studies, we still require genome-wide significance thresholds (5×10^{-8}) analogous to those used in human studies in order to minimize the chance of false positive results.

7.3 Achievements

7.3.1 Progress of Association Mapping in Soybean

So far, a series of research papers focusing on LD and association mapping have been published, spanning a number of complex traits of importance in soybean (Fig. 7.1), such as seed composition (Hwang et al. 2014; Vaughn et al. 2014; Bandillo et al. 2015), 100-seed weight (Lü et al. 2011; Zhang et al. 2015; Zhou et al. 2015; Wang et al. 2016), salt tolerance (Guan et al. 2014; Patil et al. 2016), cyst nematode (Vuong et al. 2015; Li et al. 2016), ureide concentration (Ray et al. 2015), nitrogen traits (Dhanapal et al. 2015a, b), photochemical reflectance index (Herritt et al. 2016), brown stem rot (Rincker et al. 2016), iron deficiency chlorosis (Wang et al. 2008), carbon isotope ratio (Dhanapal et al. 2015a, b), agronomic traits (Kumar et al. 2014), domestication-related traits (Zhou et al. 2015, 2016; Valliyodan et al. 2016). Many challenges still demand further study as we attempt to gain a better grasp of the various genetic and statistical aspects of association mapping. We offer our opinions on some of these opportunities and challenges in the following sections.

Soybean is an autogamous plant species that exhibits high variation in LD across its genome, indicating that a large number of markers are needed to perform GWAS. The LD pattern in soybean was identified by a study that focused on three genomic regions varying from 336 to 574 kb. The populations used were 26 accessions of the wild ancestor of soybean (*Glycine soja*), 52 Asian *G. max* landraces, 17 Asian landrace introductions that became the ancestors of North American cultivars, and 25 elite cultivars from North America. In the three cultivated *G. max* groups, LD (when r^2 declines to 0.1) extended from 90 to 574 kb (Hyten et al. 2007). Despite high LD in soybean, efforts are ongoing to sequence a number of wild soybeans, landraces, and cultivars to further understand the genetic structure and to perform a number of association mapping (Fig. 7.2). Whole-genome sequencing in soybean identified about 10 million polymorphisms among 106 PIs and showed that the soybean genome is characterized by highly divergent haplotypes (Zhou et al. 2015; Valliyodan et al. 2016). The Phase I HapMap resulted in the identification of high-quality genotyping data of 20 million SNPs and about 500 K indels. These efforts provide the foundation for dissecting the complex trait variation in soybean by uniting research efforts around the world.

Soybean oil and meal are major contributors to worldwide food production. A GWAS was performed using all 12,116 *G. max* accessions with (Fig. 7.2) and then for subsets of accessions based on world region and MG class (Bandillo et al. 2015). A total of 19 significant associations were identified for protein. Clusters of highly significant markers were present on chromosomes 15 (3.82–3.96 Mb) and 20 (29.59–31.97 Mb), which collectively explained 7% of the phenotypic variance for protein. The GWAS for oil detected 18 significantly associated SNPs, with the strongest association detected at 3.82 Mb on Chromosome 15. Collectively, the three QTL identified for oil explained 6% of the phenotypic variance. The major associations detected on chromosomes 15 and 20 were highly significant

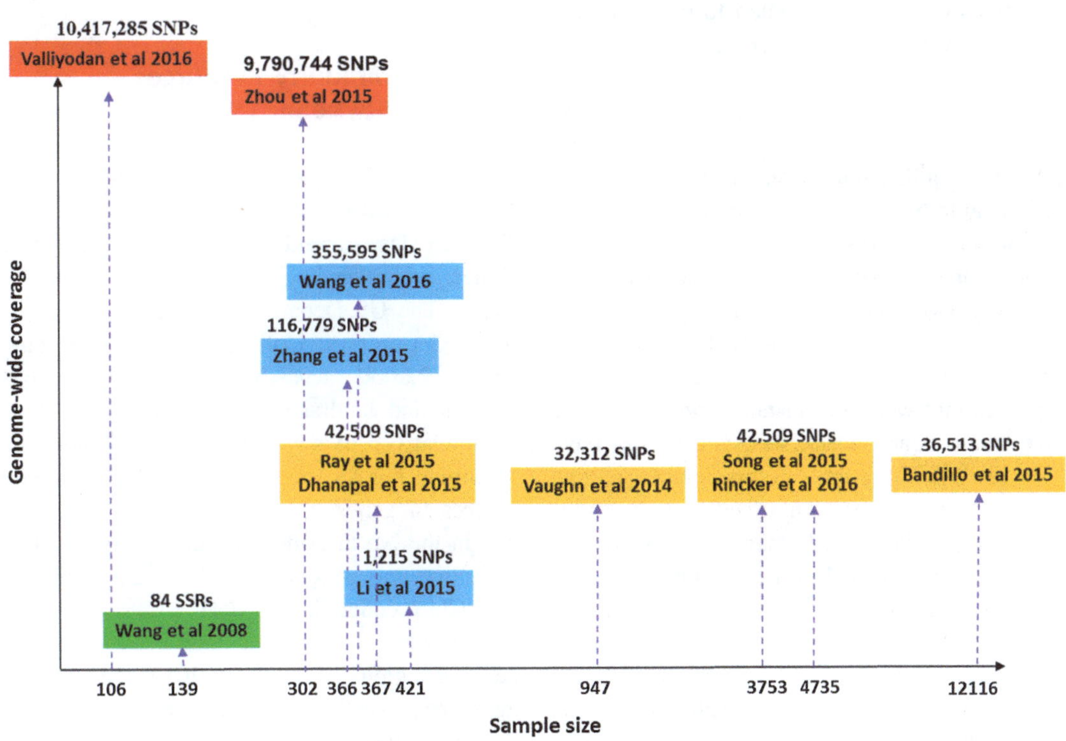

Fig. 7.2 Sample of publications on association analysis in soybean based on population size and scale of markers

for both protein and oil and have been detected repeatedly in linkage mapping studies. Haplotype analysis further narrowed down to less than 1 Mb the region spanning between 29.06 and 30.04 Mb, where encompasses only three (Glyma20 g21030, Glyma20 g21040, and Glyma20 g21080) of the original 12 potential candidate genes (Bolon et al. 2014).

The 100-seed weight is a key component of soybean yield and is generally positively correlated with yield (Mian et al. 1996). A representative sample comprising 366 accessions from the Chinese soybean landrace population (CSLRP) was tested under four growth environments to perform association analysis between 100-seed weight and a total of 116,769 SNPs, where 55 QTLs associated with seed weight were identified (Zhang et al. 2015). In another study (Wang et al. 2016), a high-throughput NJAU355K SoySNP array in 367 soybean accessions including 105 wild soybeans and 262 cultivated soybeans was developed and be used to identify genomic regions related to seed weight. Sixteen domestication genes which happened to be domestication-selective sweeps were revealed to be associated with seed weight. To further identify genomic regions related to seed weight, a GWAS was conducted across multiple environments in wild and cultivated soybeans. As a result, a strong linkage disequilibrium region on Chromosome 20 was found to be significantly correlated with seed weight in cultivated soybeans.

Soybean domestication was a complex process. Morphological features of soybeans were selected during domestication (wild soybeans vs. landrace) and improvement (landraces vs. improved cultivars). For instance, wild soybeans have small, coarse black seeds; landraces have large seeds with variable colors (black, yellow, green, or striped); and improved cultivars have large shiny yellow seeds, suggesting seed size was mainly selected during domestication, whereas seed color was uniformly selected during improvement. In addition to changes in morphology, cultivated and wild soybeans have different seed oil content. Typically, wild soybean seeds have lower oil content. These results indicated that dominant selection of different traits might have occurred at different evolutionary stages. To identify potential-selective signals during soybean domestication and improvement, Zhou et al. (2015) scanned genomic regions with extreme allele differentia ion over genomic regions using likelihood method. A total of 121 domestication-selective sweeps and 109 improvement-selective sweeps were detected. GWAS for domestication traits such as plant height, flower color, stem determinacy, seed weight, seed coat color, hilum color, pubescence form and oil content which have significant variation among different populations were performed. In the analysis, selection signals were detected in almost all reported domestication-related QTL regions, and sizes of the selected regions were smaller than the QTLs. For example, six strong GWAS signals, five of which overlapped with previously identified oil content QTLs, and one that was newly identified were detected. We determined that two of the six GWAS signals, located on Chromosome 13 and Chromosome 03, overlapped with domestication-selective sweeps.

7.3.2 Community Resources in Soybean

7.3.2.1 USDA Soybean Germplasm Collection

The US Department of Agriculture Soybean Collection initiated soybean collection project in 1970s. The US Department of Agriculture Soybean Germplasm Collection (SGC) contains nearly 22,000 accessions, including 19,648 modern and landrace cultivars (*G. max*), 1168 wild relatives of soybean (*G. soja*), and 1184 perennial. The accessions in the Collection are the sources of genes for soybean improvement not just in the North America, but worldwide. This collection was genotyped with the SoySNP50 K BeadChip containing 52,041 SNPs that were chosen from euchromatic and heterochromatic genome regions (Song et al. 2013). Meanwhile, the Germplasm Resources Information Network (GRIN, www.ars-grin.gov) provides tons of phenotypic information about soybean including seven categories or 95 traits in total. The phenotype data were originally obtained from field evaluations conducted by USDA–ARS germplasm curation staff and their collaborators. The field evaluations were conducted at various locations at which accessions from one or more maturity group (MG) classes had adaptation, and such field trials often spanned several years. A series of association studies have been conducted using these community resources, such as seed composition (Hwang et al. 2014; Vaughn et al. 2014; Bandillo et al. 2015), seed weight (Song et al. 2015), cyst nematode (Vuong et al. 2015), ureide concentration (Ray et al. 2015), nitrogen traits (Dhanapal et al. 2015a, b), photochemical reflectance index (Herritt et al. 2016), brown stem rot (Rincker et al. 2016).

7.3.2.2 Nested Association Mapping

Ideally, an association mapping population can be genotyped once but phenotyped repeatedly for the same sets of traits and new sets of traits; thus, it is advantageous to have a population that can be used by the research community for different purposes. With this in mind, the soybean community has developed a nested association mapping (NAM) population to integrate the advantages of linkage analysis and association mapping, with the ultimate goal of improving the yield potential of soybean varieties (www.soybase.org/SoyNAM). The aims of developing NAM population were to (1) capture soybean genetic diversity, (2) exploit the historical recombination events in soybean, (3) use a genetic design that can take advantage of next-generation sequencing technologies, (4) generate materials for evaluation of agronomic traits in the field locations, (5) develop a population with enough power to detect QTLs and resolve QTLs to the gene level, and (6) provide a community resource that will enable a wide range of community efforts and databases for researchers. A set of 40 diverse inbred lines were crossed to hub inbred line IA3023 to create 40 populations of 140 recombinant inbred lines (RILs) each, for a total of 5600 distinct genotypes. A total of 40 parents include 17 high yielding lines from 8 state breeders, 15 lines with diverse ancestry from Randy Nelson's program, and 8 PIs identified by Jim Specht with high yields in drought. The 5600 genotypes are called NAM recombinant inbred lines (NAM RILs). All the 41 parent inbred lines have been re-sequenced and those RILs and their parents have been genotyped with 4312 SNPs by Qijian Song and Perry Cregan at USDA-ARS, Beltsville, MD and phenotyped for yield and other key agronomic traits in performance trials conducted by a collaborative group of researchers located throughout the Midwest (Diers et al. 2015).

In essence, NAM exploits a multiple RIL population derived from crosses between a common founder line and a set of diverse founders. The strategy of NAM is to genotype common-parent-specific (CPS) markers on the founders and progenies while sequencing the founders completely or densely genotyping the founders with high-density markers. With CPS markers serving as a bridge, the genetic information obtained from genotyping the founders with high-density markers can be projected from the founders to the progenies. Projecting genetic information from the parents to the progenies also reduces genotyping costs. Notice that the concept of NAM involves the development of a population. Instead of assembling existing lines to form a population, NAM selects a diverse set of founders that are representative of the main breeding pools of the target species. Importantly, as compared with the conventional association mapping approach, NAM is characterized by higher statistical power, better mapping resolution, lower sensitivity to genetic heterogeneity, and a lower requirement of SNP markers in the progenies (Yu et al. 2008; Buckler et al. 2009).

7.4 Challenges and Opportunities

7.4.1 New Gene Identification

Identifying previously unknown genes underlying a complex trait remains to be challenging. Once association signals are detected, concerted efforts are required to identify the causal genes or polymorphisms. These follow-up studies need to be well designed given the complex genome of major crops and difficulties in choosing appropriate genetic backgrounds for genetic and transformation validation. If the complete genome sequence is unavailable while conducting association mapping, new genes with small to modest effects are difficult to be recognized and followed up in further studies. On the other hand, the candidate gene approach by definition is not designed to identify novel genes underlying a complex trait. In addition, allele frequency, multiple testing, and epistasis add to the problem of low detection rate. Compared with human genetics, the number of GWAS in plants with adequate sample size and marker density is very limited. However, advances in sequencing and GWAS methodology, completion of draft genome sequence or even multiple reference genome sequences, and continued improvements in different genetic and genomic techniques would eventually make it possible to realize the potential offered by association mapping in identifying new genes underlying complex traits.

7.4.2 Missing Heritability

Association mapping strategy is based on the assumption that the common phenotypic variation is caused by common genetic variants. As a result, array-based genotyping has been used extensively in GWAS because these SNPs were selected to represent the major genetic polymorphisms and were expected to explain a decent proportion of phenotypic variation of the trait. However, if we summarize the effect of statistically significant SNPs identified through GWAS, they explain only a small proportion of the phenotypic variation, which has led to the important and hotly debated issue of where the 'missing heritability' of complex traits might be found (Maher 2008; Manolio et al. 2009).

The causes of missing heritability can be attributed to a number of factors: genotype by environment interaction, a larger number of variants of smaller effects yet to be found; rare variants (possibly with larger effects) that are poorly detected by available genotyping arrays that focus on variants present in 5% or more of the population; structural variants poorly captured by existing arrays; low power to detect gene-by-gene interactions, inadequate accounting for shared environment among relatives, statistical issues, copy number variation, and multiple testing issues (Vineis and Pearce 2010). The power of GWAS to detect variants of modest effect and low frequency remains lacking due to the low frequency of functional alleles in the mapping population, the low influence of low-frequency alleles on the population, and/or lower detection power of the association mapping strategy. The phenotypic variation caused by numerous small-effect alleles will be difficult to detect compared with a small number of large-effect alleles; this is a challenge for any complex trait dissection studies, including association mapping.

7.4.3 Genotyping-by-Sequencing (GBS) and Next-Generation Sequencing

Rapidly evolving sequencing and genotyping technologies have fundamentally changed not only the design of specific breeding and selection strategies in crops, but also how the vast amount of available germplasm diversity can be utilized efficiently (Bernardo and Yu 2007; Heffner et al. 2009; Tester and Langridge 2010). Routine use of genomic selection in plant breeding is becoming possible because of the significantly reduced cost of obtaining molecular marker information, particularly SNPs, thanks to the development of high-throughput technology from DNA extraction, sample preparation, and array-based genotyping technologies as well as cutting-edge GBS technology (Metzker 2010). Current GBS research includes species with a sequenced genome, such as rice (Huang et al. 2009), maize (Elshire et al. 2011), and sorghum (Morris et al. 2013), and those without, such as wheat and barley (Chutimanitsakun et al. 2011). Next-generation sequencing technologies have improved output and made possible sequencing of multiple samples at the same time.

Sequencing-based high-throughput genotyping combines the advantages of cost-effectiveness, less time, and dense marker data. In the first whole-genome sequencing paper of soybean (Xu et al. 2013), a bin mapping approach for analyzing the SNPs collectively rather than individually was used on 246 RILs derived from a cross between Magellan (susceptible) and PI 438489B (resistant). The SNP calling in this method is based on a recombination break point and sliding window. With the sequence-based genetic map, three QTL for root-knot nematode were identified. Of these, one major QTL was mapped in Chromosome 10, which is 29.7 kb in size and harbors three true genes and two pseudogenes.

7.4.4 Imputation Analysis

Genome-wide association studies (GWAS) have identified over thousands SNPs associated with complex traits of agronomic importance. In soybean, to date, these studies have been carried out mainly using soySNP50K iSelect BeadChip containing about 52,041 SNPs, which has been used to genotype the USDA Soybean Germplasm Collection for 19,652 *G. max* and *G. soja* accessions (Song et al. 2013, 2015; Bandillo et al. 2015). DNA re-sequencing has emerged as a powerful new technology; two whole-genome sequencing (WGS) analyses were performed in soybean (Zhou et al. 2015; Valliyodan et al. 2016). Genotyping arrays used for GWAS are based on tagging SNPs and therefore do not directly genotype all variation in the genome. Costs to generate whole-genome sequence data are decreasing rapidly. It is expected that, in the next few years, whole-genome sequence data will be widely available for crops and livestocks, as is already the case for human studies. Despite the fact that costs of sequencing are decreasing, it is still expensive to fully sequence a core set of individuals. Hence, a feasible strategy is that lower density genotypes of the individuals are imputed to whole-genome sequence genotypes using the re-sequenced lines as reference.

Genotype imputation is the term used to describe the process of predicting or imputing genotypes that are not directly assayed in a sample of individuals. There are several distinct scenarios in which genotype imputation is desirable, but the term now most often refers to the situation in which a reference panel of haplotypes at a dense set of SNPs is used to impute into a study sample of individuals that have been genotyped at a subset of the SNPs (Marchini and Howie, 2010). As a consequence, we can assess the effect of more SNPs than those on the original micro-array. Genotype imputation hence helps tremendously in narrowing down the location of probably causal variants and leads to more powerful association analyses. Importantly, imputation has facilitated meta-analysis of datasets that have been genotyped on different arrays, by increasing the overlap of variants available for analysis between arrays.

7.4.5 Rare Variants

Because GWAS focus on the identification of common variants, it is plausible that analyses of low-frequency variants ($0.5\% \leq MAF < 5\%$) and rare variants ($MAF < 0.5\%$) are usually being ignored. However, rare variants may play a significant role in complex disease, as well as some Mendelian conditions. Rare variants may be responsible for a portion of the missing heritability of quantitative traits. The theoretical case for a significant role of rare variants is that alleles that strongly predispose an individual to disease will be kept at low frequencies in populations by purifying selection (Cirulli and Goldstein 2010). Rare variants are increasingly being studied, as a consequence of exome and whole-genome sequencing efforts. While these variants are individually infrequent in certain populations, they can be unique to specific populations. They are more likely to be deleterious than common variants, as a result of rapid population growth and weak purifying selection. They have been suspected of acting independently or along with common variants to cause disease states.

At present, GWAS is unable to detect rare variants through common SNP markers (Ott et al. 2011). The power to detect an association is a function of allele frequency, and individually, rare alleles have little influence on the population, which renders their detection difficult. Several specific statistical approaches have been developed to assess the significance of rare variants (Morgenthaler and Thilly 2007; Li and Leal 2008; Madsen and Browning 2009; Price et al. 2010a, b; Wu et al. 2011).

7.4.6 High-Throughput Phenotyping

Constraints in field phenotyping capability limit our ability to dissect the genetics of quantitative traits, particularly those related to yield and stress tolerance (e.g., yield potential as well as increased drought tolerance, flooding tolerance, and nutrient use efficiency). The development of effective field-based high-throughput phenotyping platforms (HTPPs) remains a bottleneck for future breeding advances. However, progress in sensors, aeronautics, and high-performance computing is paving the way. In recent years, there has been increased interest in HTPPs. Most HTPPs, both those run by the big transnational seed companies and the most advanced public plant research institutions around the world, such as the Australian Plant Phenomics Facility (http://www.plantphenomics.org.au/), the European Plant Phenotyping Network (www.plant-phenotyping-network.eu/eppn/structure), and the USDA (www.nifa.usda.gov/nea/plants/), are fully automated facilities in greenhouses or growth chambers with robotics, precise environmental control, and remote sensing techniques to assess plant growth and performance. Using a combination of HTPPs and GWAS, it has shown that high-throughput phenotyping has the potential to replace traditional phenotyping and serves as a novel tool for studies of plant genetics, genomics, gene characterization, and breeding (Yang et al. 2014).

7.4.7 Integrating Collective Biological Information and Analytical Tools into GWAS

Most of the GWAS in plant genetics so far were based on single common variant analyses (Manolio et al. 2009), although it has been shown that multiple rare variants together may account for a few proportions of phenotypic variation for complex traits (Bansal et al. 2010). But these studies with a focus on rare variants were the analysis of one or several candidate genes, and re-sequenced-based association studies are still not available. Pathway-based approaches have recently been developed to use prior biological knowledge on gene function to facilitate the analysis of GWAS dataset (Wang et al. 2010). A Composite Resequencing-Based Genome-Wide Association Studies (CR-GWAS) approach was recently proposed to address this issue (Zhu et al. 2011). This approach integrates next-generation sequencing, prediction of biological function of SNPs, statistical test for rare allele variants, and genome databases and gene networks. Using this strategy, the known true positives were confirmed and several new promising associations were identified. Promising genes were shown to control for flowering time through either common variants or rare variants within a diverse set of Arabidopsis accessions (Zhu et al. 2011).

References

Aulchenko YS et al (2007) GenABEL: an R library for genome-wide association analysis. Bioinformatics 23 (10):1294–1296

Bandillo N et al (2015) A population structure and genome-wide association analysis on the USDA soybean germplasm collection. Plant Genome 8 (3):1–13

Bansal V et al (2010) Statistical analysis strategies for association studies involving rare variants. Nat Rev Genet 11(11):773–785

Barsh GS et al (2012) Guidelines for genome-wide association studies. PLoS Genet 8(7):e1002812

Bernardo R, Yu JM (2007) Prospects for genomewide selection for quantitative traits in maize. Crop Sci 47 (3):1082–1090

Bolon YT et al (2014) eQTL networks reveal complex genetic architecture in the immature soybean seed. Plant Genome 7(1):1–14

Bradbury PJ et al (2007) TASSEL: software for association mapping of complex traits in diverse samples. Bioinformatics 23(19):2633–2635

Buckler ES et al (2009) The genetic architecture of maize flowering time. Science 325(5941):714–718

Chutimanitsakun Y et al (2011) Construction and application for QTL analysis of a restriction site associated DNA (RAD) linkage map in barley. BMC Genomics 12:4

Cirulli ET, Goldstein DB (2010) Uncovering the roles of rare variants in common disease through whole-genome sequencing. Nat Rev Genet 11(6):415–425

Crainiceanu CM, Ruppert D (2004) Likelihood ratio tests for goodness-of-fit of a nonlinear regression model. J Multivariate Anal 91(1):35–52

Devlin B, Roeder K (1999) Genomic control for association studies. Biometrics 55(4):997–1004

Devlin B, Bacanu SA, Roeder K (2004) Genomic control to the extreme. Nat Genet 36(11):1129–1130

Dhanapal AP et al (2015a) Genome-wide association study (GWAS) of carbon isotope ratio (delta C-13) in diverse soybean [*Glycine max* (L.) Merr.] genotypes. Theor Appl Genet 128(1):73–91

Dhanapal AP et al (2015b) Genome-wide association analysis of diverse soybean genotypes reveals novel markers for nitrogen traits. Plant Genome 8(3):1–15

Diers BW et al (2015) Nested association mapping of agronomic traits in soybeans. In: Soybean Breeders Workshop, St. Louis

Elshire RJ et al (2011) A robust, simple genotyping-by-sequencing (GBS) approach for high diversity species. PLoS ONE 6(5):e19379

Gilmour AR (2007) Mixed model regression mapping for QTL detection in experimental crosses. Comput Stat Data Anal 51(8):3749–3764

Guan RX et al (2014) Salinity tolerance in soybean is modulated by natural variation in GmSALT3. Plant J 80(6):937–950

Hardy OJ, Vekemans X (2002) SPAGEDi: a versatile computer program to analyse spatial genetic structure at the individual or population levels. Mol Ecol Notes 2(4):618–620

Heffner EL, Sorrells ME, Jannink JL (2009) Genomic selection for crop improvement. Crop Sci 49(1):1–12

Henderson CR (1984) Applications of linear models in animal breeding. University of Guelph, Guelph

Herritt M et al (2016) Identification of genomic loci associated with the photochemical reflectance index by genome-wide association study in soybean. Plant Genome 9(2):93–104

Huang XH et al (2009) High-throughput genotyping by whole-genome resequencing. Genome Res 19(6): 1068–1076

Hwang EY et al (2014) A genome-wide association study of seed protein and oil content in soybean. BMC Genomics 15:1

Hyten DL et al (2007) Highly variable patterns of linkage disequilibrium in multiple soybean populations. Genetics 175(4):1937–1944

Kang HM et al (2008) Efficient control of population structure in model organism association mapping. Genetics 178(3):1709–1723

Kang HM et al (2010) Variance component model to account for sample structure in genome-wide association studies. Nat Genet 42(4):348–354

Kariya T, Kurata H (2004) Generalized least squares. Wiley, London

Kumar B et al (2014) Population structure and association mapping studies for important agronomic traits in soybean. J Genet 93(3):775–784

Li B, Leal SM (2008) Methods for detecting associations with rare variants for common diseases: application to analysis of sequence data. Am J Hum Genet 83(3): 311–321

Li Y-H et al (2016) Dissecting the genetic basis of resistance to soybean cyst nematode combining linkage and association mapping. Plant Genome 9(2): 16–26

Lipka AE et al (2012) GAPIT: genome association and prediction integrated tool. Bioinformatics 28(18): 2397–2399

Lippert C et al (2011) FaST linear mixed models for genome-wide association studies. Nat Methods 8(10): 833–835

Loh PR et al (2015) Efficient Bayesian mixed-model analysis increases association power in large cohorts. Nat Genet 47(3):284–290

Lü H-Y et al (2011) Epistatic association mapping in homozygous crop cultivars. PLoS ONE 6(3):e17773

Madsen BE, Browning SR (2009) A groupwise association test for rare mutations using a weighted sum statistic. PLoS Genet 5(2):e1000384

Maher B (2008) Personal genomes: the case of the missing heritability. Nature 456(7218):18–21

Manolio TA et al (2009) Finding the missing heritability of complex diseases. Nature 461(7265):747–753

Marchini J, Howie B (2010) Genotype imputation for genome-wide association studies. Nat Rev Genet 11(7):499–511

Metzker ML (2010) Applications of next-generation sequencing sequencing technologies—the next generation. Nat Rev Genet 11(1):31–46

Mian MAR et al (1996) Molecular markers associated with seed weight in two soybean populations. Theor Appl Genet 93(7):1011–1016

Morgenthaler S, Thilly WG (2007) A strategy to discover genes that carry multi-allelic or mono-allelic risk for common diseases: a cohort allelic sums test (CAST). Mutat Res Fundam Mol Mech Mutagen 615(1–2): 28–56

Morris GP et al (2013) Population genomic and genome-wide association studies of agroclimatic traits in sorghum. Proc Natl Acad Sci USA 110(2):453–458

Myles S et al (2009) Association mapping: critical considerations shift from genotyping to experimental design. Plant Cell 21(8):2194–2202

Ott J, Kamatani Y, Lathrop M (2011) Family-based designs for genome-wide association studies. Nat Rev Genet 12(7):465–474

Patil G et al (2016) Genomic-assisted haplotype analysis and the development of high-throughput SNP markers for salinity tolerance in soybean. Sci Rep 6:19199

Patterson N, Price AL, Reich D (2006) Population structure and eigenanalysis. PLoS Genet 2(12): 2074–2093

Price AL et al (2006) Principal components analysis corrects for stratification in genome-wide association studies. Nat Genet 38(8):904–909
Price AL et al (2010a) Pooled association tests for rare variants in exon-resequencing studies. Am J Hum Genet 86(6):832–838
Price AL et al (2010b) New approaches to population stratification in genome-wide association studies. Nat Rev Genet 11(7):459–463
Price AL et al (2013) Mixed models can correct for population structure for genomic regions under selection response. Nat Rev Genet 14(4):300
Pritchard JK, Stephens M, Donnelly P (2000) Inference of population structure using multilocus genotype data. Genetics 155(2):945–959
Quaas RL, Pollak EJ (1981) Modified equations for sire models with groups. J Dairy Sci 64(9):1868–1872
Ray JD et al (2015) Genome-wide association study of ureide concentration in diverse maturity group IV soybean [*Glycine max* (L.) Merr.] accessions. G3 5 (11):2391–2403
Rincker K et al (2016) Genome-wide association study of brown stem rot resistance in soybean across multiple populations. Plant Genome 9(2):68–78
Risch N, Merikangas K (1996) The future of genetic studies of complex human diseases. Science 273 (5281):1516–1517
Segura V et al (2012) An efficient multi-locus mixed-model approach for genome-wide association studies in structured populations. Nat Genet 44(7): 825–830
Sham PC, Purcell SM (2014) Statistical power and significance testing in large-scale genetic studies. Nat Rev Genet 15(5):335–346
Song QJ et al (2013) Development and evaluation of SoySNP50K, a high-density genotyping array for soybean. PLoS ONE 8(1):e54985
Song QJ et al (2015) Fingerprinting soybean germplasm and its utility in genomic research. G3 5(10):1999–2006
Tang Y et al (2016) GAPIT version 2: an enhanced integrated tool for genomic association and prediction. Plant Genome 9(2):1–9
Tester M, Langridge P (2010) Breeding technologies to increase crop production in a changing world. Science 327(5967):818–822
Thornsberry JM et al (2001) Dwarf8 polymorphisms associate with variation in flowering time. Nat Genet 28(3):286–289
Valliyodan B et al (2016) Landscape of genomic diversity and trait discovery in soybean. Sci Rep 6:23598
Varshney RK, Terauchi R, McCouch SR (2014) Harvesting the promising fruits of genomics: applying genome sequencing technologies to crop breeding. PLoS Biol 12(6):e1001883
Vaughn JN et al (2014) The genetic architecture of seed composition in soybean is refined by genome-wide association scans across multiple populations. G3 4(11): 2283–2294
Vineis P, Pearce N (2010) Missing heritability in genome-wide association study research. Nat Rev Genet 11(8):589
Vuong TD et al (2015) Genetic architecture of cyst nematode resistance revealed by genome-wide association study in soybean. BMC Genomics 16:593
Wang J et al (2008) Association mapping of iron deficiency chlorosis loci in soybean (*Glycine max* L. Merr.) advanced breeding lines. Theor Appl Genet 116(6):777–787
Wang K, Li M, Hakonarson H (2010) Analysing biological pathways in genome-wide association studies. Nat Rev Genet 11(12):843–854
Wang J et al (2016) Development and application of a novel genome-wide SNP array reveals domestication history in soybean. Sci Rep 6:20728
Wu MC et al (2011) Rare-variant association testing for sequencing data with the sequence kernel association test. Am J Hum Genet 89(1):82–93
Xu X et al (2013) Pinpointing genes underlying the quantitative trait loci for root-knot nematode resistance in palaeopolyploid soybean by whole genome resequencing. Proc Natl Acad Sci 110(33):13469–13474
Yang W et al (2014) Combining high-throughput phenotyping and genome-wide association studies to reveal natural genetic variation in rice. Nat Commun 5:5087
Yu JM, Buckler ES (2006) Genetic association mapping and genome organization of maize. Curr Opin Biotechnol 17(2):155–160
Yu JM et al (2006) A unified mixed-model method for association mapping that accounts for multiple levels of relatedness. Nat Genet 38(2):203–208
Yu JM et al (2008) Genetic design and statistical power of nested association mapping in maize. Genetics 178 (1):539–551
Zhang ZW et al (2010) Mixed linear model approach adapted for genome-wide association studies. Nat Genet 42(4):355–360
Zhang YH et al (2015) Establishment of a 100-seed weight quantitative trait locus-allele matrix of the germplasm population for optimal recombination design in soybean breeding programmes. J Exp Bot 66(20):6311–6325
Zhou X, Stephens M (2012) Genome-wide efficient mixed-model analysis for association studies. Nat Genet 44(7):821–824
Zhou ZK et al (2015) Resequencing 302 wild and cultivated accessions identifies genes related to domestication and improvement in soybean. Nat Biotechnol 33(4):408–414
Zhou L et al (2016) Identification and validation of candidate genes associated with domesticated and improved traits in soybean. Plant Genome 9(2):232–248
Zhu CS, Yu JM (2009) Nonmetric multidimensional scaling corrects for population structure in association mapping with different sample types. Genetics 182(3): 875–888

Zhu CS et al (2008) Status and prospects of association mapping in plants. Plant Genome 1(1):5–20
Zhu CS et al (2011) Integrating rare-variant testing, function prediction, and gene network in composite resequencing-based genome-wide association studies (CR-GWAS). G3 1(3):233–243

Impact of Genomic Research on Soybean Breeding

8

Zenglu Li, Benjamin Stewart-Brown, Clinton Steketee and Justin Vaughn

Abstract

Soybean (*Glycine max* L. Merrill) is the leading oilseed crop in the world and a primary source of vegetable oil for human consumption and protein meal for animal feed. The USA is the world's largest soybean producer followed closely by Brazil and Argentina. Soybean breeding has been successful in developing soybean varieties with high yield, enhanced seed composition, and disease and pest resistance using conventional breeding approaches. The main challenge faced by soybean researchers has been continuously increasing genetic gain, yet increases in genetic gain have been observed over the past 80 years. Genomic technologies and DNA markers have been successfully developed and utilized in soybean over the past two decades to identify quantitative trait loci (QTL)/genes for traits of economic importance. These technologies have subsequently been utilized to introgress and select for these traits, enabling breeders to accelerate the breeding cycle and develop productive soybean cultivars. In this chapter, first we briefly review DNA marker technologies and how they have been used to characterize soybean germplasm. Then, we present examples of how genomic tools have been used for QTL discovery for traits of importance and conducting molecular breeding. To conclude, we provide our perspectives on the future of soybean breeding using DNA markers and next-generation sequencing.

8.1 Introduction

Soybean is an important source of protein meal for animal feed and oil for human consumption (Huth 1995). In addition, its nitrogen fixation capacity benefits the crops grown after soybean in crop rotations. Soybean breeding often involves introgressing desirable traits from exotic

Z. Li (✉) · B. Stewart-Brown · C. Steketee · J. Vaughn
Department of Crop and Soil Sciences, Institute of Plant Breeding, Genetics, and Genomics, University of Georgia, Athens, GA, USA
e-mail: zli@uga.edu

H.T. Nguyen and M.K. Bhattacharyya (eds.), *The Soybean Genome*, Compendium of Plant Genomes, DOI 10.1007/978-3-319-64198-0_8

germplasm, landraces, or wild relatives using conventional backcrossing approaches, which may take years to achieve. As new genomic technologies have been developed, soybean breeding programs have rapidly accepted them to help improve the yield and quality of the crop for growers. Everything from the advent of modern farming technologies and equipment to the adoption of biotech traits has contributed to the continuous increase in soybean yield over the past decades. The generation and publication of the soybean genome has likewise aided soybean researchers to develop genomic tools to support the development of improved cultivars for growers (Schmutz et al. 2010).

Soybean productivity is often affected by both biotic and abiotic stresses such as drought, insects, nematodes, and diseases. Development of soybean cultivars with both abiotic and biotic stress resistance is one of the major goals in soybean breeding programs. Although conventional breeding approaches have been successful to address these abiotic and biotic stresses, advanced genomic technologies have enabled soybean breeders to more efficiently introgress these traits into elite gene pools and consistently develop improved varieties. One way to measure improvement of a trait is to determine the amount of genetic gain obtained for it over the course of time. In soybean, yield is the most important trait for growers and, therefore, also for breeders. By comparing cultivars released over time, researchers found that genetic gains in yield have increased by about 20 kg ha^{-1} $year^{-1}$ (Fox et al. 2013; Koester et al. 2014; Rogers et al. 2015). Genetic gains in soybean yield over the last century are mainly due to physiological alteration of the plant, such as increased pod development on branches, shorter plant height, and increased node number (Suhre et al. 2014). Longer growing season, reduced lodging, increased aboveground biomass, and better allocation of energy into seed production have also aided in yield gains for modern soybean cultivars (Koester et al. 2014). There is also strong evidence that modern soybean cultivars are able to better adapt to genotype by environment (G $\times$ E) effects than older varieties, with newer cultivars consistently achieving higher yields (Suhre et al. 2014; Cober and Morrison 2015). The most recently developed soybean cultivars have had the added advantage of using molecular markers to aid in selection during the breeding process, which allowed for selection of desirable traits while limiting the negative effects of the traits. In order to utilize advanced genomic resources and empower soybean breeders, the deployment of breeder-friendly genomics and decision support tools will be crucial to accelerate breeding cycles and genetic gain. In this chapter, we provide an overview of genomic tools available for soybean research and how to apply these technologies in soybean breeding programs.

8.2 DNA Marker Discovery and Development

Advancement of genomic technologies prompted DNA marker discovery and development in soybean over the past three decades. The first generation of molecular markers was restriction fragment length polymorphism (RFLP), random amplified polymorphic DNA (RAPD), and amplified fragment length polymorphism (AFLP) markers. RFLPs were the first DNA-based markers and first studied in soybean in the late 1980s (Apuya et al. 1988). The method is based on restriction enzyme digestion and DNA hybridization. Due to the restriction enzyme digestion and hybridization steps, this method requires a large amount of high-quality DNA with radioactively labeled probes (Apuya et al. 1988) and is very time-consuming. RAPD markers were then developed which is a polymerase chain reaction (PCR)-based method and requires no prior sequence information (Welsh and McClelland 1990; Williams et al. 1990). AFLPs were developed in the early 1990s (Prabhu and Gresshoff 1994) and are based on restriction enzyme digestion, followed by selective amplification of a subset of restriction fragments. However, with this technology, developing locus-specific markers from individual fragments can be difficult (Mueller and Wolfenbarger 1999).

The second generation of DNA markers was microsatellites, which are also known as simple sequence repeats (SSRs). These markers typically are short tandem repeats of 2–6 base pairs of DNA, which are present in variable copy numbers at a given locus (Akkaya et al. 1995; Cregan et al. 1999a). SSR markers are typically codominant and abundant in plant genomes. They can be detected using unique sequences of flanking regions as primers and the PCR process (Akkaya et al. 1995). With capillary and fluorescent technologies, detection of SSRs became high-throughput and cost was reduced as only small amounts of DNA are required for the PCR amplification (Guichoux et al. 2011). However, the development of correctly functioning primers is often a tedious and costly process (Grover and Sharma 2016). In addition, the distribution of SSRs in the soybean genome is sparse, which may limit the resolution for fine mapping traits of interest.

A single-nucleotide polymorphism (SNP) is a variation of a single nucleotide that occurs at a specific locus in the genome (Zhu et al. 2003). SNPs are abundant in the soybean genome, biallelic, codominant, and have become the primary DNA markers used by soybean breeders for QTL mapping, molecular breeding, and germplasm characterization (Eathington et al. 2007; Pham et al. 2013; Song et al. 2013, 2014; Shi et al. 2015). Next-generation sequencing technologies have made SNP discovery easier and cheaper in soybean (Hyten et al. 2008). Different platforms are available for SNP detection such as Illumina and Affymetrix assays for whole-genome SNP detection (Song et al. 2013) and TaqMan, Kompetitive Allele-Specific PCR (KASP), and SimpleProbe for detection of a small number of SNPs (Pham et al. 2013; Shi et al. 2015). These SNP detection platforms are high-throughput and can be automated for the process, which has made whole-genome analysis and large-scale marker-assisted selection (MAS) possible.

By resequencing six cultivated soybean genotypes, Essex, Evans, Archer, Minsoy, Noir 1, and Peking, and two wild soybean accessions, PI 468916 and PI 479752, Song et al. (2013) identified 209,903 SNPs. Of these SNPs, over 50,000 SNPs were selected based on polymorphism and coverage to develop Illumina Infinium iSelect BeadChips for high-throughput genotyping (Song et al. 2013). The SoySNP50K BeadChips allow for rapid identification of polymorphic alleles and provide adequate saturation of most regions of the soybean genome. A 6000 SNP subset of the SoySNP50K BeadChips was further selected based on the quality, polymorphism, and distribution of SNPs, to create the SoySNP6K BeadChips (Song et al. 2014). These SoySNP6K BeadChips have provided affordable, powerful tools for germplasm characterization, rough QTL mapping, and genomic selection (Akond et al. 2014; Anderson et al. 2015; Lee et al. 2015).

8.3 Characterizing Soybean Germplasm Using DNA Markers

With the development of DNA markers in soybean, it became possible to characterize soybean germplasm lines to understand their genetic diversity. Soybean is believed to have been first domesticated in south China, and then, it spread outward to northern China and other parts of Asia (Guo et al. 2010). As with most other domesticated plants, there was a significant loss of sequence diversity during the domestication process (Hyten et al. 2006; Guo et al. 2010). In North American elite lines, based on pedigree analysis, it was found that only 80 landraces accounted for 99% of the genetic diversity, with 17 out of the 80 landraces accounting for 86% of the genetic diversity in the elite soybean lines released between 1947 and 1988 (Gizlice et al. 1994). Similar to the North American studies, researchers investigating genetic diversity in Chinese elite lines using 100 RAPD and 35 SSR markers found a very narrow genetic background being used in breeding programs (Li et al. 2008). However, unlike most other major crop species studied, soybean diversity did not significantly change during the modern breeding era (Hyten et al. 2006).

One of the first studies of RFLPs in the soybean genome noted the lack of diversity in marker polymorphism in the cultivars examined, indicating a potential breeding bottleneck during domestication (Apuya et al. 1988). Soybean genetic diversity was then evaluated with different marker systems. Using RAPD markers, Li and Nelson (2001) evaluated soybean germplasm from China, Japan, and S. Korea. They found that the mean genetic distance of lines within China is much larger than that within Japan or S. Korea, but smaller than that between China and Japan or S. Korea. Cluster and principal component analyses almost completely separated the accessions from China from those of Japan and S. Korea, but could not distinguish between the accessions from Japan and S. Korea. Similarly, the genetic diversity among 35 North American soybean ancestors, 66 North American soybean cultivars, 59 modern Chinese cultivars, and 30 modern Japanese cultivars was assessed using 332 AFLP markers. Clustering and principal component analysis grouped the cultivars into three major groups according to their geographic origins, and North American soybean ancestors overlapped with all three groups (Ude et al. 2003).

Using 496 SNP markers in North American elite lines derived from 80 Asian landraces, researchers found that elite lines had around 75% of the diversity found in the landraces used to develop the lines (Hyten et al. 2006). They suggested that although the elite lines examined maintained genetic diversity during the breeding process, there was still about 50% more diversity in the wild progenitor *Glycine soja* (Sieb. and Zucc.), which could be utilized for introgression into the elite North American lines (Hyten et al. 2006). This suggests that wild soybean is a source for genetic diversity; however, it presents a challenge for soybean breeders to introgress desirable traits into elite lines without also introducing negative wild traits. The sequencing of the cultivated and wild soybean reference genomes, and subsequent development of the SoySNP50K BeadChips, will greatly help identify the genomic regions of importance and improve the introgression of wild soybean genes into elite lines. In particular, the inclusion of *G. soja* SNP data onto the SoySNP50K BeadChips allows researchers not only to identify markers associated with desirable traits in wild soybean, but also to have a large number of markers readily available to utilize in order to reduce linkage drag with other wild genes (Song et al. 2013).

Although less diverse than *G. soja,* soybean landraces also are valuable sources of novel alleles for trait introgression. One study investigating sequence diversity in a European soybean collection found that over half of the alleles identified from SSR markers were not found in the landraces used to develop North American elite lines (Tavaud-Pirra et al. 2009). Song et al. (2013) evaluated soybean germplasm consisting of both *G. max* and *G. soja* using SoySNP50K BeadChip data. They found that 42 genomic regions (100 kb windows) between landrace and elite populations and 620 genomic regions between *G. soja* and landrace across the whole-genome contained loci with high average fixation indices ($F_{st} \geq 0.6$). These genomic regions could be those under breeding selection during the 80 years of North American soybean breeding, or could be regions associated with domestication of the wild soybean to cultivated soybean. Understanding these regions could provide insight into selection and genetic improvement of soybean.

Using SoySNP50K BeadChip data, Vaughn and Li (2016) reported on the genome-wide characterization of the structure and history of North American soybean populations and signatures of selection in these populations. Findings concluded that prehybridization line selections resulted in a clonal structure that dominated early breeding and explained many of the reductions in diversity found in the initial generations of soybean hybridization. The rate of allele frequency change does not deviate sharply from neutral expectation, yet some regions bare hallmarks of strong selection, suggesting a highly variable range of selection strengths biased toward weak effects. This article also discussed the importance of haplotypes as units of analysis when complex traits fall under novel selection regimes. The USDA Soybean Germplasm Collection contains over 20,000 domesticated and wild soybean accessions. The entire collection has been genotyped with the SoySNP50K BeadChips

(Song et al. 2015). This genotyping information has been used to identify core collections of *G. max* and *G. soja*. The *G. max* core set consists of 1418 accessions of *G. max* and 81 accessions of *G. soja*, respectively (Dr. Qijian Song, personal communication). These core collections are great resources to mine favorable alleles for yield and other traits of economic importance.

8.4 Genetic Mapping of QTL Using Genomic Tools

8.4.1 QTL Mapping Using Biparental Populations

The first soybean linkage map using RFLPs was developed in 1988 (Apuya et al. 1988). Another RFLP map developed two years later was the first mapping project to associate RFLP markers to quantitative traits in soybean (Keim et al. 1990). Two RFLP markers were identified by Concibido et al. (1994) to be significantly associated with resistance to soybean cyst nematode (SCN, *Heterodera glycines* L.) race 3. Concibido et al. (1997) also mapped the major QTL, *rhg1*, for resistance to SCN races 1, 3, and 6 from Peking, PI 88788, and PI 90763 on chromosome 18 (LG-g), which explained 35, 50, and 54 percent of phenotypic variation, respectively. Other linkage maps and QTL mapping projects quickly followed. In 1997, QTLs for resistance to three different species of root-knot nematode (*Meloidogyne incognita, Meloidogyne arenaria,* and *Meloidogyne javanica*) were reported (Tamulonis et al. 1997a, b, c). QTLs for both antibiosis and antixenosis resistance to corn earworm in soybean have also been reported (Rector et al. 1998, 1999, 2000).

The advent of PCR-based molecular markers allowed for faster development of linkage maps using larger mapping populations and increased marker density. A soybean linkage map using a combination of RFLPs, AFLPs, and RAPDs was developed in 1997 (Keim et al. 1997). At the same time, SSR markers were gaining in popularity among soybean researchers due to the increased frequency of polymorphisms observed in the highly inbred species. Several researchers had developed genetic linkage maps using combinations of the molecular markers described above (Cregan et al. 1994; Yu et al. 1994; Akkaya et al. 1995; Mansur et al. 1996), leading to numerous maps that often had conflicting marker placements due to the different pedigrees used to generate the mapping populations.

In 1999, an integrated linkage map was developed that used mapping populations and markers from three different sources to develop a common set of linkage groups and markers (Cregan et al. 1999a). This integrated map was also the first time that the number of linkage groups agreed with the chromosome number in soybean (20), leading to greater confidence that the linkage maps correlate to physical locations in the soybean genome (Cregan et al. 1999a). The integrated map was quickly adopted by soybean researchers, and a big push to convert the more time-consuming markers such as RFLPs and AFLPs to SSR markers was initiated. Using the integrated map, researchers successfully identified SSR markers that could replace RFLP markers known to associate with four insect resistance (antibiosis and antixenosis) QTLs (Narvel et al. 2001) and with QTL for southern root-knot nematode resistance (Li et al. 2001). The SSR markers used by Narvel et al. (2001) and Li et al. (2001) were often more saturated than the RFLP markers identified by previous reports (Tamulonis et al. 1997a; Rector et al. 2000) and therefore more tightly linked to the resistance QTLs. Other QTLs for traits such as seed composition (Csanádi et al. 2001; Hyten et al. 2004), Sclerotinia stem rot [*Sclerotinia sclerotiorum* (Lib.) de Bary] resistance (Arahana et al. 2001), and maturity (Molnar et al. 2003) have also been mapped using SSR markers.

At the same time that the first integrated genetic map in soybean was being developed, sequence data was starting to become publicly available for the species. Much of the data was derived from bacterial artificial chromosome (BAC) libraries or expressed sequence tag (EST) sequencing projects. These resources allowed for the identification of additional molecular markers and the publication of a second integrated soybean genetic map in 2004 (Song et al. 2004), which combined linkage maps from five different sources with even better saturation. EST data was also being utilized to develop a new class of molecular marker, which differentiated by a single nucleotide. The first SNP markers in soybean were developed in 2003 (Zhu et al. 2003) and were used in a published linkage map in 2007 (Choi et al. 2007). Using the data from the second integrated soybean genetic map (Song et al. 2004), researchers sequenced SSR marker regions linked to two southern root-knot nematode [*Meloidogyne incognita* (Kofoid and White)] QTLs, as well as BAC clones linked to the SSR markers, to look for potential SNPs (Ha et al. 2007). Polymorphic SNPs were found between the susceptible and resistant parents. The resistant and susceptible SNP haplotypes were then used to genotype other soybean lines to perform a haplotype association analysis with nematode resistance (Ha et al. 2007).

The first draft soybean sequence was publicly released in 2010 and was derived from whole-genome shotgun sequencing of 'Williams 82' (Schmutz et al. 2010). The publication of a reference sequence for soybean allowed for subsequent identification of many additional SSR and SNP markers. Researchers studying Asian soybean rust (SBR, *Phakopsora pachyrhizi* Syd.) resistance utilized the published soybean genome to identify polymorphic SNPs between a susceptible and resistant line (Monteros et al. 2010). The polymorphic SNPs were used in conjunction with SSR markers to genotype mapping populations and identify the resistance SNP haplotype for evaluation of other soybean lines (Monteros et al. 2010). The soybean reference sequence allowed for rapid development of SNP markers in the species, which quickly became the preferred molecular marker for researchers due to the vast quantity of SNPs located throughout the genome and its high frequency of biallelic polymorphisms (Song et al. 2013).

Development of the SoySNP50K BeadChips in 2013, a high-throughput array containing over 50,000 SNPs located throughout the soybean genome (Song et al. 2013), allowed for rapid identification of polymorphic alleles and good saturation of most regions of the soybean genome. The SoySNP50K BeadChips are often used to help saturate previously identified QTLs with additional markers to narrow the size of the QTL regions in numerous mapping projects. In addition, the saturation of SNPs across the soybean genome often facilitated the discovery of causative genes for many traits and allowed for the identification of markers either within the gene or in very tight linkage that are ideal for MAS. While fine mapping two closely linked QTLs for resistance to frogeye leaf spot (*Cercospora sojina* K. Hara), researchers used polymorphic SNPs identified from the SoySNP50K BeadChips to narrow the QTL region from 3–4 Mbp down to 72.6 kb and identified five candidate genes responsible for resistance (Pham et al. 2015). Additionally, four candidate genes for southern root-knot nematode resistance were identified using a similar fine mapping approach (Pham et al. 2013). Sequencing of these genes identified putative functional SNPs that were used in haplotype analysis to select for resistant genotypes (Pham et al. 2013, 2015). Additionally, researchers used a panel of 75 landraces originating from seven different countries to screen for novel sources of SBR resistance by genotyping resistant lines with the SoySNP50K BeadChips and identified three lines that may have unique resistance genes (Harris et al. 2015). One line, PI 567068A, was found to have novel resistance, that is either tightly linked or allelic to *Rpp6* (King et al. 2016a).

8.4.2 Genome-Wide Association Studies

Due to limited recombination in biparental populations, estimated QTL intervals usually have relatively low resolution. Although numerous QTLs have been mapped with biparental populations, knowledge of the QTL/gene(s) underlying traits of importance remains limited. A high-quality physical map of the soybean genome (Schmutz et al. 2010) has enabled the development of dense genotyping platforms (Song et al. 2013; Sonah et al. 2015). Development of the SoySNP50K BeadChips (Song et al.

2013) allowed for genome-wide association studies or scans (GWAS). A major benefit of any dense genotyping platform is that highly diverse germplasm can be used to localize the genetic loci underlying a trait to a very narrow genomic position (Myles et al. 2009). Many recombinations are likely to have occurred since the lineages in a GWAS population shared a common ancestor. These recombinations reduce the linkage disequilibrium (LD) to much lower levels than typically found in a biparental mapping population.

In some cases, if the LD in a GWAS population is low and the genotyping density is high, the causal mutations influencing a trait can be directly discovered (Huang et al. 2010, 2012). Even when a trait-associated region of the genome contains many genes, a small number of candidate causal genes can be identified based on either their known biochemical function or their homology to a gene implicated in another system. Thus, in soybean, which generally has the highest LD reported of any crop, a plausible list of candidate genes for traits can be identified using GWAS (Sonah et al. 2015). Because the populations used in GWAS do not have known pedigrees, unknown population structure can confound analysis. For a trait controlled by many loci, a non-causal variant can be at a high frequency in a subpopulation with an abundance of positive or negative effect alleles. In this case, the non-causal variants-associated markers can have an even higher significance of association than markers of causal variants (Platt et al. 2010). Many algorithms have been developed to overcome such artifacts. Vaughn et al. (2014) described a highly significant association between protein concentration and a genomic interval on chromosome 20 spanning from coordinates ~30.5 to ~31.7 Mb. The positive-effect allele appears to be rare in soybean landraces and concentrated in later maturing lines of Korean origin. Hwang et al. (2014) mapped the association to a large linkage block ~3 Mb upstream from this position, although the significance of this association was much lower. Using high-density SNP marker arrays and a GWAS approach, QTLs for days to flowering, days to maturity, duration of flowering-to-maturity, and plant height in early maturing soybean germplasm were detected (Zhang et al. 2015). Carbon isotope ratio (δ13C) is considered as a surrogate measure of water use efficiency (WUE) in soybean. Identification of QTL controlling improved WUE could lead to develop cultivars with increased yield under drought. Using a collection of 373 diverse soybean genotypes and 12,347 SNP markers, GWAS identified 39 SNPs associated with δ13C (Dhanapal et al. 2015).

Genes that are rare in global germplasm may be rare because, in most cases, they are either deleterious or neutral and, thus, are lost in most subpopulations. Yet, any gene with a substantial phenotypic impact is likely to find a niche use in certain subpopulations of a species that spans many latitudes and diverse environments (Carter et al. 2004). Because of their rarity, these alleles will go undetected by GWAS. Alternatively, every allele will be present at ~50% in such a biparental mapping population, regardless of whether it is rare in the global germplasm or not. For these reasons (among others), biparental populations will remain an important tool in identifying alleles for MAS.

8.4.3 Genotyping by Sequencing

Advances in next-generation sequencing technologies have driven the costs of DNA sequencing down. With reduced cost for resequencing, genotyping by sequencing (GBS) became an attractive application for discovering and genotyping SNPs for QTL mapping and for characterization of soybean germplasm and other crop species (Spindel et al. 2013; Xu et al. 2013; Bastien et al. 2014; Sonah et al. 2015).

Xu et al. (2013) applied two interesting and complementary approaches in mapping root-knot nematode resistance to almost single-gene resolution. In their study, 246 recombinant inbred lines (RILs) derived from a cross between resistant and susceptible lines were resequenced at a low coverage (0.19×). The two parent lines were resequenced at ~15× coverage. Along the entire genome, low-coverage reads for each line were used to impute the parent haplotype that was inherited by a respective line. Resultant recombination bins were associated with the resistance phenotype, and then, the expression of all genes contained within this interval was checked for response to the pest (see Sect. 8.4.4).

In another study, a GBS approach was utilized to generate more than 47,000 genome-wide SNPs across a panel of 304 short-season soybean lines (Sonah et al. 2015). Genomic loci associated with maturity, plant height, seed weight, seed oil, and protein were also identified from this panel of soybean lines and most of these loci identified with GBS-GWAS were consistent with previously reported QTLs for these traits. By optimizing GBS protocol, a total of 8397 SNPs were generated from a diverse collection of 101 soybean lines to map the QTLs underlying resistance to Sclerotinia stem rot in soybean (Iquira et al. 2015). Three genomic regions were detected that were significantly associated with resistance to Sclerotinia stem rot in their study.

Whole-genome resequencing is the ultimate manifestation of high-density genotyping, in that theoretically all polymorphisms present in an accession can be detected. Zhou et al. (2015) took this approach using 302 wild and cultivated soybean lines, noting numerous overlaps between regions under selection and QTLs as identified by GWAS.

8.4.4 Transcriptomic Analysis

The soybean sequence predicts over 60,000 putative genes, with over 40,000 genes being predicted with high confidence (Schmutz et al. 2010). High-throughput sequencing has made profiling gene expression possible in order to better understand soybean biology and develop functional genomic tools to characterize these genes and their related functions. Utilizing the Illumina sequencing platform, Libault et al. (2010) sequenced cDNA derived from 14 conditions (tissues) of the cultivar 'Williams 82.' Their results generated the transcription of 55,616 annotated genes, of which 13,529 are putative pseudogenes. This work has provided a sequence-based transcriptome atlas in soybean that will allow for further characterization of these genes. Similar results were also achieved by Severin et al. (2010) by using seeds derived from introgressing *G. soja* (PI468916) into *G. max* (A81-356022).

Transcriptome analysis could provide a way to narrow the candidate genes within known QTL intervals, assuming that causal genes generally show differential transcription. Many researchers have adopted this strategy for traits that have a complex genetic architecture. In an attempt to further refine the large-effect protein QTL on chromosome 20, Bolon et al. (2010) used both microarrays and RNA-Seq to interrogate the transcriptome at various stages of soybean seed development. By contrasting a low-protein line with a near-isogenic line into which the high-protein locus had been introgressed, they could focus on transcriptomic differences mediated by *cis* and *trans* effects of the introgressed region. There were clearly marked transcriptional differences across the introgressed region relative to the genomic background, which is expected given that expression levels are generally highly polymorphic.

In an attempt to identify genes underlying differential canopy wilting responses in soybean, Shin et al. (2015) further refined the application of transcriptomics to biparental mapping. As in the studies mentioned here and many others, two genotypes are assayed across a range of conditions. To identify differentially regulated genes, a matrix of pairwise differences in transcript abundance is created: For two genotypes assayed across four conditions, three within genotype comparisons were tested (assuming the first condition is the control) and the different genotypes are compared to one another at every condition. Thus, in the given example, 10 tests of differential expression would be made for a single gene. Yet, researchers are generally interested in more holistic trends in the data, such as 'How many genes are responding similarly in both genotypes' or 'Which genes are responding, but have a response that is conditioned by the genotype?' As with the root-knot nematode resistance mapping described above, this last question is very important with regard to identifying causal genes underlying differences in phenotypes between two lines. To facilitate a more biologically relevant analysis, Shin et al. (2015) devised an additional processing step to categorize genes based on their response profile within each genotypic background. The approach allowed them to differentiate genes involved in a generic response from those that had responses unique to a particular genotype. These latter type of genes, which they refer to as a $G \times E$ genes, can be overlaid with known QTL information to facilitate further candidate gene identification.

8.5 Molecular Breeding to Improve Efficiency of Soybean Breeding

Marker development and high-throughput genotyping technologies have led to the development of marker-trait associations and mapped QTLs for many soybean traits including protein, oil, nematode resistance, disease resistance, and other agronomic traits. In turn, this resulted in the implementation of MAS workflows in soybean breeding programs over the last two decades. These approaches have greatly improved soybean breeding efficiency. In summary, there are four molecular breeding approaches: MAS, marker-assisted backcrossing (MABC), marker-assisted recurrent selection (MARS), and genomic selection (GS). Phenotyping in either the greenhouse or the field typically is labor intensive for most traits, and some traits are not measurable until after the opportunity for crossing (i.e., seed composition and yield). DNA markers can speed up the selection process and improve the accuracy of selection.

8.5.1 Marker-Assisted Selection

MAS is an indirect breeding selection method. The first step to perform MAS in a soybean breeding program is to identify polymorphic markers between the parents at mapped QTL regions that can be traced in the segregating populations. The selection for traits of interest is based on the marker allele, which is linked to the trait of interest or a genetic variant of genes of interest, and can be performed at any growth stage. Discovery of QTL/DNA markers made MAS possible, which are widely used in soybean breeding programs in recent years. MAS has shown to have an advantage when the traits of interest are difficult and expensive to measure in the field or greenhouse compared to conventional breeding. MAS can enable selection for traits of importance and allows highly accurate selection which is unaffected by environmental factors. Therefore, MAS can be used to select these traits in the laboratory at an early stage, which could result in reduction of the total number of lines to be tested for each pedigree. QTL mapping results are the foundation to develop DNA markers for MAS. Therefore, it is important to confirm the QTL prior to MAS. Although MAS has been successful in the past two decades, it works better with qualitative traits rather than quantitative traits.

MAS is typically conducted at the F_2 or F_3 generation using leaf tissue or seed harvested from individual plants, depending on the traits to be selected. Over the past two decades, there were many successful examples in soybean for MAS. One of the most successful examples is MAS for SCN resistance. SCN resistance is a quantitative trait, but there are key genes that control a majority of the variation for this trait (Mansur et al. 1993). SCN resistance can be screened phenotypically in a greenhouse, but the process is both time and labor consuming. Using MAS has drastically improved the efficiency of selecting soybean lines resistant to SCN. There are two primary resistance genes for SCN: *rhg1* and *Rhg4* (Concibido et al. 2004). These genes offer resistance to SCN race 3, which is the most abundant race of SCN present in soybean producing areas of the USA (Jackson 2014). Resistance to SCN, specifically race 3, is commonly referred to as PI 88788-type resistance or Peking-type resistance. For PI 88788-type resistance to be effective, the *rhg1* resistance allele must be present (Concibido et al. 1997). For Peking-type resistance to be effective, the *rhg1* and *Rhg4* resistant alleles must both be present (Meksem et al. 2001; Liu et al. 2012). The *rhg1* locus has been identified as a 31.2-kb fragment containing three genes (Glyma18g02580, Glyma18g02590, and Glyma18g02610) that all contribute SCN resistance (Cook et al. 2012). With QTL discovery efforts, SSR markers have been used for selection of SCN-resistant soybean lines (Cregan et al. 1999b; Meksem et al. 2001; Suzuki et al. 2012). Based on the information reported by Cook et al. (2012) and Liu et al. (2012), Shi et al. (2015) developed SNP markers for the *Rhg1* and *Rhg4* loci utilizing the KASP assay for high-throughput genotyping. Two functional SNP markers were designed for selection of the *rhg1* resistance allele and one functional SNP marker for selection of the *Rhg4* resistance allele.

Frogeye leaf spot (FLS, *Cercospora sojina* K. Hara) is another detrimental disease of soybean, for which MAS shows promise for improving selection efficiency. FLS resistance can also be quite time/labor intensive to phenotype similar to SCN, making MAS a viable alternative for identifying lines with resistance. Screening for resistance to FLS traditionally occurs by inoculating seedling trifoliolate leaves in a greenhouse at the V2 or V3 growth stage (Fehr et al. 1971, Mian et al. 2008). The scoring of resistance involves observations of both lesion size and how well defined the lesion-center is, which can be a tedious process (Mian et al. 2008). There are five single genes that have been recognized by the Soybean Genetics Committee to convey some level of resistance to various FLS races (Athow and Probst 1952; Probst et al. 1965; Phillips and Boerma 1982; Yorinori 1987; Zou et al. 1999). Pham et al. (2015) reported fine mapping of *Rcs* genes from PI 594891 and PI 594774 on chromosome 13. Sequencing and qPCR analysis of five candidate genes in this genomic region indicated three genes (Glyma13g25320, Glyma13g25340, Glyma13g25350) with notable genetic mutations and increased gene expression when exposed to FLS that may be conferring resistance to FLS. Four SNP markers were designed utilizing the KASP assay platform for MAS of *Rcs* (PI 594774) and *Rcs* (PI 594891) in breeding programs (Pham et al. 2015). MAS has also been performed for many other soybean traits such as resistance to root-knot nematodes (Pham et al. 2013), aphids (Li et al. 2007), fatty acids (Shi et al. 2015), and other disease-resistance traits.

8.5.2 Marker-Assisted Backcrossing

Backcrossing is taking progeny created from a biparental cross and repeatedly crossing it to one of its parents, called a recurrent parent, in subsequent generations. This is done to reduce the amount of genome from the other parent (donor parent) in the line. DNA markers can help select target genes/QTL during backcrossing and can also help select the favorable genome background to minimize the genome of the donor parent and maximize the genome of the recurrent parent during repeated backcrossing of the offspring, allowing for earlier selections. MABC has become a very effective molecular breeding approach for improving elite genotypes for one or two traits, or for pyramiding of a few genes/QTLs. The efficiency of MABC depends on a number of factors including population size at each backcross generation, linkage of the markers with the target gene, and number/distribution of background markers used for backcrossing. There have been many examples of using MABC in order to introgress traits of importance such as disease resistance, insect resistance, herbicide tolerance, and seed composition traits into elite germplasm (Kim et al. 2008; Maroof et al. 2008; Landau-Ellis and Pantalone 2009; Diers et al. 2005; King et al. 2016b).

MABC has been used in the public sector to combat two important diseases of soybean, SCN and SBR. LDX01-1-65 (PI 636464) is an SCN-resistant soybean line developed by the University of Illinois and Iowa State University using MABC (Diers et al. 2005). PI 468916, a *G. soja* accession, was found to have two SCN-resistance QTLs that mapped to the genomic regions not associated with *rhg1* and *Rhg4*, the predominant alleles conferring SCN resistance in most soybean lines. Due to the fact that these novel QTLs were present in a *G. soja* line with poor agronomic traits, MABC was implemented to introgress these two QTLs into an Iowa State experimental line, A81-356022 (recurrent parent). Closely linked markers were used to select for the QTLs during the backcrossing process. This mitigated the need for extensive SCN-resistance screening throughout the process. This also allowed for ease of selecting lines throughout the process that were heterozygous for both QTLs which would be difficult to definitively determine phenotypically in a greenhouse screening. After four rounds of backcrossing, a BC_4F_1 was selected which was heterozygous for both QTLs. A population of BC_4F_3-derived lines was developed, and markers were used to identify the genotypic combinations for each QTL (Diers et al. 2005).

Resistance to SBR is conferred by *Rpp* genes. SBR is capable of causing significant yield losses in the southeastern USA (Wrather and Koenning 2009). The University of Georgia soybean breeding program used MABC to introgress several key *Rpp* genes into an elite line, G00-3213 (King et al. 2016b) to develop near-isogenic lines (NILs): G00-3213*Rpp1* (PI 676017), G00-3213*Rpp2* (PI 676018), G00-3213*Rpp3* (PI 676019), and G00-3213*Rpp4* (PI 676020). Five backcrosses were made to G00-3213 to develop each *Rpp* NIL. These high yielding, rust-resistant NILs have resistance from different PI sources and can be used as parental lines in future breeding efforts to combat various rust isolates. Markers designed for selection of each *Rpp* gene were used during the backcrossing process to select F_1 progeny heterozygous for each region. No phenotypic screening occurred during the backcrossing process and after five backcrosses and a generation of selfing, populations of BC_5F_2-derived lines were developed and screened for SBR resistance in a greenhouse using GA12, a bulk isolate of Georgia SBR spores. SNP markers ensured proper introgression of the *Rpp* genes in each screened NIL. Diers et al. (2014) utilized MABC to introgress *Rpp1*, *Rpp1-b*, *Rpp? (Hyuuga)*, and *Rpp5* into LD01-7323 and LD00-3309.

MABC has been utilized to enhance soybean lines not only for disease resistance, but also for seed composition traits as well. Landau-Ellis and Pantalone (2009) took a low-phytate germplasm line, CX1834-1-2, and backcrossed the two recessive alleles for low phytate into an elite soybean cultivar, '5601T.' During each stage of the backcrossing process, SSR markers were used to select the two recessive alleles, as well as the backcross progeny incorporating the highest amount of the recurrent parent genome. Using background marker selection in each generation is beneficial, because one can achieve donor genome contribution equivalent to six backcrosses after as little as three backcrosses. This is more tedious in soybean relative to other crops because of the difficulty in producing F_1s on a large scale (Visscher et al. 1996; Hospital and Charcosset 1997; Frisch et al. 1999). Rawal et al. (2014) looked to develop a soybean line with reduced off-flavor through introgression of a null allele for lipoxygenase-2 from PI 596540. This null allele was introgressed into 'JS 97-52' using SSR markers linked to the null allele as well as markers across the genome for background selection.

8.5.3 Marker-Assisted Recurrent Selection

MARS is another molecular breeding approach. The objective of this approach is to identify, select, and accumulate favorable alleles from several genomic regions for quantitative traits such as yield within a single population. The approach requires estimates of the marker allele effect for the target trait and then recombines the selected individuals to pyramid multiple favorable alleles in that population (Eathington et al. 2007; Bernardo 2010). The recombining typically takes two to three cycles. Eathington et al. (2007) provided basic information for how Monsanto goes about performing MARS in soybean and how successful they have been compared to conventional selection. MARS is most effective for complex traits with many genes controlling small amounts of variation in the trait that cannot easily be selected for in a single generation. MARS is also useful because markers can assist in making selections in off-season nurseries where the heritability for a trait is low compared to a target environment. For target populations in target environments, a selection model is developed that estimates marker allele effects for the trait of interest. This model weights markers with more significant effects more heavily than markers with minimal effects. After genotyping the population in question with DNA markers, the breeding values are determined for each line (often RILs) in the population. Superior lines with higher breeding values are selected for crossing to produce new populations and the cycle starts over. The key is that the continuous selection and controlled pollination of superior lines lead to an accumulation of favorable alleles within a population for a trait of interest such as yield which cannot be easily selected for using simpler MAS schemes involving a lesser number of markers. Multiple traits can be combined in selection models that weight the importance of certain traits over others.

Eathington et al. (2007) compared advances made performing MARS for 1 year versus conventional selection in 43 soybean breeding populations and found a 37.6 kg ha^{-1} benefit for yield from MARS. Bernardo (2010) proposed a breeding scheme for MARS that can be implemented, while simultaneously minimizing the amount of crosses that need to be made by hand. This scheme utilizes off-season nurseries and greenhouses to continue advancement year-round. This scheme assumes that a selection model is already in place built upon previous yield data obtained from similar genetic materials grown in similar environments. This process begins with yield testing RILs from a target population in target environments. The superior RILs are cross-pollinated to produce F_1 seeds. These F_1 plants are then allowed to self-pollinate and develop F_2 seed. The F_2 plants are then grown out and genotyped using a panel of molecular markers related to the selection model. The model allows for the selection of favorable F_2 plants to recombine and create another round of F_1 seeds. This process continues cyclically for several rounds until favorable genetic diversity becomes fixed and no further considerable genetic gain can be made.

8.5.4 Genomic Selection

Conventional breeding approaches have led to the development of a large number of improved soybean cultivars from both private and public soybean breeding programs. Over the past two decades, DNA markers have been successfully used to select for traits such as SCN and root-knot nematode resistance (Pham et al. 2013; Shi et al. 2015). However, the use of MAS has generally not been effective for the improvement of quantitative traits, such as yield. Genomic selection (GS or 'genome-wide selection') is, in effect, MAS scaled to the entire genome (Meuwissen et al. 2001; Goddard and Hayes 2009). By genotyping across the genome, a breeder can estimate the genetic effects of each locus and track all yield QTL alleles, because they are likely to be in high LD with at least one marker. Still, for GS to be effective, the combined genetic effect of all QTL must be estimated with fairly high accuracy.

Decreased genotyping costs and new statistical models enable simultaneous estimations of all marker effects. GS is a new form of MAS that estimates all marker effects across the whole-genome to calculate genomic estimated breeding value (GEBV). Markers are not tested for significance—all markers are used in selection. The SoySNP6K BeadChips developed by Song et al. (2013) are a subset of the SNPs selected from SoySNP50K BeadChips, which have been used to genotype the entire USDA Soybean Germplasm Collection. Utilizing these SoySNP6K BeadChips for GS projects will allow soybean breeders to reference the SNP alleles in projects back to the Soybean Germplasm Collection. GS has been shown to be an effective tool to predict and select quantitative traits such as yield. However, it has yet to be applied to large-scale public breeding efforts in soybean. Many opportunities are available to start and implement a GS workflow in soybean breeding: (1) high density of SNP markers available—50 and 6 K SNP Infinium Chips, which provide low cost per data point; (2) LD in soybean tends to be quite high; (3) pedigree information is accurate and extensive; (4) high-throughput genotyping for MAS is already routinely implemented in numerous programs; (5) numerous SNP markers have been developed for key traits and can be used for selection along with the GS model for yield.

Jarquín et al. (2014) produced one of the first genomic prediction modeling publications in soybean. In this study, the researchers utilized a population of 301 experimental lines from the University of Nebraska-Lincoln soybean breeding program. These lines were composed of 34 biparental families spanning maturity groups I, II, and III and were genotyped via GBS. A G-BLUP model was used for prediction. Prediction accuracy for grain yield was assessed using cross-validation and found to be 0.64 under conditions they considered optimal based on factors they modified in an attempt to improve the model. The researchers modeled additive-by-additive epistasis and saw no significant difference in prediction accuracy. They also found that their prediction accuracy showed no significant improvement once the training population size approached 100.

Xavier et al. (2016) performed GS for six soybean traits (plant height, days to maturity, number of reproductive nodes, pods per node, number of nodes, and grain yield) on the Soybean Nested Association Mapping (SoyNAM) population which contained 5555 RILs with maturity groups ranging from II to IV. IA3023 was the common parent, and 40 founder parents were used to develop the SoyNAM population. Lines were genotyped using a 5-K SNP chip designed specifically for SoyNAM. Cross-validation was performed while experimenting with different training population sizes, genotyping densities, and prediction models. Training population size was found to be the most significant factor affecting prediction accuracy, which began to plateau around 2000 individuals. Ma et al. (2016) looked to evaluate how prediction accuracy for plant height and grain yield were affected by the number of markers used in the prediction model as well as if preselecting markers based on certain criteria could improve prediction accuracy. This study developed a population of 235 soybean varieties courtesy of the National Key Facility for Crop Gene Resources and Genetic Improvement (NFCIR) in China. One hundred and eighty-five of these soybean lines were North Spring soybean, while the remaining 50 were Huang-Huai summer soybean. These lines were genotyped using the SoySNP6K BeadChips and analyzed using ridge regression best linear unbiased prediction (RR BLUP). Results showed that for plant height, different marker densities and differences in marker preselection strategies had little impact on prediction accuracy. For grain yield, results indicated that haplotype block analysis for marker preselection improved prediction accuracy.

Models for genomic prediction have been shown to be effective for quantitative disease resistance in soybean as well. Bao et al. (2014) used a panel of 282 accessions from the University of Minnesota breeding program to see whether GS could be implemented to select for quantitative resistance to SCN. A genomic prediction model was developed using RR BLUP, and major effect QTLs for SCN resistance were included as fixed effects. This model was found to be more accurate in selecting resistant materials than simply performing MAS. Bao et al. (2015) also looked to improve selection for resistance to sudden death syndrome (SDS, *Fusarium virguliforme*). The same panel of 282 soybean accessions was evaluated for root lesion severity (RLS), foliar symptom severity (FSS), root retention (RR), and dry matter reduction (DMR) after being soil inoculated with SDS. Cross-validation of an RR BLUP model was performed. Using single-trait models, prediction accuracy was 0.64 for RLS, 0.20 for FSS, 0.18 for RR, and 0.16 for DMR. Several multi-trait models were tested, but none of them improved prediction accuracy primarily because these traits were weakly correlated.

8.6 Prospects on Utilization of DNA Markers and Next-Generation Sequencing in Soybean Breeding

Soybean is an important source of protein and oil in the world. With the world population continuously increasing, sustained or increased demands for soybeans for animal feed, vegetable oil, and biodiesel are expected (Wilson 2010). To meet such demands, it is critical to accelerate the rate of genetic gain for soybean yield by developing and deploying improved soybean varieties with disease and pest resistance, and abiotic stress tolerance (Specht et al. 2014). New genomic technologies and tools developed in recent years have been integrated into soybean breeding programs, which has accelerated breeding cycles and improved the rate of genetic gain.

Cost-effective next-generation sequencing technologies have brought genomics research and soybean breeding into a new era. Completion of the reference genome of the soybean cultivar 'Williams 82' was assembled to capture approximately 975 Mb of the 1100 Mb soybean genome (Schmutz et al. 2010). About 66,000 protein-coding loci were predicted, and are available at http://www.phytozome.net/soybean. In addition, several hundred soybean lines and milestone cultivars have been resequenced (Lam et al. 2010; Li et al. 2013; Zhou et al. 2015). With these sequence resources, soybean researchers are able to locate the genome sequence corresponding to QTL regions and search the soybean gene annotation in these QTL regions to pinpoint the genes responsible for disease and pest resistance, as well as other value-added and agronomic traits. The identification of these candidate genes for the traits of importance could allow soybean researchers to develop functional or user-friendly DNA markers to perform effective MAS without screening parents of populations. These functional markers will also help breeders to quickly identify the germplasm carrying desirable target genes for germplasm introgression. When introgressing new favorable alleles into elite germplasm, flanking regions from non-elite sources or wild relatives are always a concern and could potentially cause linkage drag for yield. High-density DNA markers and next-generation sequencing information could enable soybean breeders to reduce the flanking regions of genes of interest to avoid or minimize linkage drag during the trait introgression.

SoySNP6K Infinium Chips have provided a cost-effective way for breeders to rough map QTL, build reference populations, and perform GS. Moreover, the SoySNP6K data generated from breeding programs could be used to reference back to the USDA Soybean Germplasm Collection to understand allele variation and diversity. In recent years, several emerging genomic technologies have been developed such as GWAS and GS in soybean. These technologies have been applied in soybean research and breeding programs. Coupled with high-throughput phenotyping technologies in soybean, the genomics information generated could allow soybean breeders to conduct effective predictive breeding. In addition, important information generated through other new technologies such as transcriptomics, proteomics, metabolomics, epigenomics, and CRISPR/Cas9 and other targeted genome editing will further help develop new genomic tools and breeding strategies for productive soybean breeding to increase the rate of genetic gain in soybean.

Acknowledgements We thank Drs. Donna Harris and Rebecca Tashiro for the initial participation.

References

Akkaya MS, Shoemaker RC, Specht JE, Bhagwat AA, Cregan PB (1995) Integration of simple sequence repeat DNA markers into a soybean linkage map. Crop Sci 35:1439–1445

Akond M, Liu S, Boney M, Kantartzi SK, Meksem K, Bellaloui N, Lightfoot DA, Kassem MA (2014) Identification of quantitative trait loci (QTL) underlying protein, oil, and five major fatty acids' contents in soybean. Am J Plant Sci 5:158–167

Anderson J, Akond M, Kassem MA, Meksem K, Kantartzi SK (2015) Quantitative trait loci underlying resistance to sudden death syndrome (SDS) in MD96-5722 by 'Spencer' recombinant inbred line population of soybean. Biotechnology 5:203–210

Apuya N, Frazier B, Keim P, Roth EJ, Lark K (1988) Restriction fragment length polymorphisms as genetic markers in soybean, *Glycine max* (L.) Merrill. Theor Appl Genet 75:889–901

Arahana VS, Graef GL, Specht JE, Steadman JR, Eskridge KM (2001) Identification of QTLs for resistance to *Sclerotinia sclerotiorum* in soybean. Crop Sci 41:180–188

Athow K, Probst AH (1952) The inheritance of resistance to frogeye leaf spot of soybeans. Phytopathology 42:660–662

Bao Y, Vuong T, Meinhardt C, Tiffin P, Denny R, Chen S, Nguyen HT, Orf JH, Young ND (2014) Potential of association mapping and genomic selection to explore PI 88788 derived soybean cyst nematode resistance. Plant Genome. doi:10.3835/plantgenome2013.11.0039

Bao Y, Kurle JE, Anderson G, Young ND (2015) Association mapping and genomic prediction for resistance to sudden death syndrome in early maturing soybean germplasm. Mol Breed 35:128

Bastien M, Sonah H, Belzile F (2014) Genome wide association mapping of *Sclerotinia sclerotiorum* resistance in soybean with a genotyping-by-sequencing approach. Plant Genome. doi:10.3835/plantgenome2013.10.0030

Bernardo R (2010) Genome-wide selection with minimal crossing in self-pollinated crops. Crop Sci 50:624–627

Bolon YT, Joseph B, Cannon SB, Graham MA, Diers BW, Farmer AD, May CD, Muehlbauer GL, Specht JE, Tu Z, Weeks N, Xu WW, Shoemaker RC, Vance CP (2010) Complementary genetic and genomic approaches help characterize the linkage group I seed protein QTL in soybean. BMC Plant Biol 10:41

Carter TE, Nelson RL, Sneller CH, Cui Z (2004) Genetic diversity in soybean. In: Boerma HR, Specht JE (eds) Soybeans: improvement, production, and uses. Agron Monogr 16. American Society of Agronomy, Crop Science Society of America, and Soil Science Society of America, Madison, WI, pp 303–416

Choi IY, Hyten DL, Matukumalli LK, Song Q, Chaky JM, Quigley CV, Chase K, Lark KG, Reiter RS, Yoon MS, Hwang EY, Yi SI, Young ND, Shoemaker RC, Tassell CPV, Specht JE, Cregan PB (2007) A soybean transcript map: gene distribution, haplotype and single-nucleotide polymorphism analysis. Genetics 176: 685–696

Cober ER, Morrison MJ (2015) Genetic improvement estimates, from cultivar × crop management trials, are larger in high-yield cropping environments. Crop Sci 55:1425–1434

Concibido VC, Denny RL, Boutin SR, Hautea R, Orf JH, Young ND (1994) DNA marker analysis of loci underlying resistance to soybean cyst nematode (*Heterodera glycines Ichinohe*). Crop Sci 34:240–246

Concibido VC, Lange DA, Denny RL, Orf JH, Young ND (1997) Genome mapping of soybean cyst nematode resistance genes in 'Peking', PI 90763, and PI 88788 using DNA markers. Crop Sci 37:258–264

Concibido VC, Diers BW, Arelli PR (2004) A Decade of QTL mapping for cyst nematode resistance in soybean. Crop Sci 44:1121–1131

Cook DE, Lee TG, Guo X, Melito S, Wang K, Bayless AM, Wang J, Hughes TJ, Willis DK, Clemente TE, Diers BW, Jiang J, Hudson ME, Bent AF (2012) Copy number variation of multiple genes at *Rhg1* mediates nematode resistance in soybean. Science 338:1206–1209

Cregan PB, Bhagwat AA, Akkaya MS, Rongwen J (1994) Microsatellite fingerprinting and mapping of soybean. Methods Mol Cell Biol 5:49–61

Cregan PB, Jarvik T, Bush AL, Shoemaker RC, Lark KG, Kahler AL, Kaya N, VanToai TT, Lohnes DG, Chung J, Specht JE (1999a) An integrated genetic linkage map of the soybean genome. Crop Sci 39:1464–1490

Cregan PB, Mudge J, Fickus EW, Danesh D, Denny R, Young ND (1999b) Two simple sequence repeat markers to select for soybean cyst nematode resistance conditioned by the rhg1 locus. Theor Appl Genet 99:811–818

Csanádi G, Vollman J, Stift G, Lelley T (2001) Seed quality QTLs identified in a molecular map of early maturing soybean. Theor Appl Genet 103:912–919

Dhanapal AP, Ray JD, Singh SK, Hoyos-Villegas V, Smith JR, Purcell LC, King CA, Cregan PB, Song Q, Fritschi FB (2015) Genome-wide association study (GWAS) of carbon isotope ratio (δ13C) in diverse soybean [*Glycine max* (L.) Merr.] genotypes. Theor Appl Genet 128:73–91

Diers BW, Arelli PR, Carlson SR, Fehr WR, Kabelka EA, Shoemaker RC, Wang D (2005) Registration of LDX01-1-65 soybean germplasm with soybean cyst nematode resistance derived from *Glycine soja*. Crop Sci 45:1671–1672

Diers BW, Kim K, Frederick RD, Hartman GL, Unfried J, Schultz S, Cary T (2014) Registration of eight soybean germplasm lines resistant to soybean rust. J Plant Reg 8:96–101

Eathington SR, Crosbie TM, Edwards MD, Reiter RS, Bull JK (2007) Molecular markers in a commercial breeding program. Crop Sci 47:S-154–S-163

Fehr WR, Caviness CE, Burmood DT, Pennington JS (1971) Stage of development descriptions for soybeans, *Glycine max* (L.) Merrill. Crop Sci 11:929–931

Fox CM, Cary TR, Colgrove AL, Nafziger ED, Haudenshield JS, Hartman GL, Specht JE, Diers BW (2013) Estimating soybean genetic gain for yield in the northern United States—influence of cropping history. Crop Sci 53:2473–2482

Frisch M, Bohn M, Melchinger AE (1999) Comparison of selection strategies for marker-assisted backcrossing of a gene. Crop Sci 39:1295–1301

Gizlice Z, Carter TE, Burton JW (1994) Genetic base for North American public soybean cultivars released between 1947 and 1988. Crop Sci 34:1143–1151

Goddard ME, Hayes BJ (2009) Mapping genes for complex traits in domestic animals and their use in breeding programs. Nat Rev Genet 10:381–391

Grover A, Sharma PC (2016) Development and use of molecular markers: past and present. Crit Rev Biotechnol 36:290–302

Guichoux E, Lagache L, Wagner S, Chaumeil P, Léger P, Lepais O, Lepoittevin C, Malausa T, Revardel E, Salin F, Petit R (2011) Current trends in microsatellite genotyping. Mol Ecol Resour 11:591–611

Guo J, Wang Y, Song C, Zhou J, Qiu L, Huang H, Wang Y (2010) A single origin and moderate bottleneck during domestication of soybean (*Glycine max*): implications from microsatellites and nucleotide sequences. Ann Bot 106:505–514

Ha B, Hussey RS, Boerma HR (2007) Development of SNP assays for marker-assisted selection of two southern root-knot nematode resistance QTL in soybean. Crop Sci 47:S-73–S-82

Harris DK, Kendrick MD, King ZR, Pedley KF, Walker DR, Cregan PB, Buck JW, Phillips DV, Li Z, Boerma HR (2015) Identification of unique genetic sources of soybean rust resistance from the USDA soybean germplasm collection. Crop Sci 55:2161–2176

Hospital F, Charcosset A (1997) Marker-assisted introgression of quantitative trait loci. Genetics 147:1469–1485

Huang X, Wei X, Sang T, Zhao Q, Feng Q, Zhao Y, Li C, Zhu C, Lu T, Zhang Z, Li M, Fan D, Guo Y, Wang A, Wang L, Deng L, Li W, Lu Y, Weng Q, Liu K, Huang T, Zhou T, Jing Y, Li W, Lin Z, Buckler ES, Qian Q, Zhang Q-F, Li J, Han B (2010) Genome-wide association studies of 14 agronomic traits in rice landraces. Nat Genet 42:961–967

Huang X, Zhao Y, Wei X, Li C, Wang A, Zhao Q, Li W, Guo Y, Deng L, Zhu C, Fan D, Lu Y, Weng Q, Liu K, Zhou T, Jing Y, Si L, Dong G, Huang T, Lu T, Feng Q, Qian Q, Li J, Han B (2012) Genome-wide association study of flowering time and grain yield traits in a worldwide collection of rice germplasm. Nat Genet 44:32–39

Huth PJ (1995) Nutritional aspects of soybean oil and protein. In: Erickson DR (ed) Practical handbook of soybean processing and utilization. AOCS Press, Urbana, pp 460–482

Hwang EY, Song Q, Jia G, Specht JE, Hyten DL, Costa J, Cregan PB (2014) A genome-wide association study of seed protein and oil content in soybean. BMC Genomics 15:1

Hyten DL, Pantalone VR, Sams CE, Saxton AM, Landau-Ellis D, Stefaniak TR, Schmidt ME (2004) Seed quality QTL in a prominent soybean population. Theor Appl Genet 109:552–561

Hyten DL, Song Q, Zhu Y, Choi I-Y, Nelson RL, Costa JM, Specht JE, Shoemaker RC, Cregan PB (2006) Impacts of genetic bottlenecks on soybean genome diversity. Proc Natl Acad Sci USA 103:16666–16671

Hyten DL, Song Q, Choi I-Y, Yoon M-S, Specht JE, Matukumalli LK, Nelson RL, Shoemaker RC, Young ND, Cregan PB (2008) High-throughput genotyping with the GoldenGate assay in the complex genome of soybean. Theor Appl Genet 116:945–952

Iquira E, Humira S, François B (2015) Association mapping of QTLs for sclerotinia stem rot resistance in a collection of soybean plant introductions using a genotyping by sequencing (GBS) approach. BMC Plant Biol 15:1

Jackson G (2014) Soybean cyst nematode. Mississippi State University Extension Service Publication 1293

Jarquín D, Crossa J, Lacaze X, Cheyron PD, Daucourt J, Lorgeou J, Piraux F, Guerreiro L, Pérez P, Calus M, Burgueño J, Campos GDL (2014) A reaction norm model for genomic selection using high-dimensional genomic and environmental data. Theor Appl Genet 127:595–607

Keim P, Diers BW, Olson TC, Shoemaker RC (1990) RFLP mapping in soybean: association between marker loci and variation in quantitative traits. Genetics 126:735–742

Keim P, Schupp JM, Travis SE, Clayton K, Zhu T, Shi L, Ferreira A, Webb DM (1997) A high-density soybean genetic map based on AFLP markers. Crop Sci 37:537–543

Kim KH, Kim MY, Van K, Moon JK, Kim DH, Lee SH (2008) Marker-assisted foreground and background selection of near isogenic lines for bacterial leaf pustule resistant gene in soybean. J Crop Sci Biotechnol 11:263–268

King ZR, Harris DK, Pedley KF, Song Q, Wang D, Wen Z, Buck JW, Li Z, Boerma HR (2016a) A novel *Phakopsora pachyrhizi* resistance allele (Rpp) contributed by PI 567068A. Theor Appl Genet 129:517–534

King ZR, Harris DK, Wood ED, Buck JW, Boerma HR, Li Z (2016b) Registration of four near-isogenic soybean lines of G00-3213 for resistance to Asian soybean rust. J Plant Reg 10:189–194

Koester RP, Skoneczka JA, Cary TR, Diers BW, Ainsworth EA (2014) Historical gains in soybean (*Glycine max* Merr.) seed yield are driven by linear increases in light interception, energy conversion, and partitioning efficiencies. J Exp Bot 65:3311–3321

Lam HM, Xu X, Liu X, Chen W, Yang G, Wong FL, Li M-W, He W, Qin N, Wang B, Li J, Jian M, Wang J, Shao G, Wang J, Sun SS-M, Zhang G (2010) Resequencing of 31 wild and cultivated soybean genomes identifies patterns of genetic diversity and selection. Nat Genet 42:1053–1059

Landau-Ellis D, Pantalone VR (2009) Marker-assisted backcrossing to incorporate two low phytate alleles into the Tennessee soybean cultivar 5601T. In: Shu QY (ed) Induced plant mutations in the genomics era. Food and Agriculture Organization of the United Nations, Rome, pp 316–318

Lee S, Freewalt KR, Mchale LK, Song Q, Jun TH, Michel AP, Dorrance AE, Mian MAR (2015) A high-resolution genetic linkage map of soybean based on 357 recombinant inbred lines genotyped with BARCSoySNP6K. Mol Breed 35:58

Li Z, Nelson RL (2001) Genetic diversity among soybean accessions from three countries measured by RAPDs. Crop Sci 41:1337–1347

Li Z, Jakkula L, Hussey RS, Tamulonis JP, Boerma HR (2001) SSR mapping and confirmation of QTL from PI96354 conditioning soybean resistance to southern root-knot nematode. Theor Appl Genet 103:1167–1173

Li Y, Hill CB, Carlson SR, Diers BW, Hartman GL (2007) Soybean aphid resistance genes in the soybean cultivars Dowling and Jackson map to linkage group M. Mol Breed 19:25–34

Li W, Han Y, Zhang D, Yang M, Teng W, Jiang Z, Qiu L, Sun G (2008) Genetic diversity in soybean genotypes from north-eastern China and identification of candidate markers associated with maturity rating. Plant Breed 127:494–500

Li Y, Zhao S, Ma J, Li D, Yan L, Li J, Qi X, Guo X, Zhang L, He W, Chang R, Liang Q, Guo Y, Ye C, Wang X, Tao Y, Guan R, Wang J, Liu Y, Jin L, Zhang X, Liu Z, Zhang L, Chen J, Wang K, Nielsen R, Li R, Chen P, Li W, Reif JC, Purugganan M, Wang J, Zhang M, Wang J, Qiu L (2013) Molecular footprints of domestication and improvement in soybean revealed by whole genome re-sequencing. BMC Genomics 14:579

Libault M, Farmer A, Joshi T, Takahashi K, Langley RJ, Franklin LD, He J, Xu D, May G, Stacey G (2010) An integrated transcriptome atlas of the crop model *Glycine max*, and its use in comparative analyses in plants. Plant J 63:86–99

Liu S, Kandoth PK, Warren SD, Yeckel G, Heinz R, Alden J, Yang C, Jamai A, El-Mellouki T, Juvale PS, Hill J, Baum TJ, Cianzio S, Whitham SA, Korkin D, Mitchum MG, Meksem K (2012) A soybean cyst nematode resistance gene points to a new mechanism of plant resistance to pathogens. Nature 492:256–260

Ma YS, Reif JC, Jiang Y, Wen ZX, Wang DC, Liu ZX, Guo Y, Wei SH, Wang SM, Yang CM, Wang HC, Yang CY, Lu WG, Xu R, Zhou R, Wang RZ, Sun ZD, Chen HZ, Zhang WH, Wu J, Hu GH, Liu CY, Luan XY, Fu YS, Guo T, Han TF, Zhang MC, Sun BC, Zhang L, Chen WY, Wu CX, Sun S, Yuan BJ, Zhou XN, Han DZ, Yan HR, Li WB, Qiu LJ (2016) Potential of marker selection to increase prediction accuracy of genomic selection in soybean (*Glycine max* L.). Mol Breed 36:113

Mansur LM, Carriquiry AL, Rao-Arelli AP (1993) Generation mean analysis of resistance to race 3 of soybean cyst nematode. Crop Sci 33:1249–1253

Mansur L, Orf JH, Chase K, Jarvik T, Cregan PB, Lark KG (1996) Genetic mapping of agronomic traits using recombinant inbred lines of soybean. Crop Sci 36:1327–1336

Maroof MAS, Jeong SC, Gunduz I, Tucker DM, Buss GR, Tolin SA (2008) Pyramiding of soybean mosaic virus resistance genes by marker-assisted selection. Crop Sci 48:517–526

Meksem K, Ruben E, Hyten DL, Schmidt ME, Lightfoot DA (2001) High-throughput genotyping for a polymorphism linked to soybean cyst nematode resistance gene Rhg4 by using TaqMan (TM) probes. Mol Breed 7:63–71

Meuwissen THE, Hayes BJ, Goddard ME (2001) Prediction of total genetic value using genome-wide dense marker maps. Genetics 157:1819–1829

Mian MAR, Missaoui AM, Walker DR, Phillips DV, Boerma HR (2008) Frogeye leaf spot of soybean: a review and proposed race designations for isolates of *Cercospora sojina* Hara. Crop Sci 48:14–24

Molnar SJ, Rai S, Charette M, Cober ER (2003) Simple sequence repeat (SSR) markers linked to E1, E3, E4, and E7 maturity genes in soybean. Genome 46:1024–1036

Monteros MJ, Ha BK, Phillips DV, Boerma HR (2010) SNP assay to detect the 'Hyuuga' red-brown lesion resistance gene for Asian soybean rust. Theor Appl Genet 121:1023–1032

Mueller UG, Wolfenbarger LL (1999) AFLP genotyping and fingerprinting. Trends Ecol Evol 14:389–394

Myles S, Peiffer J, Brown PJ, Ersoz ES, Zhang ZW, Costich DE, Buckler ES (2009) Association mapping: critical considerations shift from genotyping to experimental design. Plant Cell 21:2194–2202

Narvel JM, Walker DR, Rector BG, All JN, Parrott WA, Boerma HR (2001) A retrospective DNA marker assessment of the development of insect resistant soybean. Crop Sci 41:1931–1939

Pham AT, McNally K, Abdel-Haleem H, Boerma HR, Li Z (2013) Fine mapping and identification of candidate genes controlling the resistance to southern root-knot nematode in PI 96354. Theor Appl Genet 126:1825–1838

Pham AT, Harris DK, Buck J, Hoskins A, Serrano J, Abdel-Haleem H, Cregan P, Song QJ, Boerma HR, Li Z (2015) Fine mapping and characterization of candidate genes that control resistance to *Cercospora sojina* K. Hara in two soybean germplasm accessions. PLoS ONE 10(5):e0126753

Phillips DV, Boerma HR (1982) Two genes for resistance to race 5 of *Cercospora sojina* in soybeans. Phytopathology 72:764–766

Platt A, Vilhjálmsson BJ, Nordborg M (2010) Conditions under which genome-wide association studies will be positively misleading. Genetics 186:1045–1052

Prabhu RR, Gresshoff PM (1994) Inheritance of polymorphic markers generated by DNA amplification fingerprinting and their use as genetic-markers in soybean. Plant Mol Biol 26:105–116

Probst AH, Athow KL, Laviolette FA (1965) Inheritance of resistance to race 2 of *Cercospora sojina* in soybeans. Crop Sci 5:332

Rawal R, Kumar V, Rani A, Gokhale SM, Husain SM (2014) Marker assisted backcross selection for genetic removal of lipoxygenase-2 from popular soybean (*Glycine max*) variety JS 97-52: parental polymorphism survey and hybridity validation. Indian J Agric Sci 84:414–417

Rector BG, All JN, Parrott WA, Boerma HR (1998) Identification of molecular markers linked to quantitative trait loci for soybean resistance to corn earworm. Theor Appl Genet 96:786–790

Rector BG, All JN, Parrott WA, Boerma HR (1999) Quantitative trait loci for antixenosis resistance to corn earworm in soybean. Crop Sci 39:531–538

Rector BG, All JN, Parrott WA, Boerma HR (2000) Quantitative trait loci for antibiosis resistance to corn earworm in soybean. Crop Sci 40:233–238

Rogers J, Chen PY, Shi AN, Zhang B, Scaboo A, Smith SF, Zeng AL (2015) Agronomic performance and genetic progress of selected historical soybean varieties in the southern USA. Plant Breed 134:85–93

Schmutz J, Cannon SB, Schlueter J, Ma JX, Mitros T, Nelson W, Hyten DL, Song QJ, Thelen JJ, Cheng JL, Xu D, Hellsten U, May GD, Yu Y, Sakurai T, Umezawa T, Bhattacharyya MK, Sandhu D, Valliyodan B, Lindquist E, Peto M, Grant D, Shu SQ, Goodstein D, Barry K, Futrell-Griggs M, Abernathy B, Du JC, Tian ZX, Zhu LC, Gill N, Joshi T, Libault M, Sethuraman A, Zhang XC, Shinozaki K, Nguyen HT, Wing RA, Cregan P, Specht J, Grimwood J, Rokhsar D, Stacey G, Shoemaker RC, Jackson SA (2010) Genome sequence of the palaeopolyploid soybean. Nature 463:178–183

Severin AJ, Woody JL, Bolon YT, Joseph B, Diers BW, Farmer AD, Muehlbauer GJ, Nelson RT, Grant D, Specht JE, Graham MA, Cannon SB, May GD, Vance CP, Shoemaker RC (2010) RNA-Seq Atlas of *Glycine max*: a guide to the soybean transcriptome. BMC Plant Biol 10:160

Shi Z, Bachleda N, Pham AT, Bilyeu K, Shannon G, Nguyen H, Li Z (2015) High-throughput and functional SNP detection assays for oleic and linolenic acids in soybean. Mol Breed 35:1–10

Shin JH, Vaughn JN, Abdel-Haleem H, Chavarro C, Abernathy B, Do Kim K, Jackson SA, Li Z (2015) Transcriptomic changes due to water deficit define a general soybean response and accession-specific pathways for drought avoidance. BMC Plant Biol 15:26

Sonah H, O'Donoughue L, Cober E, Rajcan I, Belzile F (2015) Identification of loci governing eight agronomic traits using a GBS-GWAS approach and validation by QTL mapping in soya bean. Plant Biotechnol J 13:211–221

Song QJ, Marek LF, Shoemaker RC, Lark KG, Concibido VC, Delannay X, Specht JE, Cregan PB (2004) A new integrated genetic linkage map of the soybean. Theor Appl Genet 109:122–128

Song QJ, Hyten DL, Jia GF, Quigley CV, Fickus EW, Nelson RL, Cregan PB (2013) Development and evaluation of SoySNP50K, a high-density genotyping array for soybean. PLoS ONE 8(1):e54985. doi:10.1371/journal.pone.0054985

Song QJ, Jia G, Quigley CV, Fickus EW, Hyten D, Nelson RL, Cregan PB (2014) Soybean BARC-SoySNP6K Beadchip—a tool for soybean genetics research. Poster presented at: plant and animal genome XXII conference, San Diego, CA, USA, 10–15 January 2014, Abstract no. P306

Song QJ, Hyten DL, Jia GF, Quigley CV, Fickus EW, Nelson RL, Cregan PB (2015) Fingerprinting soybean germplasm and its utility in genomic research. Genes Genomes Genet 5:1999–2006

Specht JE, Diers BW, Nelson RL, Francisco J, de Toledo F, Torrion JA, Grassini P (2014) Soybean. In: Smith S, Diers B, Specht J, Carver B (eds) Yield gains in major U.S. field crops. American Society of Agronomy, Crop Science Society of America, and Soil Science Society of America, Madison, WI, USA, Crop Science Society of America Special Publication 33, pp 311–356

Spindel J, Wright M, Chen C, Cobb J, Gage J, Harrington S, Lorieux M, Ahmadi N, McCouch S (2013) Bridging the genotyping gap: using genotyping by sequencing (GBS) to add high-density SNP markers and new value to traditional bi-parental mapping and breeding populations. Theor Appl Genet 126:2699–2716

Suhre JJ, Weidenbenner NH, Rowntree SC, Wilson EW, Naeve SL, Conley SP, Casteel SN, Diers BW, Esker PD, Specht JE, Davis VM (2014) Soybean yield partitioning changes revealed by genetic gain and seeding rate interactions. Agron J 106:1631–1642

Suzuki C, Tanaka Y, Takeuchi T, Yumoto S, Shirai S (2012) Genetic relationships of soybean cyst nematode resistance originated in Gedenshirazu and PI84751 on rhg1 and rhg4 loci. Breed Sci 61:602–607

Tamulonis JP, Luzzi BM, Hussey RS, Parrott WA, Boerma HR (1997a) RFLP mapping of resistance to southern root-knot nematode in soybean. Crop Sci 37:1903–1909

Tamulonis JP, Luzzi BM, Hussey RS, Parrott WA, Boerma HR (1997b) DNA markers associated with resistance to Javanese root-knot nematode in soybean. Crop Sci 37:783–788

Tamulonis JP, Luzzi BM, Hussey RS, Parrott WA, Boerma HR (1997c) DNA marker analysis of loci conferring resistance to peanut root-knot nematode in soybean. Theor Appl Genet 95:664–670

Tavaud-Pirra M, Sartre P, Nelson R, Santoni S, Texier N, Roumet P (2009) Genetic diversity in a soybean collection. Crop Sci 49:895–902

Ude GN, Kenworthy WJ, Costa JM, Cregan PB, Alvernaz J (2003) Genetic diversity of soybean cultivars from China, Japan, North America, and North American ancestral lines determined by amplified fragment length polymorphism. Crop Sci 43:1858–1867

Vaughn JN, Li Z (2016) Genomic signatures of North American soybean improvement inform diversity enrichment strategies and clarify the impact of hybridization. G3(6):2693–2705

Vaughn JN, Nelson RL, Song QJ, Cregan PB, Li Z (2014) The genetic architecture of seed composition in soybean is refined by genome-wide association scans across multiple populations. Genes Genomes Genet 4:2283–2294

Visscher PM, Haley CS, Thompson R (1996) Marker-assisted introgression in backcross breeding programs. Genetics 144:1923–1932

Welsh J, Mcclelland M (1990) Fingerprinting genomes using PCR with arbitrary primers. Nucl Acids Res 18:7213–7218

Williams JGK, Kubelik AR, Livak KJ, Rafalski JA, Tingey SV (1990) DNA polymorphisms amplified by arbitrary primers are useful as genetic-markers. Nucl Acids Res 18:6531–6535

Wilson RF (2010) Outlook for soybeans and soybean products in 21st century markets. Lipid Technol 22:199–202

Wrather JA, Koenning SR (2009) Effects of diseases on soybean yields in the United States 1996 to 2007. Plant Health Progress. doi:10.1094/PHP-2009-0401-01-RS

Xavier A, Muir WM, Rainey KM (2016) Assessing predictive properties of genome-wide selection in soybeans. Genes Genomes Genet 6:2611–2616

Xu XY, Zeng L, Tao Y, Vuong T, Wan JR, Boerma R, Noe J, Li Z, Finnerty S, Pathan SM, Shannon JG, Nguyen HT (2013) Pinpointing genes underlying the quantitative trait loci for root-knot nematode resistance in palaeopolyploid soybean by whole genome resequencing. Proc Natl Acad Sci USA 110:13469–13474

Yorinori JT (1987) Frogeye leaf spot of soybeans (*Cercospora sojina Hara*). In: Shibbles R (ed) World soybean production and utilization conference IV. Westview Press Inc., Boulder, pp 1275–1283

Yu YG, Maroof MAS, Buss GR, Maughan PJ, Tolin SA (1994) RFLP and microsatellite mapping of a gene for soybean mosaic-virus resistance. Phytopathology 84:60–64

Zhang JP, Song QJ, Cregan PB, Nelson RL, Wang XZ, Wu JX, Jiang GL (2015) Genome-wide association study for flowering time, maturity dates and plant height in early maturing soybean (*Glycine max*) germplasm. BMC Genomics 16:217

Zhou ZK, Jiang Y, Wang Z, Gou ZH, Lyu J, Li WY, Yu YJ, Shu LP, Zhao YJ, Ma YM, Fang C, Shen YT, Liu TF, Li CC, Li Q, Wu M, Wang M, Wu YS, Dong Y, Wan WT, Wang X, Ding ZL, Gao YD, Xiang H, Zhu BG, Lee SH, Wang W, Tian ZX (2015) Resequencing 302 wild and cultivated accessions identifies genes related to domestication and improvement in soybean. Nat Biotechnol 33:408–414

Zhu YL, Song QJ, Hyten DL, Van Tassell CP, Matukumalli LK, Grimm DR, Hyatt SM, Fickus EW, Young ND, Cregan PB (2003) Single-nucleotide polymorphisms in soybean. Genetics 163:1123–1134

Zou JJ, Dong W, Yang QK, Cao YP, Chen SY (1999) Inheritance of resistance to race 7 of *Cercospora sojina* in soybeans and RAPD tagging of the resistance gene. Chin Sci Bull 44:452–456

Soybean Genomic Libraries, TILLING, and Genetic Resources

9

Liu Shiming, Naoufal Lakhssassi, Zhou Zhou, Vincent Colantonio, My Abdelmajid Kassem and Khalid Meksem

Abstract

Soybean (*Glycine max* (L.) Merr.) is one of the most important crops worldwide, providing a sustainable source of protein and oil. Development and utilization of large-scale chemical mutagenesis in soybean is a promising strategy to develop new soybean genetic resources (germplasm) without the regulatory hurdles of genetic modification. Mutagenized soybean populations can be used with high throughput screening by Targeted Induced Local Lesions IN Genomes (TILLING) to identify mutations within genes of interest. By correlating an altered phenotype to the occurrence of mutations within a corresponding gene, protein function can be elucidated without the requirement of genetic transformation. Mutagenized soybean populations and their genomic libraries have been successfully applied to the identification of a soybean cyst nematode (SCN) resistance gene by correlating mutations within *GmSHMT* to a loss of SCN resistance, demonstrating that *GmSHMT* is the *Rhg4* gene conferring SCN resistance. Additionally, by screening for mutations within genes involved in fatty acid biosynthesis, germplasm that accumulates high levels of oleic acid, stearic acid, or palmitic acid were discovered. Chemical mutagenesis as a forward genetics approach when coupled with TILLING as a reverse genetics approach has been proven to be a valuable tool for soybean researchers in the discovery and development of agronomically important traits.

L. Shiming · N. Lakhssassi · Z. Zhou · V. Colantonio · K. Meksem (✉)
Department of Plant, Soil and Agricultural Systems, Southern Illinois University, Carbondale, IL 62901, USA
e-mail: meksem@siu.edu

M.A. Kassem
Department of Biological Sciences, Fayetteville State University, Fayetteville, NC 28301, USA

9.1 Introduction

Soybean [*Glycine max* (L.) Merr.] is a major commercial crop grown throughout the world, providing a sustainable source of protein and oil. Soybean has become the most widely grown protein/oilseed crop in the USA since the 1930s

H.T. Nguyen and M.K. Bhattacharyya (eds.), *The Soybean Genome*, Compendium of Plant Genomes, DOI 10.1007/978-3-319-64198-0_9

and contributes an estimated $38 billion to the current US economy (NAICS 11111, 2014). Soybean has been extensively used for food, livestock feed, and industrial applications. With an emerging trend toward renewable energy resources in the market, the demand for soybean continues to rise. Hence, a sustainable soybean supply raises a serious challenge. In particular, improvement of the soybean germplasm, breeding of new genetic soybean resources, and control of diseases are of top priority.

The completion of whole-genome shotgun sequencing of the soybean cultivar (cv.) Williams 82 predicted over 46,000 genes within the soybean genome (Schmutz et al. 2010). Since then, the challenge facing soybean researchers has become how to efficiently associate genomic sequences with phenotypic variation in order to identify the functions of all those genes. Due to two duplication events which occur 59 and 13 million years ago, soybean has a polyploidy genome, and 70–80% of the genes are duplicated and present under multiple copies, which make difficult the identification of genes involved in important agronomical traits (i.e., oil, protein, yield, and resistant to biotic and abiotic stresses) and complicate the development of soybean breeding programs (Yin et al. 2013; Zhu et al. 2014; Singh and Jain 2015; Lakhssassi et al. 2017b). Several approaches have been used for gene functional annotation and each approach has its own advantages. For instance, RNA interference (RNAi) is a means of tissue-specific silencing that allows a knockdown of various genes, and this has been broadly applied for post-transcriptional gene silencing of target genes in soybean (Nunes et al. 2006; Flores et al. 2008; Zhang et al. 2014). Because soybean is known to be recalcitrant to stable transformation, virus induced gene silencing (VIGS) is another method that has been used to knock down genes in soybean; thus, rapidly identifying gene functions without the requirement of stable transformation (Senda et al. 2004; Nagamatsu et al. 2007, 2009; Zhang et al. 2009; Liu et al. 2012). In addition, T-DNA has widely been implemented to characterize the functional role of genes in plants, and new DNA sequencing technology can be used to facilitate transposon tagging in plants (Henikoff and Comai 2003). Recently, CRISPR/Cas9 technology has also been gradually developed in soybean (Cai et al. 2015; Jacobs et al. 2015; Sun et al. 2015).

Transformation is required in using approaches/systems such as T-DNA, VIGS, RNAi, or CRISPR/Cas9 for functional gene analysis. However, it is hard to generate transgenic plants in soybean. Hairy root transformation is a method usually employed for transient soybean transformations, but this system is unstable, prone to a low transformation efficiency, and only phenotypes limited to roots are likely to be tested. Because of this, chemically and physically mutagenized soybean populations are promising. Chemical mutagens such as ethyl methanesulfonate (EMS) can cause random mutations (predominately C to T and G to A transversions in EMS) that produce single nucleotide polymorphisms (SNPs) and insertion–deletions (InDels) at the whole-genome level. Using chemical mutagenesis, a collection of mutants with significant phenotypes can be easily produced, and mutations within genes of interest can be discovered by Targeted Induced Local Lesions IN Genomes (TILLING). TILLING is a reverse genetic tool for high throughput mutant screening and can be used for functional gene analysis without genetic transformation. Fast neutron (FN) radiation is a reliable approach to create knockout mutant populations and is more efficient for null allele identification due to the majority of mutations resulting from large deletions (Men et al. 2002). Coupling FN mutagenesis with high throughput screening technologies like comparative genomic hybridization (CGH) and next-generation sequencing, FN could have a promising prospect for plant functional genomics (Bolon et al. 2011). Large-scale mutagenized soybean populations can be used as forward genetic resources for mutation breeding of novel germplasms without passing through genetic modification (GM). Taken together, it is of significant importance to develop mutagenized soybean populations as genetic resources, construct their genomic libraries, and screen mutants by TILLING for novel germplasm development and functional gene analysis.

9.1.1 Genetic Soybean Resources—Mutagenized Soybean Populations

Single base pair mutations caused by chemical mutagens are classified into missense, nonsense, and silent mutations. Missense mutations result in a codon that encodes a different amino acid in the protein sequence. Nonsense mutations result in a premature stop codon, terminating the protein early. Both missense and nonsense mutations potentially cause an alteration of gene functions and phenotypes. Silent mutations, on the other hand, do not lead to any amino acid changes and usually do not display a phenotypic alteration. Chemical and physical mutagens such as EMS and FN are commonly used to develop large-scale mutagenized crop populations. Mutagenized populations are a collection of germplasm with varying phenotypes and can be used as novel genetic resources in reverse genetics (functional genomics, etc.) and forward genetics (selection of mutant traits, mutation breeding, etc.). So far, many large-scale mutagenized populations of crops such as Arabidopsis, barley, maize, rice, soybean, tomato, watermelon, and wheat have been developed using chemical mutagens (Greene et al. 2003; Perry et al. 2003; Till et al. 2004; Cooper et al. 2008; Dalmais et al. 2008; Porceddu et al. 2008; Suzuki et al. 2008; Talamè et al. 2008; Wang et al. 2008; Xin et al. 2008; Dong et al. 2009).

Soybean cyst nematode (SCN, *Heterodera glycines* Ichinohe) is the most devastating pest in soybean production and causes an annual yield loss of over $1 billion in the USA alone (Koenning and Wrather 2010). Genetic resistance of host soybean varieties is the most effective, economical, and environmentally friendly means to managing SCN. The soybean *cv.* Forrest is resistant to SCN races 1 and 3 and root-knot nematode (*Meloidogyne incognita*) (Hartwig and Epps 1973). Forrest SCN resistance is mainly controlled by two quantitative trait loci (QTL), *rhg1* and *Rhg4* (Meksem et al. 2001), and is derived from the cultivar 'Peking'. To identify the functions of genes by reverse genetics, such as genes underlying SCN resistance, and to use mutation breeding to develop desired soybean genetic resources, such as germplasm with high levels of seed stearic and oleic acids, we used Forrest as the genetic background to develop large-scale EMS-mutagenized populations and statistically analyzed major morphological and agronomic phenotypes.

9.1.1.1 Development of Large-Scale Forrest Mutant Populations

To develop a mutagenized population, a mutagenesis pretest is conducted at a series of ethylmethane sulfonate (EMS) concentrations ranging from 0 to 1.2% (v/v): 0, 0.3, 0.5, 0.55, 0.6, 0.65, 0.7, 0.8, 1.0, and 1.2%. This allows for a determination of an appropriate EMS concentration for use in the development of large-scale mutant populations. One hundred Forrest seeds are treated with ranging EMS concentrations in a 500 ml bottle at room temperature overnight ($\sim$15 h) in the fume hood. Afterward, the seeds are washed thoroughly thrice with tap water (300 ml per wash) to remove EMS and the washed EMS solution is neutralized by 10% (w/v) sodium thiosulfate solution. The mutagenized seeds (M1) are then sowed immediately in a greenhouse under 16 h/8 h (light/dark) photoperiod at 28–30 °C. After about 10 days, the germination rate of each treatment is recorded. The concentration of EMS used to treat the large-scale Forrest population was chosen from the treatment that had an approximately 50% germination rate compared to the wild-type Forrest (Meksem et al. 2008). The results indicated that the appropriate concentration of EMS was 0.57% (v/v).

In 2011, a total of about 4000 Forrest seeds were soaked in a 0.57% (v/v) EMS solution overnight (15 h) in the fume hood at room temperature followed by three thorough washes with tap water. The mutagenized M1 seeds were then planted in Fafard Canadian Growing Mix

No. 2 soil using 48-cell trays in the greenhouse at the Horticulture Research Center (HRC) of Southern Illinois University Carbondale (SIUC). M1 seedlings were grown under 16 h/8 h photoperiod at 28–30 °C. After 3–4 weeks, the seedlings were transplanted to the field at the HRC and M1 plants produced M2 seeds through self-fertilization. M2 seeds were harvested and stored in the seed storage room at Agronomy Research Center of SIUC.

In 2012, two M2 seeds from each M1 family were sowed in the HRC greenhouse, and seedlings were transplanted from greenhouse to the field after about 4 weeks. Young leaf tissues from each M2 plant were collected in the field for DNA extraction to construct the genomic libraries of the mutagenized population and later M3 seeds were harvested and stored long-term at −20 °C. In total, 1588 Forrest M2 families were developed from 2011 to 2012.

9.1.1.2 Morphological and Phenotypic Variation of the Mutagenized Forrest Population

For each mutagenized population, a variety of morphological and phenotypic data should be collected from three successive soybean generations. This allows for the mutant variation to be characterized as mutations become affixed. In the chimeric M1 generation, the germination rate of the mutagenized seeds is scored and the growth stages of M1 seedlings are recorded in the greenhouse. In the heterozygous M2 generation, morphological traits of mutant plants become more visible. For this, pictures of the plants are taken and documented in an archive. At harvest, the total number of M3 seeds from each M2 plant is counted. In addition, the color and size of M3 seeds are screened, and the weight (g) of 10 M3 seeds from each M2 family is measured.

The major morphological phenotypes of the mutagenized Forrest population are summarized in Fig. 9.1. Some detailed morphological phenotypes are included as follows. (1) Chlorophyll deficiency. The symptoms of chlorophyll deficiency are fairly common among the M2 plants due to mutations in genes of chlorophyll biosynthesis or related pathways. Yellow-green to white leaves from the first trifoliate to full flowering are typical phenotypes of chlorophyll deficiency in mutant soybeans. EMS is an efficient chemical mutagen to generate chlorophyll deficient variants in soybean M2 populations (Carroll et al. 1986). In the whole population, 1.46% of the mutants showed a chlorotic phenotype and 8.35% of mutants exhibited other leaf color phenotypes (reddish, dark hue, or strange color). (2) Leaf shape. The wild-type Forrest plants exhibited odd, weird and ovate leaves in each growth and development stage while various leaf shapes were found in M2 plants. There were 4.38% of mutants displaying oblong leaves (elliptical shaped, long and skinny) and 2.51% of mutants displaying other leaf shape phenotypes (rough, bumpy, chordate, etc.). An individual mutant (F761) presented at least three different leaf shapes compared to the wild type. (3) Leaflet number. Most of the leaves of wild-type soybeans are trifoliate. However, a mutant showed a cluster of leaflets at a single node in the field. Instead of two unifoliate leaves, two small branches attached opposite to each other at the same node above the cotyledons were exhibited on another mutant. (4) Branch architecture. In the field, angled, stiff, or twisted branches were observed from mature M2 plants at harvest. The rate of lateral branching ×2 (2 branches), ×3 (3 branches), and ×4 (4 branches) mutants accounted for 10.44, 6.58, and 0.94% of the population, respectively. A bushy mutant (F1257) grew thickly with 3–4 branches and produced a number of pods. (5) Stem length. Stem length is regarded as a parameter of growth habit for soybean plants. Although there were no significant differences in stem lengths between the wild type and mutants, some dwarf M2 plants were found in the field (0.3%). (6) Seed coat color. Several loci, including *I*, *T*, *W1*, *R*, and *O*, control soybean seed coat colors ranging from black to light yellow due to deposition of different anthocyanin pigments. The pigmentation of the wild-type Forrest seed coat is yellow. However, a total of 22 soybean mutant lines having black seeds were recorded from the whole M2 population. The number of black seeds and yellow seeds of the mutant lines is shown in

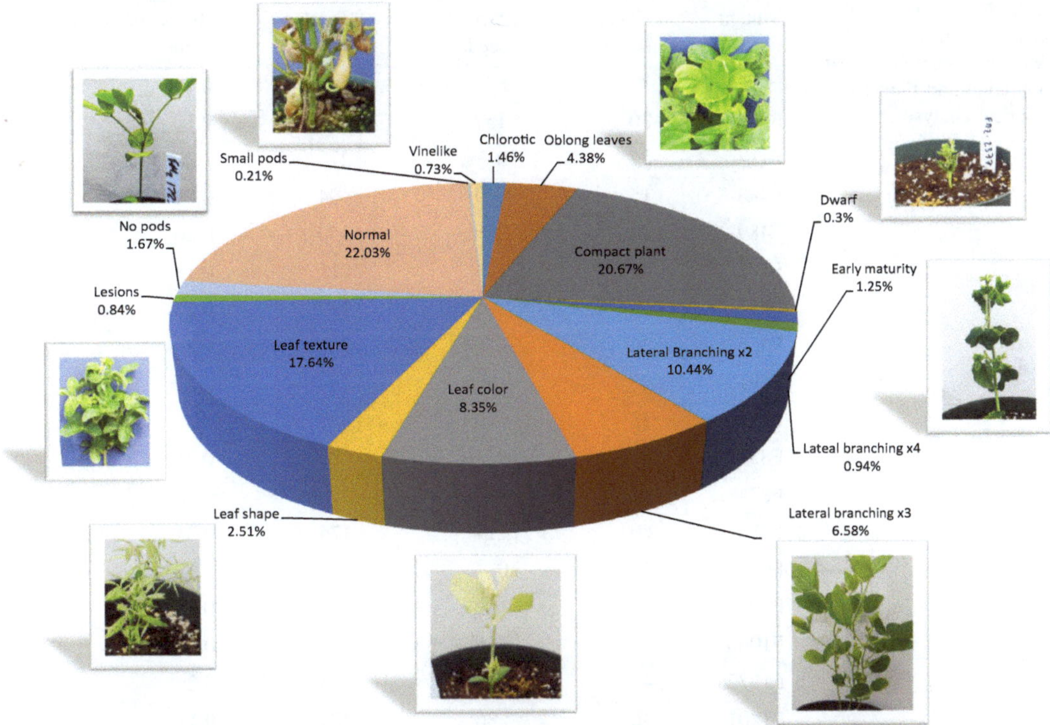

Fig. 9.1 Major morphological phenotypes of the developed EMS-mutagenized Forrest-mutagenized population

Table 9.1. The ratio of black seeds to total seeds from each mutant ranged from 0.02 to 1.00, and 0.37 on average. There were 4 mutants (F414, F874, F995, and F1516) that only had black seeds and one mutant (F1496) that had 99% black seeds. (7) Seed weight. Seed weight is a productivity trait to evaluate soybean yield. After threshing, 10 M3 seeds were randomly selected from each M2 family and weighted with a balance. The full scale of seed weight ranged from 0.40 to 2.38 g per 10 seeds. The average weight per 10 seeds of mutant was 1.45 g, while that of wild type was 1.11 g. In addition, the seed weight of 98 M2 families accounted for 6% of total M2 population's seed weight and was above 1.9 g (Fig. 9.2). (8) Seed size. Seed size is considered as an important yield component in soybean production. M3 seeds from each M2 family were screened by hand at ARC of SIUC. Some mutants such as F976 displayed a smaller size of M3 seeds compared to the wide type. (9) Pod development. Pod number per plant imposes a limitation on the soybean yield. 1.67% of mutants did not produce any pods and 0.21% of mutants produced small pods. (10) Plant height. Plant height determines the pod number on a plant at some degree, which indirectly influences the yield potential. A horizontal or erect stature of mutant plants was seen in the field, while compact and miniature mutant plants were also present. (11) Maturity. In general, the wild-type soybean reaches full maturity in approximate 128 days. Based on the record of days after planting (DAP) at emergence stage, unifoliate leaf stage and the first trifoliate leaves stage, the growth of mutant plants was usually getting slower at the early growth stages; however, 1.25% of mutants matured earlier in the end, compared to the wild type.

9.1.2 Genomic Libraries of Soybean Mutant Populations

To construct genomic libraries, young leaf tissue of each Forrest M2 plant is collected and genomic DNA is extracted. Next, DNA samples are

Table 9.1 Mutants with black seeds in the whole EMS-mutagenized Forrest population and their seed statistical analyses. Twenty-two mutant lines were found in the whole population (1558 lines)

Mutant ID	Black seeds	Yellow seeds	Total seeds	Black seeds: Total seeds
F174	7	82	89	0.08
F375	30	152	182	0.16
F414	27	0	27	1.00
F448	15	679	694	0.02
F489	92	358	450	0.20
F502	40	181	221	0.18
F557	26	101	127	0.20
F558	26	67	93	0.28
F691	4	13	17	0.24
F692	33	650	683	0.05
F831	4	9	13	0.31
F874	32	0	32	1.00
F988	60	114	174	0.34
F995	28	0	28	1.00
F1023	25	112	137	0.18
F1060	21	565	586	0.04
F1072	45	128	173	0.26
F1201	36	286	322	0.11
F1381	29	271	300	0.10
F1489	26	41	67	0.39
F1496	334	4	338	0.99
F1516	31	0	31	1.00
Average	–	–	–	0.37

The ratio of black seeds to total seeds from each mutant ranged from 0.02 to 1.00, 0.37 on average

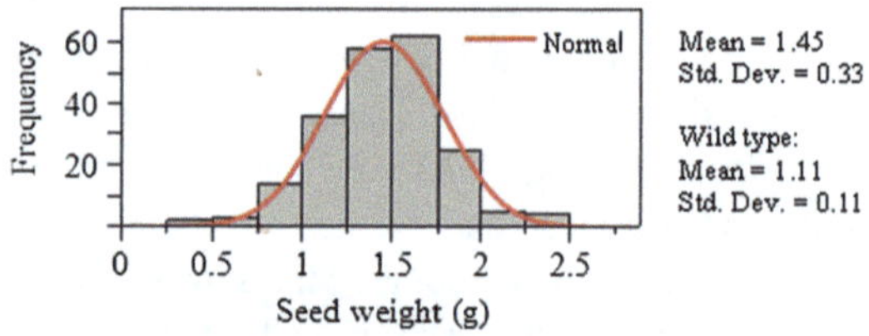

Fig. 9.2 Statistical analyses of weight (g) per 10 seeds from 98 mutant families of the developed EMS-mutagenized Forrest population

pooled into 96-well plates with DNA of 8 mutant families in each well. These are then used as genomic libraries of the mutagenized soybean populations.

9.1.2.1 DNA Extraction and Quantification

The high quality and yield of DNA samples are critical to mutation screening by TILLING, genomic sequencing, etc. Roughly the same amount (about 50–100 mg) of leaf tissue from each mutant line is used to extract DNA using DNeasy 96 Plant Kit (QIAGEN, Valencia, CA, USA). Agarose gel electrophoresis is used to evaluate the DNA quality and DNA samples are then stored at −20 °C for quantification.

The quantity of DNA is estimated and then normalized. A pentahydrate bis-Benzimide stain, known as Hoechst 33258, binds specifically to A–T base pairs in dsDNA and has a fluorescence proportional to the total nucleic acid concentration of each sample (BioTek Instruments Inc., Winooski, VT, USA). The concentration of DNA samples can be read through Gen5 Microplate Data Collection and Analysis Software on a PC using Synergy 2 Multi-Mode Microplate Reader (BioTek Instruments Inc., Winooski, VT, USA). λ-DNA stock at 100 ng/μl is used to make serial concentrations of a standard solution (10, 8, 4, 2 ng/μl) with 1× TNE buffer (0.2 M NaCl, 10 mM Tris-HCl, 1 mM EDTA, pH 7.4) to generate standard curves. Before DNA samples are added, concentrated Hoechst 33258 stock (1 mg/ml) is diluted to 0.11 μg/ml in working buffer using 1xTNE. Aliquots from each DNA sample are transferred to a new 96-well plate and diluted 100-fold prior to quantification. Eighty-eight DNA samples (20 μl each), five standard concentrations (20 μl each), and 3 blanks can be loaded in a 96-well plate added with 180 μl (samples and standards) or 200 μl (blanks) of dye in each well for quantification. The plate layout is then set up through Gen5, and the matrix, standard curves, and statistical information can be read by running the DNA quantification protocol. Finally, DNA of each sample is quantified and diluted to 100 ng/μl.

9.1.2.2 Construction of Genomic Libraries by Pooling Genomic DNA and Identification of Individuals

Quantified DNA samples (100 ng/μl) are diluted 20 times for eightfold pooling. The DNA samples from eight 96-well plates are pooled using a bi-dimensional arraying strategy for high throughput screening (Fig. 9.3a). In row pooling, DNA samples arrayed from the same column of one plate are pooled in one row of a new 96-well plate. Each well of a row in the new plate contains eight samples from the same column of one plate. Correspondingly, in column pooling, DNA samples arrayed from each column of eight plates are placed vertically in one column of a new 96-well plate. Each well of a column in the new plate contains eight samples from the same column of eight plates. Thus, each pooled plate

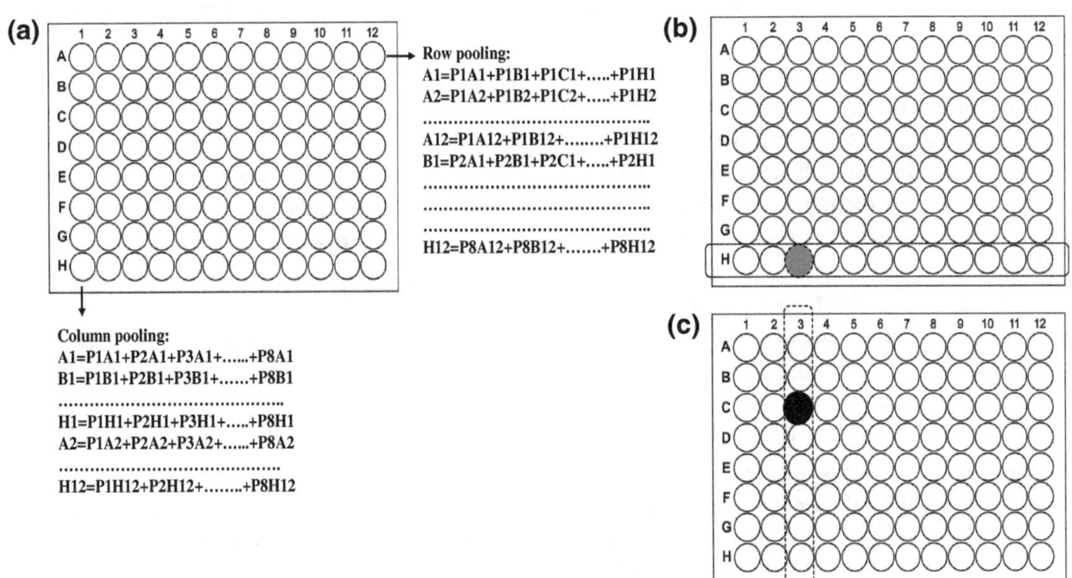

Fig. 9.3 Construction of genomic libraries by pooling of genomic of DNA and identification of screened individuals. **a** Pooling methods, **b** Rowing pooling screening, **c** Column pooling screening. An example of an individual identification: in (**b**) screened H3 is a mixture of P1A3, P1B3, P1C3, P1D3, P1E3, P1F3, P1G3, and P1H3; in (**c**) screened C3 is a mixture of P1C3, P2C3, P3C3, P4C3, P5C3, P6C3, P7C3, and P8C3, so, in both (**b**) and (**c**), only P1C3 is overlapped, namely the positive screened individual is P1C3

represents a total of 768 DNA samples (8 × 96) from M2 families. All pooled plates from one EMS-mutagenized Forrest population are stored at −20 °C as a genomic library for mutation screening, genotyping, and other uses. For individual screening and identification, the overlapped sample positively screened from the corresponding row and column pooled plates is identified as the individual mutant (Fig. 9.3b, c).

9.1.3 Soybean TILLING

TILLING is a reverse genetic tool for high throughput screening of induced mutations by combining traditional chemical mutagenesis and PCR (McCallum et al. 2000). In the initial TILLING protocol, denaturing HPLC (dHPLC) was described to detect mutations in genes of interest. In order to improve the efficiency of screening, gel-based TILLING uses endonucleases such as ENDO I and CEL I to specifically cleave mismatches in heteroduplexes formed between wild type and mutants. This method is more reliable and rapid for genome-scale screening of chemically induced mutations in plants (Colbert et al. 2001). With the development and application of next-generation sequencing, TILLLING by sequencing has recently been developed and also applied in plants (Tsai et al. 2011). During the past decade, TILLING has been widely used in a number of species including *Arabidopsis*, barley, grass, *Lotus japoniucs*, maize, rice, sorghum, soybean, and wheat (Greene et al. 2003; Perry et al. 2003; Till et al. 2004; Cooper et al. 2008; Dalmais et al. 2008; Suzuki et al. 2008; Talamè et al. 2008; Xin et al. 2008; Dong et al. 2009; Weil 2009). A model legume *Lotus japonicus* was used to establish the first legume-TILLING platform. The phenotype of mutant plants can be accessed via a Web-based database (http://data.jic.ac.uk/cgi-bin/lotusjaponicus). Mutations in the *SYMRK* gene were screened by TILLING for the study of root nodule symbiosis (Perry et al. 2003). A few TILLING databases consisting of information on mutant populations are accessible online and several TILLING public services are available in Europe and the USA (Kurowska et al. 2011) including for three legumes: soybean, *M. truncatula*, and *L. japonicus* (http://www.soybeantilling.org; Tadege et al. 2009).

9.1.3.1 Conventional TILLING

A TILLING system was successfully extended to soybean even with its highly duplicated, allotetraploid genome (http://www.soybeantilling.org). In general, high throughput soybean TILLING includes the following steps, (1) M1 seeds are mutagenized by chemical mutagens such as EMS and then sowed to generate M1 plants (mutagenized populations); (2) M1 plants are self-fertilized to produce M2 seeds; (3) M2 plants are grown, leaf tissue from each M2 line is collected, and M3 seeds are harvested; (4) DNA samples from M2 leaf tissue collected are extracted, quantified and pooled eightfold in 96-well plates to construct the genomic libraries; (5) Genes of interest are amplified by PCR with fluorescent tailed primers using the constructed genomic libraries, and heteroduplexes are formed after reannealing; (6) An endonuclease such as CEL I or ENDO I is used to cleave mismatches of heteroduplexes; (7) Endonuclease-cleaved PCR amplicons are cleaned and electrophoresed on polyacrylamide gels using LI-COR 4300 DNA Analyzer System; (8) Putative mutations are detected with bands shown in the gel images; (9) Mutant individuals are identified through deconvolution of pools and then sequenced to be confirmed; (10) Seeds from the individual mutant are planted for mutation confirmation and phenotyping (Cooper et al. 2008; Meksem et al. 2008).

In two soybean cultivars, Forrest and Williams 82, four mutagenized populations were developed using EMS and *N*-nitroso-*N*-methylurea (NMU) for TILLING. Seven targets were screened in each population and discovered a total of 116 mutations with a mutation rate from 1/140 to 1/550 kb (Cooper et al. 2008). Using this soybean TILLING platform, mutations in the *GmCLV1A* gene were identified to study the function of this gene in soybean (Meksem et al. 2008). For soybean seed meal and oil composition traits, EMS-induced mutations in both raffinose synthase gene *RS2* and omega-6 fatty acid desaturase gene *FAD2-1A* were screened and identified by TILLING. The novel mutant alleles can be deployed to assist soybean breeding for desired seed phenotypes (Dierking and Bilyeu 2009).

More recently, TILLING was applied to investigate the functional roles of candidate genes conferring resistance to SCN. In a pilot experiment, TILLING results indicated that no causal correlations exist between an *Rhg4 LRR-RLK* gene and SCN resistance. The use of Eco-TILLING, a modified TILLING technology for discovery of natural nucleotide polymorphisms, was integrated with other approaches to do functional genomics research (Till et al. 2006; Liu et al. 2011). Using TILLING, two missense mutations in the *GmSHMT* gene were identified in Forrest mutant populations and resulted in an alteration of the SCN resistance phenotype. The *SHMT* gene was isolated from soybean cv. Forrest by map-based positional cloning and validated to confer SCN resistance through a series of reverse genetic, genomic, and molecular studies (Liu et al. 2012).

To thoroughly investigate the connection between gene sequence and phenotype, the emergence of TILLING overcomes restraint on other reverse genetic strategies in several aspects. An allelic series of point mutations can be produced simultaneously. Compared to transgenic plants, there is no restriction on the utilization of mutants generated by chemical mutagenesis. Most importantly, it allows for an identification of mutations in specific regions of interest through large-scale screening in a relatively short period of time. To date, TILLING is widely used to screen mutant populations of model and crop plants, whereas extra outcrossing workload is required due to the background mutations from random mutagenesis (Bilyeu 2008; Anai 2012).

9.1.3.2 TILLING by Sequencing

On the basis of conventional TILLING, TILLING by sequencing was developed by applying next-generation sequencing (NGS) technology to TILLING. In TILLING by sequencing, the genes of interest are amplified by normal PCR using the constructed genomic libraries of the mutagenized populations, and then amplicons are cleaned and new libraries are constructed to do NGS. Afterward, the obtained raw sequences are filtered and the cleaned sequences are aligned to a reference genome. This results in SNPs and InDels detected within the aligned consensus genome sequences (Tsai et al. 2011, 2013, 2015). TILLING by sequencing is highly efficient, powerful, and, as the costs of NGS decreases, is being increasingly utilized to discover rare mutations within genes of interest in mutagenized populations.

In *Drosophila melanogaster*, an EMS-induced point mutation of the gene controlling eggshell morphology, *encore*, was identified by Illumina whole-genome sequencing (Blumenstiel et al. 2009). In plants, a novel mutation screening method was developed in mutagenized populations of rice (*Oryza sativa*) and wheat (*Triticum durum*) using a combination of Illumina GA platform sequencing, multidimensional pooling strategies, and a bioinformatics pipeline (Tsai et al. 2011). TILLING by sequencing has also been applied successfully in high throughput mutant screening of canola and peanut (Gilchrist et al. 2013; Guo et al. 2015).

9.1.4 Application of Soybean Genetic Resources

The mutagenized soybean populations produce many new genetic resources, which can be used in functional genomics and mutation breeding of germplasms with desired agronomical traits.

9.1.4.1 Genetic Resources for Functional Genomics-Identification of Soybean Genes Conferring Resistance to SCN

In 1954, SCN was first discovered in North Carolina. To date, SCN is widely distributed in most countries of the world where soybean is produced. In the family Heteroderidae, the cyst nematodes belonging to the genera *Heterodera* and *Globodera* are the most economically important species for agricultural systems. SCN is an obligate sedentary endoparasite that enters host roots, migrates to the vascular tissue, and starts feeding. It is globally regarded as the most economically destructive pathogen in soybean production and causes over $1 billion in annual losses in the USA (Koenning and Wrather 2010). Planting resistant cultivars and nonhost crop rotation are the primary management strategies of controlling SCN. Nevertheless, the use of limited commercial cultivars for SCN resistance will eventually lead to SCN population shifts and the loss of resistance (Vuong et al. 2010). Hence, the study of the nature of genetic resistance is indispensable to discover the crop's mechanism of action against the pathogen and ultimately improve resistances.

With the advent of concerted efforts to investigate the inheritance of resistance to SCN, several genes were reported in soybean cultivars to confer SCN resistance including *rhg1*, *rhg2*, *rhg3* (Caldwell et al. 1960), and *Rhg4* (Matson and Williams 1965). Molecular marker technology has been applied to localize and characterize quantitative trait loci (QTL) underlying SCN resistance in the soybean genome (Concibido et al. 2004; Kassem et al. 2014). QTL associated with SCN resistance to different races in numerous soybean cultivars have been identified through genetic mapping. Although several studies have shown inconsistencies in the mapping of SCN resistance QTL, *rhg1* on chromosome 18 and *Rhg4* on chromosome 8 have had relatively consistent mapping results (Concibido et al. 2004). Since the 2000s, many in-depth studies focused on the genetic resistance mechanism of *rhg1* and *Rhg4*. A study showed that resistance to SCN race 3 requires both *rhg1* and *Rhg4* in soybean *cv.* Forrest and that *Rhg4* is dominant (Meksem et al. 2001). A Forrest bacterial artificial chromosome (BAC)-based physical map was constructed to identify candidate genes within QTL (Ruben et al. 2006). Two candidate genes for resistance to SCN, each encoding a leucine-rich repeat receptor-like kinase protein (*LRR-RLK*), were identified at the *rhg1* and *Rhg4* loci (Hauge et al. 2001; Lightfoot and Meksem 2002). However, the function of the *LRR-RLK* genes at the *rhg1* and *Rhg4* loci was not identified until recently. Using TILLING to screen mutants of the *LRR-RLK* at *Rhg4* from an EMS-mutagenized Forrest M2 population, a nonsense mutant Q263* was identified. However, no matters the zygosity (heterozygote or homozygote) of the progeny of mutant Q263*, their SCN-infection phenotype was not altered, indicating that the *Rhg4 LRR-RLK* gene did not confer resistance to SCN (Liu et al. 2011). More recently, by employing the EMS-mutagenized Forrest M2 population, two missense mutants (E61K and M125I) of *GmSHMT* (serine hydroxymethyltransferase) were identified by TILLING. With map-based positional cloning, high-density genetic mapping, and genomic sequencing, *GmSHMT* was hypothesized to be the sole candidate gene at the *Rhg4* locus conferring resistance to SCN. By correlating the *GmSHMT* mutants to a loss of SCN resistance, *GmSHMT* was shown to be the *Rhg4* gene, pointing to a novel mechanism of plant resistance against pathogens (Liu et al. 2012).

9.1.4.2 Genetic Resources for Mutation Breeding of Germplasms with Altered Fatty Acid Profiles

In plant breeding, genetic variety of desired traits is needed for crop improvement. Unfortunately, the practical use of varieties from spontaneous mutations is encumbered due to low mutation rates and extensive selection processes. On the other hand, induced mutations have been increasingly used in plant breeding programs since the discovery of the genetic effect of X-rays on *Drosophila* (Muller 1927) and maize (Stadler 1928). For the past nine decades, physical and chemical mutagens like gamma rays and EMS have been widely applied to major crops such as barley, rice, wheat, and beans for novel varieties. Many new crop cultivars from mutation breeding projects have been released to increase yield and quality traits, as mutation breeding is not restricted by the same regulations of genetically-modified organisms (GMOs) (Parry et al. 2009).

During the 1930s, tobacco (*Nicotiana tabacum*) was mutagenized by X-ray irradiation to create the first commercial mutant cultivar '*chlorina*' (Goodspeed 1929; Tollenaar 1938). The application of mutation breeding has since been enhanced to economically improve important traits in crop plants (Ahloowalia and Maluszynski 2001), and many mutant plant varieties have been released throughout the world (http://www.mvgs.iaea.org). Since the 1980s, mutation breeding of soybean has attracted interest as a part of efforts to improve grain legume production. Several novel mutant soybean cultivars have been released with improved agronomic traits including seed yield, disease resistance, and symbiotic nitrogen fixation (Micke 1993).

As the most consumed vegetable oil in the world, soybean oil has been used substantially in the food industry (http://www.soystats.com). Commodity soybean oil typically contains 11% palmitic acid (16:0), 4% stearic acid (18:0), 25% oleic acid (18:1), 52% linoleic acid (18:2), and 8% linolenic acid (18:3) (Fehr 2007). Due to negative effects of high linoleic acid on nutrition and food production, recent studies have aimed to genetically increase the contents of palmitic acid, stearic acid, and oleic acid in soybean seeds (Pham et al. 2011). High oleic acid cultivars have especially been sought after. Using TILLING in soybean, three missense mutations were identified in the omega-6 fatty acid desaturase gene *FAD2-1A*, of which one mutation resulted in an increase in oleic acid content and a decrease in linoleic acid content in the seed oil (Dierking and Bilyeu 2009). From a NMU-mutagenized William 82 M3 population of 4566 families, 52 mutants were identified with remarkably high or low content of one of 5 fatty acids (Hudson 2012). Three of these mutants had a high level of seed palmitic acid caused by mutations of *KASII* (3-ketoacyl-ACP synthase II) (Head et al. 2012), and 6 of these mutants contained high levels of seed stearic acid caused by mutations of *SACPD-C* (delta-9-stearoyl-acyl-carrier protein desaturase) (Carrero-Colón et al. 2014).

In the developed Forrest M2 population, total fatty acid content of 484 families of M3 seed and percentages of 5 types of fatty acids (palmitic acid, stearic acid, oleic acid, linoleic acid and linolenic acid) were measured by gas chromatography and statistically analyzed (Fig. 9.4). The contents of those 5 fatty acids in seeds of the wild-type Forrest were 10.93, 3.25, 18.89, 53.71, and 6.89%, respectively ($n = 21$ plants) (Fig. 9.5). The contents in the seed of those 484 mutant families ranged from 5.11 to 31.89% with a mean of 11.97; 2.23–11.0% with a mean of

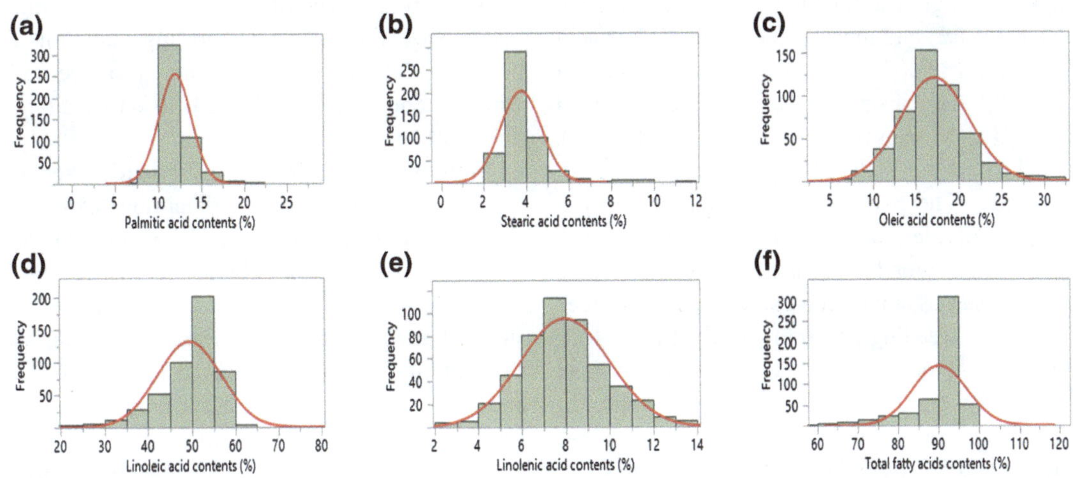

Fig. 9.4 Measurement and statistical analyses of the contents of 5 fatty acids in seed of 484 mutant families of the developed EMS-mutagenized Forrest mutant population

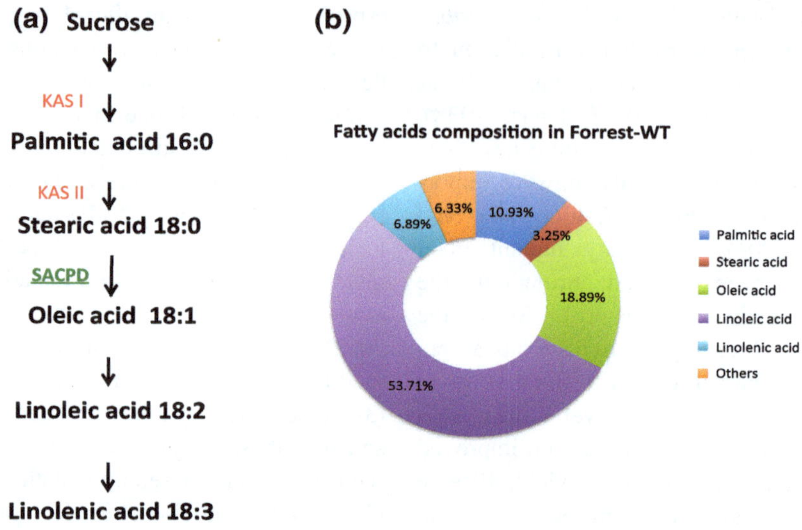

Fig. 9.5 **a** The biosynthetic pathway of fatty acids. Predominant biosynthetic pathway in soybean seed. **b** Fatty acid seed composition in the Forrest wild type in soybean. *KAS* ketoacyl synthase, *FAD2* fatty acid desaturase 2, *SACPD* Stearoyl-Acyl Carrier Protein Desaturase (Lakhssassi et al. 2017c)

3.77%; 6.9–41.14% with a mean of 17.18%; 8.86–62.63% with a mean of 17.18%; and 2.08–16.31% with a mean of 7.96%, respectively. The contents of palmitic acid, stearic acid, and/or oleic acid in seeds of many mutants were significantly higher than those in seeds of wild-type Forrest, and the content of linolenic acid in seeds of some mutants was lower than that in seeds of wild-type Forrest. For example, F367 seeds contained significantly higher contents of oleic acid (41.14%) and lower contents of linoleic acid (30.02%). In F403, the content of seed oleic acid (30.24%) is close to that of linoleic acid (35.97%). Moreover, one mutant (F848) was found to contain only 2.08% of linolenic acid in its seeds. These mutants identified with high contents of oleic acid, stearic acid, and/or palmitic acid and/or low contents of linolenic acid can be used as novel genetic soybean resources for further breeding programs. Interestingly, four mutations of omega-6 fatty acid desaturase gene *FAD2-1A* were identified (one silent mutation in F1274 and three missense mutations in F784, F1235, and F1284) (Fig. 9.6), and one missense mutation of *FAD2-1B* in F812 were recently identified (Fig. 9.7) (Anderson et al. 2017; Lakhssassi et al. 2017c). From the developed Forrest mutant populations, all three mutants resulted in an increase in oleic acid content. Combination of mutant *FAD2-1A* and *FAD2-1B* and combination of *FAD2* and *FAD3* could produce high oleic acid content soybean lines (>80%), combining high oleic acid and low linolenic acid content soybean lines, respectively (Pham et al. 2010, 2012). Thus, making crosses between two screened Forrest mutants with high contents of seed oleic acid carrying mutations at both FAD2-1A and FAD2-1B may produce lines with contents of seed oleic acid close to or higher than 80%.

Moreover, employing Forward genetic approach, EMS mutants containing increased seed stearic acid content has been also characterized in two different cultivars. In Williams 82, six missense mutants carrying mutations in the stearoyl-acyl carrier protein desaturase (SACPD-C) have been obtained (Carrero-Colón et al. 2014). In Forrest, a nonsense mutant and four missense mutants presenting increased seed, nodule, and leaf stearic acid have been identified to carry mutations in the SACPD-C isoform (Fig. 9.8) (Lakhssassi et al. 2017a). SACPD

(a)

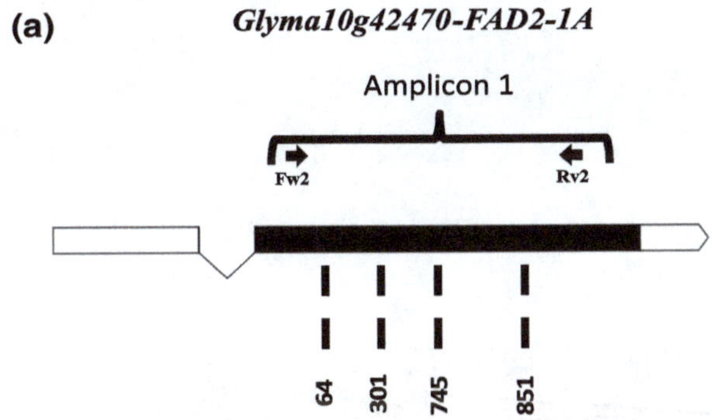

(b)

Forrest-WT	C	C	C	C

Mutant line	Nucleic acid changes				Oleic acid %	Effect
F784	G	C	C	C	**35.10**	R22G
F1235	C	T	C	C	**50.86**	L101F
F1274	C	C	T	C	**47.20**	L249=
F1284	C	C	C	T	**45.97**	P284L
Forrest-WT	C	C	C	C	**18.04±3.92**	-

Fig. 9.6 *FAD2-1A* mutations of the screened four high oleic acid mutants. **a** *FAD2-1A* gene model in the wild-type Forrest. **b** Predicted FAD2-1A protein showing amino acid differences between the Forrest-WT and the four high oleic mutants identified by forward genetics, three missense mutations (R22G, L101F, and P284L), and one silent mutation (L249=) within the *FAD2-1A*. The oleic acid level represents the highest content obtained from each line. The primers used for TILLING and gDNA sequencing are indicated by *arrows* (Lakhssassi et al. 2017c)

(a)

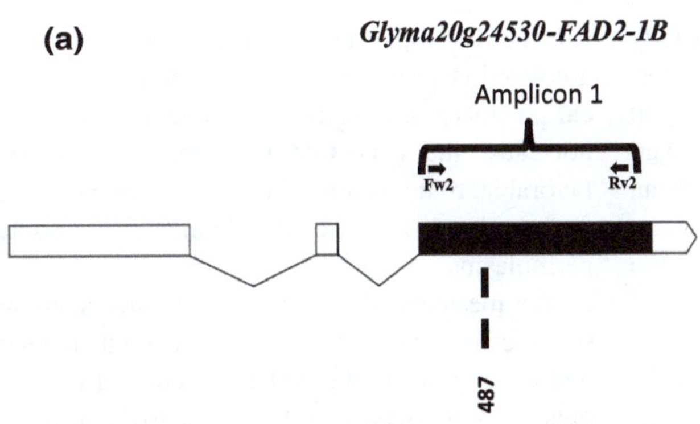

(b)

Forrest-WT	C

Mutant line no.		Oleic acid %	Effect	Provean	Prediction	FI %
F812	T	**30.51**	P163S	-7.291	Deleterious	5.41
Forrest-WT	C	**18.89±2.34**	-			3.9

Fig. 9.7 *FAD2-1B* mutations of the screened high oleic acid mutant. **a** *FAD2-1B* gene model in wild-type Forrest. **b** Predicted FAD2-1B protein showing amino acid differences between the Forrest-WT and the high oleic mutant identified by forward genetics; one missense mutations (P163S) within the *FAD2-1B*. The oleic acid level represents the highest content obtained. The primers used for PCR genotyping are indicated by *arrows* (Lakhssassi et al. 2017c)

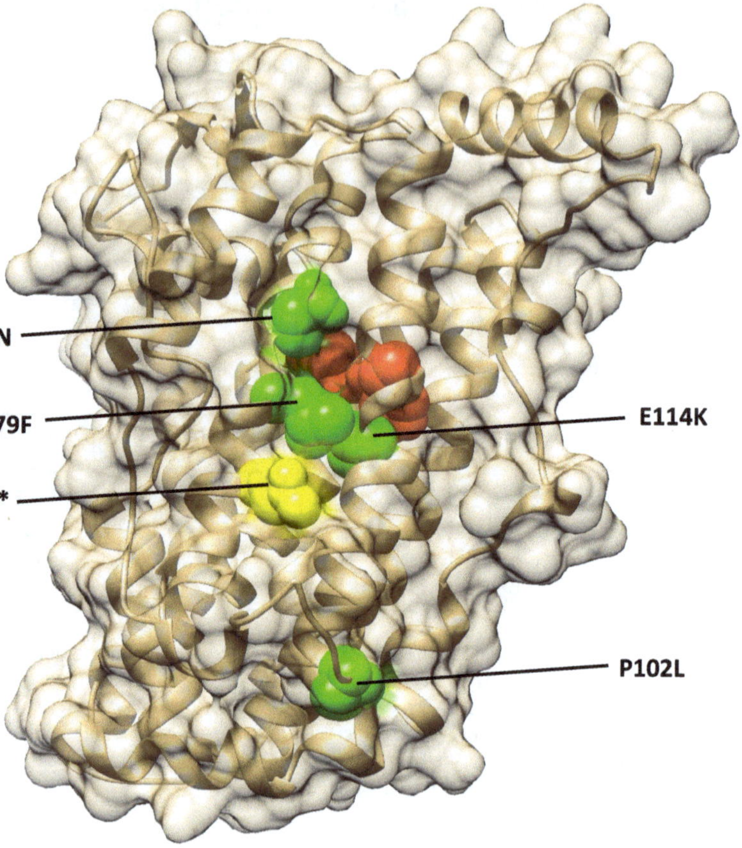

Fig. 9.8 Homology modeling of the SACPD-C from Forrest with important catalytic residues and the five identified *sacpd-c* mutations mapped. Figure present one subunit containing a di-iron coordination complex (*red*). Amino acid differences between the five mutants identified by forward genetics: four missense mutations (D77N, L79F, P102L, and E114K) in *green* and one nonsense mutation (Q83*) in *yellow* within the *SACPD-C* (Lakhssassi et al. 2017a)

proteins are known for their critical roles in fatty acid synthesis, biotic and abiotic stresses. Interestingly, it has been discovered recently that mutations at SACPD-C are involved also in plant development. In fact, EMS mutations in soybean at *GmSACPD-C* uncover an impact of stearic acid content in leaf and nodule structure and morphology (Lakhssassi et al. 2017a).

9.2 Discussion

As shown in Fig. 9.1, many kinds of morphological phenotypes of soybean at each growth and development stage can be observed from developed EMS-mutagenized Forrest populations. Still about 22.03% of mutants look like the wild-type Forrest without morphological changes indicating that on one hand, EMS introduces SNPs and InDels mutations in Forrest genome and/or alterations of many agronomical traits, and the developed Forrest population is actually a collection of mutants with different agronomical phenotypes. Because EMS mutagenesis does not cause the same GMO regulatory problems, favorable mutants may be directly released or further bred for the development of novel germplasms.

We measured the contents of 5 fatty acids in seeds of 484 mutants, and each content ranged widely (Fig. 9.4). The higher the content of seed oleic acid of soybean is, the better soybean is for human consumption and health. Some mutants such as F367 had up to 41.14% of seed oleic acid content, an increase of 139.5% from the wild-type Forrest (17.18%). There are reports of over 80% oleic acid content soybean lines developed by breeding mutant alleles of *FAD2-1A* and *FAD2-1B* together (Pham et al. 2010). Soybean lines with high oleic acid contents and low linolenic acid contents were also developed by breeding mutant *FAD2* and *FAD3* genes

(Pham et al. 2012). A high level of seed linolenic acid is bad for human health and soybean breeders are aiming to breed soybeans with contents of linolenic acid at 3% or less. One mutant (F848) was screened to have the lowest content of seed linolenic acid (2.08%), qualifying for those breeding goals. Moreover, those mutants with high levels of oleic acid and/or low level of linolenic acid were derived (mutagenized) from the SCN resistant soybean cultivar Forrest. In soybean, the high oleic acid mutant soybean lines are usually derived from Williams 82, a cultivar susceptible to SCN. All high-level oleic acid and/or low-level linolenic acid Forrest mutants were found to retain resistance to SCN (unpublished). So, those mutants not only contained high levels of oleic acid and/or low level of linolenic acid but also conferred SCN resistance. All aspects show that mutagenized soybean populations can be useful as novel genetic resources.

TILLING is a reverse genetic tool and was initially developed in *Arabidopsis* (McCallum et al. 2000) and then TILLING platforms were established rapidly in many crops such as soybean (Cooper et al. 2008; Meksem et al. 2008). In the beginning, TILLING mutations were detected using dHPLC. Later, gel-based TILLING employed DNA analyzers (for example, LI-COR 4300 DNA analyzer) to detect mutations of fluorescence-dyed PCR amplicons digested by endonuclease such as ENDO I after the formation of heteroduplexes. However, it has been reported recently through the characterization of the *FAD2* gene family the limitations of Gel-based TILLING technique, in genes with high copy number, paralogs, and similarities within the soybean genome (Lakhssassi et al. 2017c). With the broad application and low cost of next-generation sequencing, TILLING by sequencing was developed in plants (Tsai et al. 2011). TILLING by sequencing is a powerful and effective tool to detect rare mutations (SNPs and InDels) by employing genomic libraries of mutagenized populations, amplifying genomic fragments of interest to construct DNA libraries, and then direct sequencing. Chemical mutagenesis and TILLING have many advantages compared to other molecular approaches such as RNAi, VIGS, and T-DNA. Primarily, TILLING may be used to directly identify gene functions by screened mutagenized populations without transformation. This is particularly suitable for crops such as soybean that are recalcitrant to transformation. However, mutagens such as EMS are random, and sometimes mutations of genes of interest may not be screened due to the absence of mutations. On the other hand, the screened mutations may be silent, or even missense mutations are screened, not meaning that the mutations will cause phenotype alterations. Increase of mutagenized population size and development of integrated approaches can advance functional genomics research.

Using soybean genomic libraries constructed with genomic DNAs of the EMS-mutagenized Forrest population, 2 missense mutants of *GmSHMT* (E61K and M125I) were screened by TILLING. SCN-infection phenotype of those two mutants was altered from resistant to moderate susceptibility (Liu et al. 2012), which provided an important and direct evidence that *GmSHMT* is the soybean *Rhg4* gene conferring SCN resistance. A nonsense mutant of *Rhg4 LRR-RLK* (Q263*) was also screened by TILLING but no alterations of SCN resistance was observed, no matter the zygosity, clearly indicating that the *Rhg4 LRR-RLK* had no function in SCN resistance (Liu et al. 2011). Furthermore, EMS-mutagenized soybean populations were applied in the screening of mutants of fatty acid biosynthesis. Dierking and Bilyeu (2009) reported that they identified three missense mutations of omega-6 fatty acid desaturase gene *FAD2-1A* from three soybean mutants and one of the mutants contained an increase of seed oleic acid and a decrease of seed linoleic acid.

The soybean genomic libraries constructed can also be used in genotyping by sequencing. NGS, SNPs, and InDels are efficiently used to construct high-resolution genetic maps. Furthermore, QTL can be identified with the trait-associated SNPs and InDels. Compared to classic genetic approaches to localize and identify QTL such as crossing and bulking selection, the use of mutagenized soybean populations and genotyping by sequencing may be more efficient. We can use the mutagenized soybean populations to forward genetically screen mutants with significantly phenotypic traits of interest, and then make crosses of these screened mutants with wild-type soybeans to produce recombinant inbred line (RILs) populations. The RILs with significant favorable phenotypes can be selected and whole genome sequenced by NGS, and SNPs and InDels are then identified. Afterward, a MutMap can be constructed and SNPs and InDels associated with the traits are identified. Finally, the genes controlling the traits are identified. The MutMap approach has been successfully applied in identification of genes underlying traits in rice (Abe et al. 2012; Fekih et al. 2013; Takagi et al. 2015). The utilization of mutagenized populations and their genomic libraries as genetic resources will promote soybean genomics research and breeding through TILLING and NGS.

Acknowledgements We thank all student workers who helped to develop the EMS-mutagenized mutants.

References

Abe A, Kosugi S, Yoshida K, Natsume S, Takagi H, Kanzaki H, Matsumura H, Yoshida K, Mitsuoka C, Tamiru M, Innan H, Cano L, Kamoun S, Terauchi R (2012) Genome sequencing reveals agronomically important loci in rice using MutMap. Nat Biotechnol 30:174–178

Ahloowalia BS, Maluszynski M (2001) Induced mutations—a new paradigm in plant breeding. Euphytica 118:167–173

Anderson J, Lakhssassi N, Kantartzi SK, Meksem K (2017) Non-hypothesis analysis of a Mutagenic Soybean (Glycine max (L.)) Population for Protein and Fatty Acid Composition. J Am Soc. In press

Anai T (2012) Potential of a mutant-based reverse genetic approach for functional genomics and molecular breeding in soybean. Breed Sci 61:462–467

Bilyeu KD (2008) Forward and reverse genetics in soybean. In: Stacey G (ed) Genetics and Genomics of Soybean. Springer, New York, pp 135–139

Blumenstiel JP, Noll AC, Griffiths A, Perera AG, Walton KN, Gilliland WD, Hawley RS, Staehling-Hampton K (2009) Identification of EMS-induced mutations in *Drosophila melanogaster* by whole-genome sequencing. Genetics 182:25–32

Bolon Y, Haun WJ, Xu WW, Grant D, Stacey MG, Nelson RT, Gerhardt DJ, Jeddeloh JA, Stacy G, Muehlbauer GJ, Orf JH, Naeve SL, Stuper RM, Vance CP (2011) Phenotypic and genomic analyses of a fast neutron mutant population resource in soybean. Plant Physiol 156:240–253

Cai Y, Chen L, Liu X, Sun S, Wu C, Jiang B, Han T, Hou W (2015) CRISPR/Cas9-mediated genome editing in soybean hairy roots. PLoS ONE 10:e0136064

Caldwell BE, Brim CA, Ross JP (1960) Inheritance of resistance of soybeans to the cyst nematode, *Heterodera glycines*. Agron J 52:635–636

Carrero-Colón M, Abshire N, Sweeney D, Gaskin E, Hudson K (2014) Mutations in SACPD-C result in a range of elevated stearic acid concentration in soybean seed. PLoS ONE 9:e97891

Carroll BJ, McNeil DL, Gresshoff PM (1986) Mutagenesis of soybean (*Glycine max* (L.) Merr.) and the isolation of non-nodulating mutants. Plant Sci 47:109–114

Colbert T, Till BJ, Tompa R, Reynolds S, Steine MN, Yeung AT, McCallum CM, Comai L, Henikoff S (2001) High-throughput screening for induced point mutations. Plant Physiol 126:480–484

Concibido VC, Diers BW, Arelli PR (2004) A decade of QTL mapping for cyst nematode resistance in soybean. Crop Sci 44:1121–1131

Cooper JL, Till BJ, Laport RG, Darlow MC, Kleffner JM, Jamai A, El-Mellouki T, Liu S, Ritchie R, Nielsen N, Bilyeu KD, Meksem K, Comai L, Henikoff S (2008) TILLING to detect induced mutations in soybean. BMC Plant Biol 8:9

Dalmais M, Schmidt J, Signor CL, Moussy F, Burstin J, Savois V, Aubert G, Brunaud V, Oliveira YD, Guichard C, Thompson R, Bendahmane A (2008) UTILLdb, a *Pisum sativum in silico* forward and reverse genetics tool. Genome Biol 9(2):R43

Dierking EC, Bilyeu KD (2009) New sources of soybean seed meal and oil composition traits identified through TILLING. BMC Plant Biol 9:89

Dong C, Dalton-Morgan J, Vincent K, Sharp P (2009) A modified TILLING method for wheat breeding. Plant Genome 2:39–47

Fehr WR (2007) Breeding for modified fatty acid composition in soybean. Crop Sci 47:S72–S87

Fekih R, Takagi H, Tamiru M, Abe A, Natsume S, Yaegashi H, Sharma S, Sharma S, Kanzaki H, Matsumura H, Saitoh H, Mitsuoka C, Utsushi H, Uemura A, Kanzaki E, Kosugi S, Yoshida K, Cano L, Kamoun S, Terauchi R (2013) MutMap+: genetic mapping and mutant identification without crossing in rice. PLoS ONE 8:e68529

Flores T, Karpova O, Su X, Zeng P, Bilyeu K, Sleper DA, Nguyen HT, Zhang ZJ (2008) Silencing of *GmFAD3* gene by siRNA leads to low α-linolenic acids (18:3) of *fad3*-mutant phenotype in soybean [*Glycine max* (Merr.)]. Transgen Res 17:839–850

Gilchrist EJ, Sidebottom CH, Koh CS, Macinnes T, Sharpe AG, Haughn GW (2013) A mutant *Brassica napus* (canola) population for the identification of new genetic diversity via TILLING and next generation sequencing. PLoS ONE 8:e84303

Goodspeed TH (1929) The effects of X-rays and radium on species of the genus *Nicotiana*. J Hered 20:243–259

Greene EA, Codomo CA, Taylor NE, Henikoff JG, Till BJ, Reynolds SH, Enns LC, Burtner C, Johnson JE, Odden AR, Comai L, Henikoff S (2003) Spectrum of chemically induced mutations from a large-scale reverse-genetic screen in *Arabidopsis*. Genetics 164:731–740

Guo Y, Abernathy B, Zeng Y, Ozias-Akins P (2015) TILLING by sequencing to identify induced mutations in stress resistance genes of peanut (*Arachis hypogaea*). BMC Genomics 16:157

Hartwig EE, Epps JM (1973) Registration of 'Forrest' soybeans. Crop Sci 13:287

Hauge BM, Wang ML, Parsons JD, Parnell LD (2001) Nucleic acid molecules and other molecules associated with soybean cyst nematode resistance. US Pat App Pub No. 20030005491

Head K, Galos T, Fang Y, Hudson K (2012) Mutations in the soybean 3-ketoacyl-ACP synthase gene are correlated with high levels of seed palmitic acid. Mol Breed 30:1519–1523

Henikoff S, Comai L (2003) Single-nucleotide mutations for plant functional genomics. Annu Rev Plant Biol 54:375–401

Hudson K (2012) Soybean oil-quality variants identified by large-scale mutagenesis. Int J Agron. doi:10.1155/2012/569817

Jacobs TB, LaFayette PR, Schmitz RJ, Parrott WA (2015) Targeted genome modifications in soybean with CRISPR/Cas9. BMC Plant Biotechnol 15:16

Kassem MA, Ramos L, Hyten D, Bond J, Bendahmane A, Arelli PR, Njiti VN, Cianzio S, Kantartzi SK, Meksem K (2014) Quantitative trait loci that underlie SCN resistance in soybean [*Glycine max* (L.) Merr.] PI438489B by 'Hamilton' recombinant inbred line population. Atlas J Plant Biol 1(3):29–38

Koenning SR, Wrather JA (2010) Suppression of soybean yield potential in the continental United States from plant diseases estimated from 2006 to 2009. Plant Health Prog. doi:10.1094/PHP-2010-1122-01-RS

Kurowska M, Daszkowska-Golec A, Gruszka D, Marzec M, Szurman M, Szarejko I, Maluszynski M (2011) TILLING—a shortcut in functional genomics. J Appl Genet 52:371–390

Lakhssassi N, Colantonio V, Flowers ND, Zhou Z, Henry JS, Liu S, Meksem K (2017a) Stearoyl-acyl carrier protein desaturase mutations uncover an impact of stearic acid in leaf and nodule structure. Plant Physiol 174:1531–1543

Lakhssassi N, Liu S, Bekal S, Zhou Z, Colantonio V, Lambert K, Barakat A, Meksem K (2017b) Characterization of the soluble NSF attachment protein gene family identifies two members involved in additive resistance to a plant pathogen. Sci Rep 7:45226

Lakhssassi N, Zhou Z, Liu S, Colantonio V, Abughazaleh A, Meksem K (2017c) Characterization of the FAD2 gene family in soybean reveals the limitations of gel-based TILLING in genes with high copy number. Front Plant Sci 8:324

Lightfoot DA, Meksem K (2002) Isolated polynucleotides and polypeptides relating to loci underlying resistance to soybean cyst nematode and soybean sudden death syndrome and methods employing same. US Pat App Pub No 2002144310

Liu X, Liu S, Jamai A, Bendahmane A, Lightfoot DA, Mitchum MG, Meksem K (2011) Soybean cyst nematode resistance in soybean is independent of the *Rhg4* locus *LRR-RLK* gene. Funct Integr Genom 11:539–549

Liu S, Kandoth PK, Warren SD, Yeckel G, Heinz R, Alden J, Yang C, Jamai A, El-Mellouki T, Juvale PS, Hill J, Baum TJ, Cianzio S, Whitham SA, Korkin D, Mitchum MG, Meksem K (2012) A soybean cyst nematode resistance gene points to a new mechanism of plant resistance to pathogens. Nature 492:256–260

Matson AL, Williams LF (1965) Evidence of a fourth gene for resistance to the soybean cyst nematode. Crop Sci 5:477

McCallum CM, Comai L, Greene EA, Henikoff S (2000) Targeting induced local lesions in genomes (TILLING) for plant functional genomics. Plant Physiol 123:439–442

Meksem K, Pantazopoulos P, Njiti VN, Hyten DL, Arelli PR, Lightfoot DA (2001) 'Forrest' resistance to the soybean cyst nematode is bigenic: saturation mapping of the *Rhg1* and *Rhg4* loci. Theor Appl Genet 103:710–717

Meksem K, Liu S, Liu X, Jamai A, Mitchum MG, Bendahmane A, EI-Mellouki T (2008) TILLING: A reverse genetics and a functional genomics tool in soybean. In: Kahl G, Meksem K (eds) The handbook of plant functional genomics: concepts and protocols. Wiley-VCH, Weinheim, pp 251–266

Men AE, Laniya TS, Searle IR, Iturbe-Ormaetxe I, Gresshoff I, Jiang Q, Carroll BJ, Gresshoff PM (2002) Fast neutron mutagenesis of soybean (*Glycine soja* L.) produces a supernodulating mutant containing a large deletion in linkage group H. Genome Lett 3:147–155

Micke A (1993) Mutation breeding of grain legumes. Plant Soil 152:81–85

Muller HJ (1927) Artificial transmutation of the gene. Science 66:84–87

Nagamatsu A, Masuta C, Senda M, Matsuura H, Kasai A, Hong J, Kitamura K, Abe J, Kanazawa A (2007) Functional analysis of soybean genes involved in flavonoid biosynthesis by virus-induced gene silencing. Plant Biotechnol J 5:778–790

Nagamatsu A, Masuta C, Matsuura H, Kitamura K, Abe J, Kanazawa A (2009) Down-regulation of flavonoid 3′-hydroxylase gene expression by virus-induced gene silencing in soybean reveals the presence of a threshold mRNA level associated with pigmentation in pubescence. J Plant Physiol 166:32–39

Nunes AC, Vianna GR, Cuneo F, Amaya-Farfán J, Capdeville G, Rech EL, Aragao FJ (2006) RNAi-mediated silencing of the *myo*-inositol-1-phosphate synthase gene (*GmMIPS1*) in transgenic soybean inhibited seed development and reduced phytate content. Planta 224:125–132

Parry MAJ, Madgwick PJ, Bayon C, Tearall K, Lopez AH, Baudo M, Rakszegi M, Hamada W, Al-Yassin A, Ouabbou H, Labhilili M, Phillips AL (2009) Mutation discovery for crop improvement. J Exp Bot 60:2817–2825

Perry JA, Wang TL, Welham TJ, Gardner S, Pike JM, Yoshida S, Parniske M (2003) A TILLING reverse genetics tool and a web-accessible collection of mutants of the legume *Lotus japonicus*. Plant Physiol 131:866–871

Pham AT, Lee JD, Shannon JG, Bilyeu KD (2010) Mutant alleles of FAD2-1A and FAD2-1B combine to produce soybeans with the high oleic acid seed oil trait. BMC Plant Biol 10:195

Pham AT, Lee JD, Shannon JG, Bilyeu KD (2011) A novel *FAD2-1A* allele in a soybean plant introduction offers an alternate means to produce soybean seed oil with 85% oleic acid content. Theor Appl Genet 123:793–802

Pham AT, Shannon JG, Bilyeu KD (2012) Combinations of mutant FAD2 and FAD3 genes to produce high oleic acid and low linolenic acid soybean oil. Theor Appl Genet 125:503–515

Porceddu A, Panara F, Calderini O, Molinari L, Taviani P, Lanfaloni L, Scotti C, Carelli M, Scaramelli L, Bruschi G, Cosson V, Ratet P, Larembergue HD, Duc G, Piano E, Arcioni S (2008) An Italian functional genomic resource for *Medicago truncatula*. BMC Res Notes 1:129

Ruben E, Jamai A, Afzal J, Njiti VN, Triwitayakorn K, Iqbal MJ, Yaegashi S, Bashir R, Kazi S, Arlli P, Town CD, Ishihara H, Meksem K, Lightfoot DA (2006) Genomic analysis of the *rhg1* locus: candidate genes that underlie soybean resistance to the cyst nematode. Mol Genet Genomics 276:503–516

Schmutz J, Cannon SB, Schlueter J, Ma J, Mitros T, Nelson W et al (2010) Genome sequence of the palaeopolyploid soybean. Nature 463:178–183

Senda M, Masuta C, Ohnishi S, Goto K, Kasai A, Sano T, Hong JS, MacFarlane S (2004) Patterning of virus-infected *Glycine max* seed coat is associated with suppression of endogenous silencing of chalcone synthase genes. Plant Cell 16:807–818

Singh VK, Jain M (2015) Genome-wide survey and comprehensive expression profiling of Aux/IAA gene family in chickpea and soybean. Front Plant Sci 6:918

Stadler LJ (1928) Genetic effects of X-rays in maize. Proc Natl Acad Sci USA 14:69–75

Sun X, Hu Z, Chen R, Jiang Q, Song G, Zhang H, Xi Y (2015) Targeted mutagenesis in soybean using the CRISPR-Cas9 system. Sci Rep 5:10342

Suzuki T, Eiguchi M, Kumamaru T, Satoh H, Matsusaka H, Moriguchi K, Nagato Y, Kurata N (2008) MNU-induced mutant pools and high performance TILLING enable finding of any gene mutation in rice. Mol Genet Genomics 279:213–223

Tadege M, Wang TL, Wen J, Ratet P, Mysore KS (2009) Mutagenesis and beyond! Tools for understanding legume biology. Plant Physiol 151:978–984

Takagi H, Tamiru M, Abe A, Yoshida K, Uemura A, Yaegashi H, Obara T, Oikawa K, Utsushi H, Kanzaki E, Mitsuoka C, Natsume S, Kosugi S, Kanzaki H, Matsumura H, Urasaki N, Kamoun S, Terauchi R (2015) MutMap accelerates breeding of a salt-tolerant rice cultivar. Nat Biotechnol 33:445–449

Talamè V, Bovina R, Sanguineti MC, Tuberosa R, Lundqvist U, Salvi S (2008) TILLMore, a resource for the discovery of chemically induced mutants in barley. Plant Biotechnol J 6:477–485

Till BJ, Reynolds SH, Weil C, Springer N, Burtner C, Young K, Bowers E, Codomo CA, Enns LC, Odden AR, Greene EA, Comai L, Henikoff S (2004) Discovery of induced point mutations in maize genes by TILLING. BMC Plant Biol 4:12

Till BJ, Colbert T, Tompa R, Enns L, Codomo C, Johnson J, Reynolds SH, Henikoff JG, Greene EA, Steine MN, Comai L, Henikoff S (2006) High-throughput TILLING for *Arabidopsis*. Methods Mol Biol 323:127–135

Tollenaar D (1938) Untersuchungen über mutation bei Tabak. II. Einige künstlich erzeugte chromosommutanten. Genetica 20:285–294

Tsai H, Howell T, Nitcher R, Missirian V, Watson B, Ngo KJ, Lieberman M, Fass J, Uauy C, Tran RK, Khan AA, Filkov V, Tai TH, Dubcovsky J, Comai L (2011) Discovery of rare mutations in populations: TILLING by sequencing. Plant Physiol 156:1257–1268

Tsai H, Missirian V, Ngo KJ, Tran RK, Chan SR, Sundaresan V, Comai L (2013) Production of a high-efficiency TILLING population through polyploidization. Plant Physiol 161:1604–1614

Tsai H, Ngo K, Lieberman M, Missirian V, Comai L (2015) Tilling by sequencing. Methods Mol Biol 1284:359–380

Vuong TD, Sleper DA, Shannon JG, Nguyen HT (2010) Novel quantitative trait loci for broad-based resistance to soybean cyst nematode (*Heterodera glycines* Ichinohe) in soybean PI 567516C. Theor Appl Genet 121:1253–1266

Wang N, Wang Y, Tian F, King GJ, Zhang C, Long Y, Shi L, Meng J (2008) A functional genomics resource for *Brassica napus*: development of an EMS mutagenized population and discovery of FAE1 point mutations by TILLING. New Phytol 180:751–765

Weil CF (2009) TILLING in grass species. Plant Physiol 149:158–164

Xin Z, Wang ML, Barkley NA, Burow G, Franks C, Pederson G, Burke J (2008) Applying genotyping (TILLING) and phenotyping analyses to elucidate gene function in a chemically induced sorghum mutant population. BMC Plant Biol 8:103

Yin G, Xu H, Xiao S, Qin Y, Li Y, Yan Y, Hu Y (2013) The large soybean (*Glycine max*) WRKY TF family expanded by segmental duplication events and subsequent divergent selection among subgroups. BMC Plant Biol 13:148

Zhang C, Yang C, Whitham SA, Hill JH (2009) Development and use of an efficient DNA-based viral gene silencing vector for soybean. Mol Plant Micr Interact 22:123–131

Zhang L, Yang XD, Zhang YU, Yang J, Qi GX, Guo DQ, Xing GJ, Yao Y, Xu WJ, Li HY, Li QY, Dong YS (2014) Changes in oleic acid content of transgenic soybeans by antisense RNA mediated posttranscriptional gene silencing. Int J Genomics. doi:10.1155/2014/921950

Zhu Y, Wu N, Song W, Yin G, Qin Y, Yan Y, Hu Y (2014) Soybean (*Glycine max*) expansin gene superfamily origins: segmental and tandem duplication events followed by divergent selection among subfamilies. BMC Plant Biol 14:93

Soybean Functional Genomics: Bridging the Genotype-to-Phenotype Gap

10

Jamie A. O'Rourke, Michelle A. Graham and Steven A. Whitham

Abstract

Technological advances coupled with the economic importance of soybean have led to increased efforts to understand gene function and associate genes with phenotypes of agronomic and fundamental interest. Functional genomics approaches aim to develop sufficient understanding needed to bridge the genotype-to-phenotype gap. In general terms, functional genomics approaches begin by using highly parallelized methods to analyze genomes, transcriptomes, proteomes, and metabolomes to generate hypotheses about genes that control phenotypes. Candidate genes are then tested for their contributions to phenotypes through various methods such as RNA silencing, genetic mutation, or overexpression. In this chapter, we review the current approaches, tools, and resources that are being applied for functional genomics research in soybean.

10.1 Introduction

The current assembly of the Williams 82 soybean reference genome (Wm82.a2.v1) contains 56,044 protein-coding loci and 88,647 mRNA transcripts (Schmutz et al. 2010); http://phytozome.jgi.doe.gov/pz/portal.html#!info?alias=Org_Gmax). In addition, hundreds of loci encoding regulatory non-coding RNAs, including microRNAs (miRNAs) and phased small interfering RNAs (phasiRNAs), are present in the soybean genome

J.A. O'Rourke · M.A. Graham
United States Department of Agriculture—Agricultural Research Service, Corn Insects and Crop Genetics Research Unit, Ames, IA 50011, USA

J.A. O'Rourke · M.A. Graham
Department of Agronomy, Iowa State University, Ames, IA 50011, USA

S.A. Whitham (✉)
Department of Plant Pathology and Microbiology, Iowa State University, Ames, IA 50011, USA
e-mail: swhitham@iastate.edu

H.T. Nguyen and M.K. Bhattacharyya (eds.), *The Soybean Genome*, Compendium of Plant Genomes, DOI 10.1007/978-3-319-64198-0_10

(Arikit et al. 2014). Genome sequences of hundreds of accessions of *Glycine max* and allied species are available, which is expected to identify potentially useful genes that are not present in the reference genome (Kim et al. 2010a, b; Lam et al. 2010; Li et al. 2013, 2014b; Chung et al. 2014; Zhou et al. 2015). A major challenge facing soybean researchers is determining the functions of protein-coding and non-coding genes and understanding their contributions to phenotypes that are of agronomic importance and fundamental interest.

Functional genomics is an approach that seeks to bridge the gap between genotype and phenotype by assigning functions to genes based on a variety of experimental evidence that builds upon the availability of the genome sequence. Functional genomics encompasses many types of experiments and datasets that include highly parallelized analyses of gene expression at the mRNA (transcriptome) and protein (proteome) levels, or accumulation of small RNA species [microRNAs (miRNAs) and short-interfering RNAs (siRNAs)], and metabolites (metabolome). Such analyses seek to correlate genes with traits/phenotypes. However, they typically cannot establish causality, which requires altering the function of the gene(s) of interest. Therefore, perturbing the function of candidate genes is another main component of functional genomics studies.

In this chapter, we highlight recent studies and current approaches for soybean functional genomics encompassing genomics, transcriptomics, and gene knockout and knockdown strategies. We focus on recent developments related to tools, resources, and approaches, and we discuss how these may be used to investigate genes underlying phenotypes of interest. For recent in-depth reviews on functional genomics applied to specific topics such as root hair cells and interactions with the environment (abiotic and biotic factors), we refer readers to other reviews (Tran and Mochida 2010; Hossain et al. 2015; Liu et al. 2015; Whitham et al. 2016b).

10.2 Genome-Based Resources for Soybean Improvement

The completion of the *G. max* soybean genome in 2009 (Schmutz et al. 2010) helped facilitate the integration of vast amounts of genetic, phenotypic, and genomics data. Websites such as SoyBase (Cannon et al. 2012), SoyKB (Joshi et al. 2012), and Phytozome (Goodstein et al. 2011) allow users to access data by sequence, gene names, markers and traits of interest, expression patterns, or homology to known genes in other species. For a full description of these databases, we refer readers to the corresponding publications and websites.

Initially, the reference genome was used to target relatively simple traits controlled by single dominant genes or major quantitative trait loci (QTL). Gene silencing or mutagenesis would yield easily measurable phenotypic differences. For example, functional genomics approaches were used to target candidate genes for resistance (Meyer et al. 2009; Pandey et al. 2011; Liu et al. 2012; Cook et al. 2014), abiotic stress tolerance (Atwood et al. 2014), silencing (Curtin et al. 2011), or chlorophyll content (Zhang et al. 2009). The reference genome was also used to develop a comparative genome hybridization (CGH) array for soybean that could be used to identify structural variants within (Haun et al. 2011) and between soybean accessions (McHale et al. 2012; Anderson et al. 2014). Interestingly, structural variants between accessions of soybean were often associated with biotic stress response genes.

The soybean reference genome, combined with the cost-effectiveness of high-throughput sequencing, accelerated the use of genome-wide mRNA transcriptome sequencing (RNA-Seq) analyses. Transcriptome studies provide powerful datasets for functional genomics studies, particularly in combination with whole genome re-sequencing and methylome sequencing. Depending on the type of sequencing platform used, RNA-Seq data can provide information ranging from the basic gene sequence, alternative splice variants, gene expression profiles, and polymorphic sites that can be used to develop simple sequence repeat (SSR) and single nucleotide polymorphism (SNP) markers. RNA-Seq technology can facilitate both forward and reverse genetic screens, as candidate genes underlying a phenotype of interest can be uncovered and/or the impact of a specific sequence mutation on gene expression profiles can be examined.

Transcripts can be mapped to an existing reference genome or used to develop a de novo transcriptome assembly. Though there are a wide variety of alignment programs available, the TopHat (Trapnell et al. 2009) program has proven particularly useful in aligning RNA-Seq reads to complex genomes while accounting for splice junctions. These programs count the number of reads from each sample or replicate that align to each predicted gene. Counts can then be utilized by programs including Cufflinks (Trapnell et al. 2012), EdgeR (Robinson et al. 2010; McCarthy et al. 2012), and DESeq (Anders and Huber 2010) to identify suites of differentially expressed genes between samples. Severin et al. (2010) and Libault et al. (2010) developed gene atlases providing soybean gene expression measurements in different tissues. The Severin et al. (2010) atlas focuses largely on aboveground plant tissues including leaves, flowers, pods, and seven stages of seed development as well as roots and nodules. The Libault et al. (2010) atlas focuses largely on belowground tissues including root hair cells, root tips, roots, mature nodules, and it includes aboveground tissues—leaves, shoot apical meristems, flowers, and green pods. Arikit et al. (2014) generated an atlas of small RNAs developed from 69 libraries representing vegetative and reproductive tissues. These studies paved the way for soybean researchers studying gene expression changes in response to abiotic stress (Peiffer et al. 2012; Atwood et al. 2014; Moran Lauter et al. 2014), biotic stress (Kim et al. 2011; Tremblay et al. 2013; Lin et al. 2014; Wong et al. 2014), and throughout soybean development (Zabala et al. 2012; Cho et al. 2013; Jones and Vodkin 2013; Kour et al. 2014).

For many non-model species, RNA-Seq technologies have facilitated the development of a de novo transcriptome sequence prior to, or instead of, the genome sequence. This is a particularly attractive approach for outcrossing species such as alfalfa (Yang et al. 2011; O'Rourke et al. 2015), or those with large, complicated genomes such as lentil, pigeon pea, chickpea, and lupine (Dubey et al. 2011; Kaur et al. 2011; O'Rourke et al. 2013a; Kudapa et al. 2014). For soybean, which has a high-quality genome sequence, the utilization of RNA-Seq reads to develop de novo transcriptomes may seem counterintuitive. However, reference genomes have limitations. Genes may not be present in the genotype used as the reference, but may confer important traits of interest. Developing de novo assemblies for a species with a reference genome sequence can be done by either aligning all reads from the RNA-Seq experiment to each other, developing a new transcriptome, or by mapping the reads to the reference genome then using unaligned reads to assemble novel transcripts not present in the reference. Most de novo assembly programs including Velvet/Oases, Trinity, and SOAPdenovo (Zerbino and Birney 2008; Grabherr et al. 2011; Luo et al. 2012; Schulz et al. 2012) utilize de Brujin graphs to scaffold sequence reads and re-constitute transcript sequences. This usually results in far more sequences than are reasonably expected. Programs to collapse redundant transcripts include CD-Hit (Fu et al. 2012) and CAP3 (Huang and Madan 1999); however, they often result in the loss of alternative splice variants. It is important to ensure a de novo transcriptome accurately reflects the biological reality. The N50 statistic is used to judge the quality of an assembly; the higher the N50, the better the assembly. Programs like DETONATE can assist in determining the best de novo assembly program, optimizing parameters, and determining sequencing support for an assembled contig (Li et al. 2014a). To determine the completeness or coverage of the transcriptome, Gongora-Castillo and Buell (2013) have proposed the following three parameters: the proportion of assembled reads, the saturation of improvements in N50 contig size, and the number of contigs that can be annotated. These analyses help ensure that data generated by using the assembly is biologically sound and informative.

The regulation of gene expression is complex, and it includes regulating transcription factor expression, DNA methylation, and histone modifications. While functional genomics tools traditionally focus on correlating DNA sequences and phenotypes, methylation patterns also play an important role in phenotypic expression. Methylation patterns are one facet of the rapidly expanding field of epigenetics. Broadly, epigenetics is defined as the study of heritable phenotypic changes not due to changes in the DNA sequence. Thus, epialleles are alleles with the same DNA sequence, but different methylation patterns. There are three types of epialleles: obligate, the methylation pattern is completely dependent on a DNA sequence variant; pure, the epiallele is maintained independent of the DNA sequence; and facilitated, the genetic variant can influence the genetic state, but not as reliably as the obligate epialleles (Schmitz 2014). Much like identification of suites of genes differentially expressed in plants exposed to different conditions, researchers can also identify differentially methylated regions (DMRs), which are heritable over multiple generations. Methylation is usually highest in heterochromatic regions composed of tandem or inverted repeats and transposons (Song et al. 2013)

Methylation patterns have been investigated in soybean since 1994 when researchers performed Southern blots using hypomethylated genomic DNA as probes to investigate the genomic structure and methylation patterns of duplicated regions of the soybean genome (Zhu et al. 1994). In 1999, researchers were able to systematically map the centromeric regions of soybean by comparing the mapping profiles of two amplified fragment length polymorphism (AFLP) markers targeting genomic regions with dissimilar methylation properties (methylation insensitive vs. methylation sensitive) and combining this with known genomic properties (Young et al. 1999). The methylation patterns between the annual wild soybean (*Glycine soja*) lines and cultivated soybean (*G. max*) lines were identified using methylation-sensitive polymorphism (MASP) analysis (Zhong et al. 2009). Both methylation polymorphism types were more frequently observed in *G. max* than in *G. soja*, suggesting methylation polymorphisms have been selected during domestication (Zhong et al. 2009). An AFLP analysis of the same plants revealed no correlation between genetic and methylation polymorphisms, suggesting these two polymorphisms were generated and/or maintained by separate and independent mechanisms (Zhong et al. 2009).

More recently, the genomes, methylomes, and transcriptomes of 83 recombinant inbred lines (RILs) and their parents were sequenced to determine how epialleles impact phenotypic variation (Schmitz et al. 2013). This study found that recently duplicated genomic features were more highly methylated than those arising from earlier duplications. Specifically, analyzing pairs of paralogs revealed that while CG methylation was similar for both pairs, CHG and CHH methylation (where H represents A, C, or T) was statistically enriched in one paralog compared to the other. This corresponds to the differential expression patterns measured between the paralog pairs. Since CHG and CHH methylation are a result of RNA-directed DNA methylation (RdDM), these results suggest this pathway is highly active in the soybean genome. The authors proposed these methylation patterns and subsequent impacts on gene expression are a mechanism utilized by the plant to regulate gene expression through the gene loss or subfunctionalization processes (Schmitz et al. 2013). In a separate study, DNA methylation patterns were identified and compared between roots, stems, leaves, and cotyledons of developing seeds. Between these tissue types, 2162 DMRs (each unique to a single tissue type) were identified (Song et al. 2013). As expected, hypomethylated regions significantly correlated with increased expression of flanking genes, though there was no correlation with downregulation of gene expression near hypermethylated DMRs.

Additionally, the results from this study suggest high CHG and CHH methylation within a soybean gene may induce gene silencing (Song et al. 2013).

Methylation also plays an important role in phenotypic traits. Specific seed coat variations are a result of increased methylation of a transposon, which inhibits the binding of a transposase, thus facilitating mRNA processing and subsequent anthocyanin biosynthesis (Zabala and Vodkin 2014). *Rhgl*, which mediates soybean cyst nematode (SCN) resistance, has 8 DMRs between SCN susceptible (containing a single *Rhgl* locus repeat) and SCN resistant lines. Interestingly, hypermethylated regions exhibit increased transcript abundance, though expression may be even higher if methylation is reduced. Methylation of *Rhgl* adds to the complexity of phenotypic control and was likely important in *Rhgl* evolution (Cook et al. 2014). Song et al. (2012) identified ten transcription factors whose expression is methylation dependent and are induced by salinity stress, confirming the critical role of methylation in soybean salinity stress tolerance.

10.3 Forward Genetic Resources for Soybean Improvement

Forward genetics studies, identifying a mutant phenotype then uncovering the genetic sequence causing the phenotype of interest, have been widely adopted in soybean. There are many ways to perturb the functions of plant genes, and mutant populations have been developed through the use of random mutagens including chemicals, radiation, or DNA elements such as transposons and *Agrobacterium* T-DNA. Random mutagenesis is heritable and stable, but it requires intensive screening to identify mutant phenotypes. Following identification of a mutant of interest, the responsible mutation must be fine mapped before it can be cloned and sequenced. Alternatively, mutagenized populations can be indexed and characterized by PCR-based sequencing methods. Individual mutants carrying mutations in the genes of interest can be tested for altered phenotypes. These approaches essentially generate random mutant populations, and then methods are required to de-convolute the mutations in order to identify those that are of interest. Once the population is indexed, it becomes extremely valuable in functional genomics applications. Prime examples of this are *Arabidopsis thaliana* T-DNA lines (Alonso et al. 2003). Alternatively, methods may be developed to enable rapid identification of mutations of interest from populations like targeting induced local lesions in genomes (TILLING) (Cooper et al. 2008).

Soybean has undergone several genetic bottlenecks resulting in low genetic diversity, particularly among elite cultivars (Hyten et al. 2006). While a number of spontaneous mutation events have been identified and characterized in soybean, induced mutagenesis provides a mechanism to increase heritable diversity at a faster rate than seen in nature. Unlike transgenic plants, there are no regulatory restrictions on the handling and transfer of mutant plants, making these ideal to incorporate into breeding programs. Additionally, while transgenic approaches result in few transgenic plants, each mutagenesis event results in a large number of independent mutants. Despite these advantages, mutagenesis does have limitations: Different genotypes are more tractable to mutagenesis than others and even among mutable genotypes, environmental effects can severely impact phenotypes between replicate experiments. Recent years have seen the development and utilization of multiple soybean mutant populations, each with their own advantages, which can be utilized in both forward and reverse genetic studies. In this section, we will present each population type and explore its utility as a functional genomics tool. As soybean is prized for its seeds, it is no surprise that the majority of the studies involving mutant populations focus on altered seed compositions.

10.3.1 Ionizing Radiation

Ionizing radiation includes X-rays, fast neutrons, and gamma rays. In soybean, mutant populations developed from X-ray and fast neutron radiation are publicly available. Until recently, ionizing radiation was believed to only induce large deletions. Through the adaptation of high-throughput genomics tools including CGH arrays and high-throughput genomic re-sequencing, researchers have determined ionizing radiation induces deletions, duplications, inversions, and translocations of all sizes (Bolon et al. 2011; Belfield et al. 2012; O'Rourke et al. 2013b). Ionizing radiation has been used to develop soybeans with increased yield, earlier flowering times, increased germination rates, decreased dehiscence, increased biotic and abiotic stress resistance, decreased anti-nutritional components, and altered protein and oil compositions.

X-ray mutagenesis was used to develop the earliest mutant populations. In 1955, L.F. Williams identified L67-3483, a tan saddle seed coat mutant, from X-ray irradiated Clark (Rode and Bernard 1975). The locus underlying this trait was subsequently mapped by Kato and Palmer (2003) to the top of molecular linkage group D1b +W (chromosome 2). Additional X-ray irradiation populations have been produced at Saga University in Japan resulting in soybean lines with high oleic acid, high linolenic acid, and high stearic acid (Takagi et al. 1989; Rahman et al. 1994, 1995). Similar to William's research, these early studies provided excellent material for breeding programs, but underlying sequences and genes were not identified. More recently, Anai et al. (2012) have used X-ray irradiated plants to identify two high palmitic acid mutants. Sequence analysis confirmed the mutation of *GmKASIIA* and *GmKASIIB* in two mutants. However, sequence analysis of a third mutant, with a similar phenotype, showed that it was wild type for both *GmKASII*, suggesting additional genes are involved in regulating the palmitic acid content of soybean seeds. X-ray mutagenesis was also employed at the University of Missouri where a forward genetic screen of X-ray mutants identified three mutant lines with increased stearic acid (Gillman et al. 2014). Further genetic analyses, including a reverse genetic screen of an ethylmethane sulfonate (EMS) mutant population, determined the phenotype resulted from mutations in *SACPD-C* (Gillman et al. 2014).

Mutants resulting from gamma irradiation have proven particularly useful in developing plants with improved nutritional quality including reduced phytate levels, Kunitz trypsin inhibitor activity, and lipoxygenase-free seeds (Kim et al. 2010a; Lee et al. 2011, 2014; Yuan et al. 2012). These mutants are easily integrated into large breeding programs. In Japan, almost 10% of the total acreage of soybean is from gamma ray-induced mutants (Nakagawa 2009). Nine of the commercially available cultivars are 'indirect' mutant cultivars (Nakagawa 2009), descended from gamma-irradiated parental lines. Similarly, in China, over 16.33×10^6 ha are planted with soybean varieties derived from gamma irradiation (Kharkwal and Shu 2009). India and Thailand also utilize gamma-irradiated soybeans in their breeding programs. Breeders in India identified over 55 gamma-irradiated mutants with a variety of altered traits, seven of which were publicly released (D'Souza et al. 2009; Kharkwal and Shu 2009).

While the majority of mutant plants are identified and pursued for altered seed properties, multiple groups have used gamma-irradiated populations to identify plants with increased abiotic stress tolerance. Researchers in Thailand have identified mutants with increased resistance to soybean crinkle leaf virus and increased germination rates in extreme dry or wet climates, increasing germination rates from 30 to 70% in the dry season and from 41 to 83% in the wet season (Srisombun et al. 2009). While the specific changes to DNA resulting in flood tolerance have not yet been identified, gene expression analyses of mutant plants grown in flooded and control conditions determined that genes involved in cell wall loosening and proteolysis are not upregulated in the flooded mutant. Additionally, anaerobic metabolism is more efficient in flood-tolerant mutants (Komatsu et al. 2013). All these traits provide interesting avenues for future research on improving abiotic stress tolerance in soybean and other legume species.

Fast neutron (FN) mutagenesis involves bombarding plants with neutrons to induce mutations. Until recently, these mutations were all assumed to be large deletions. Genomics studies have determined that small (<100 base pair) deletions, genome duplications (both large and small), and various translocation and genomic re-arrangements are also induced by FNs (Bolon et al. 2011; Belfield et al. 2012; Anderson et al. 2014). One of the most notable soybean FN mutants is FN37, which lacks the ability to regulate nodule production and thus forms 10× the number of nodules of the wild-type parent (Men et al. 2002). Genetic and physical mapping determined this mutant contained a 460 Kbp deletion on molecular linkage group H (chromosome 12). A dwarf plant resulting from FN mutagenesis was identified by Hwang et al. (2014). Combining forward genetic screens with next-generation sequencing, Hwang et al. (2014) identified an 803 base pair deletion, including part of a peroxidase homolog Glyma15g05831, which likely results in the dwarf phenotype, a beneficial trait in extreme climates. Similarly, high-throughput genomics approaches including CGH arrays, exon capture, and next-generation sequencing have been used to identify and confirm changes in the genome likely underlying changes in plant architecture or seed quality (Bolon et al. 2011, 2014b). The entirety of the data assembled from this FN population including seeds, photographs of obvious phenotypic changes, and NIR data on seed composition has been collected and is publicly available at www.soybase.org/mutants/. Here, sresearchers can search by sample name, genes of interest, trait, images, or phenotypes to identify a suite of mutants for which they can request seed for use in their own research projects.

10.3.2 Insertional Mutagenesis

Insertional mutagenesis has also been used to develop mutant populations. In soybean, individual transformation events are both time and labor intensive. Using transposon and retrotransposon tagging, a multitude of mutations can be derived from a single transformation event. Soybean currently has three transposon tagging mutant populations developed from three unique transposon systems (Chap. 12): *Ac/Ds*, *mPING*, and *Tnt1*. The *Ac/Ds* transposon system, originally identified by Barbara McClintock in maize, has two major advantages: (1) the preference of these transposons to insert into exonic regions, increasing the likelihood of interrupting gene function and (2) the ability to determine the location of insertion via PCR, based on the known sequence of the transposons. This system has been used to create activation tagging, gene, and enhancer trap element mutants (Mathieu et al. 2009). The miniature inverted-repeat transposable element (*mPING*), originally identified in rice, has a high transposition frequency in soybean, though transposition may be developmentally regulated and plants may continue to experience transposition events, further complicating pheno/genotyping (Hancock et al. 2011). Unlike the *Ac/Ds* transposon system which tends to insert near their original point of origin (Mathieu et al. 2009), *mPING* transposes to unlinked genomic positions, usually within 2.5 Kbp of a gene sequence (Hancock et al. 2011). *Tnt1*, a retrotransposon from tobacco, can be induced to transpose in species (including soybean) through tissue culture (Cui et al. 2013). *Tnt1* mutagenesis requires fewer initial transgenic events than the *Ac/Ds* system, and the insertions induce stable and heritable changes to the genome. The data associated with all three types of insertional mutants have been deposited at the University of Missouri soybean genome database available at the following Web page digbio.missouri.edu/gmgenedb/index.php. These mutant populations are valuable functional genomics tools for deciphering gene function in soybean and other legumes. Recently, an endogenous CACTA-type transposon *Tgm9* has been identified and utilized in creating a soybean mutant population (Chap. 12).

10.3.3 Chemical Mutagenesis

Chemically mutagenized soybean populations have been developed from both EMS (ethyl methanesulfonate) and NMU (*N*-nitroso-*N* methylurea) mutagenesis. EMS and NMU mutagenesis induce single nucleotide polymorphisms (SNPs) either A to T or G to C transitions. Such mutants are often utilized in reverse genetic studies such as TILLING where DNA from mutagenized plants is screened using primers developed for a specific DNA sequence of interest. The first successful use of TILLING in soybean identified mutations in the omega-6 fatty acid desaturase gene *FAD2-1A*. Further analyses determined these were missense mutations, one of which increased the oleic acid content of seeds, a positive change for soybean seed profiles (Dierking and Bilyeu 2009). Subsequent studies used TILLING in two novel EMS mutant populations identified plants with mutations in the *GmFAD2-1b* genes (Hoshino et al. 2010). Crossing *GmFAD2-1a* and *GmFAD2-1b* mutants produced soybeans with seed oleic acid content >80% (compared to $\sim$18% of normal soybeans) (Hoshino et al. 2010).

In addition to the protein and oil compositions of soybean seeds, improved disease resistance is an important area of research. The SCN is one of the most devastating pathogens affecting soybean production. To identify genes conferring SCN resistance, researchers performed a TILLING screen on an EMS mutagenized soybean population generated from the SCN resistant cultivar 'Forrest' (Liu et al. 2011, 2012). Screening this population for mutations in the obvious LRR-RLK gene identified mutants, but the mutants did not exhibit altered SCN resistance (Liu et al. 2011). However, screening the EMS population for the neighboring serine hydroxymethyltransferase (SHMT) gene identified two mutants with missense mutations, one of which was predicted to be deleterious. Upon SCN infection, the deleterious mutation correlated with the loss of SCN resistance phenotype. Since SHMT is not a canonical resistance gene, additional genetic analyses were employed to confirm its role in SCN resistance. The SHMT gene is involved in the one-carbon metabolism pathway and is conserved across species, making it a unique resistance mechanism.

10.4 Reverse Genetic Resources for Soybean Improvement

Reverse genetic studies start with a gene sequence of interest whose function is investigated utilizing various genetic and genomics tools. In this section, we discuss RNA-silencing techniques that can be used to downregulate or silence the expression of soybean genes. There are several RNA-silencing-based technologies that use double-stranded RNAs (dsRNAs), inverted-repeat RNAs (irRNAs), or miRNAs to knock down the expression of plant genes. DsRNAs are delivered using viral vectors or transgenes, while irRNAs and miRNAs are delivered using transgenes. RNA-silencing-based approaches are not equivalent to null mutations, because it is probable that some of the target gene product(s) is still produced, and there is variability in the extent to which RNA silencing occurs between lines.

Plant RNA-silencing systems can be manipulated, so that plants specifically shut down expression of their own genes. RNA-silencing technologies enable researchers to induce loss-of-function of the targeted gene(s) of interest. This is possible because RNA-silencing systems are programmed by small RNAs that are produced from dsRNA precursors. DsRNAs can be introduced into plant cells in the form of recombinant viruses that replicate through double-stranded RNA intermediates, form secondary structures, or by expression of irRNAs and precursors of miRNAs. Double-stranded viral RNA and irRNAs are cleaved by Dicer-like (DCL) enzymes into siRNAs ranging from 21 to 24 nucleotides in length, depending on the DCL that did the processing. The siRNAs are then bound by argonaute (AGO) proteins, and the RNA-induced silencing complex (RISC) is formed. The siRNAs program AGO to specifically cleave complementary RNA sequences present in the cell. Therefore, if a siRNA is complementary to a sequence within a plant mRNA, that mRNA will be targeted for degradation, and the expression of that gene will be reduced or possibly abolished. In short, recombinant viruses or inverted-repeat RNAs program RISC to target plant mRNAs by inducing the accumulation of pools of overlapping siRNAs that are complementary to plant mRNAs, resulting in decreased expression of target genes. Gene-silencing strategies that utilize miRNAs also exploit DCL and AGO to process the microRNA precursor, but they result in the production of single, specific miRNAs or the production of phasiRNAs that can act *in trans* on complementary mRNA targets (Schwab et al. 2006; Felippes et al. 2012; Carbonell et al. 2014, 2015).

10.4.1 Virus-Induced Gene Silencing (VIGS)

VIGS is a transient method for reducing the expression of targeted plant genes to create loss-of-function phenotypes (Becker 2013). VIGS relies on the development of infectious clones of viruses that are suitable for gene-silencing applications. The ideal VIGS virus possesses a relatively weak suppressor of RNA silencing, and its genome tolerates and stably maintains foreign inserts over the infection time course in experimental plants. For VIGS applications, the viral genome is modified to contain a cloning site that enables the insertion of fragments of plant genes to be silenced. The recombinant viruses are inoculated into plants using techniques such as biolistics, rub-inoculation, or *Agrobacterium* infiltration. As the recombinant virus spreads systemically throughout the plant, it carries the fragment of the plant gene, which programs the RNA-silencing system to degrade mRNAs that contain complementary sequences. While silencing may persist through the life of the plant or even into the next generation, the technique is considered transient, because the viruses do not insert into the plant genome.

Infectious clones of five viruses have been used for VIGS in soybean with varying degrees of efficacy, and they have been reviewed in detail elsewhere (Whitham et al. 2015). *Bean pod mottle virus* (BPMV) has been used extensively by us and others for VIGS applications, and detailed methods on two different BPMV systems are available (Kachroo and Ghabrial 2012; Zhang et al. 2013). The main difference between the two BPMV systems is how the initial inoculum is generated—one uses in vitro transcription to produce infectious RNA transcripts that are rubbed onto soybean plants (Kachroo and Ghabrial 2012), and the other uses biolistics or rub-inoculation to directly inoculate plants with plasmid DNA carrying the infectious clones (Zhang et al. 2013; Whitham et al. 2016a). After inoculation, BPMV produced from the two systems moves systemically and induces gene silencing to similar extents.

BPMV is a bipartite virus that has two components to its genome, RNA1 and RNA2. Both RNAs contain 5′ and 3′ untranslated regions that flank single, large open reading frames encoding polyproteins that are processed into the individual mature proteins by protease functions encoded by RNA1. RNA1 also encodes replication functions, and the shorter RNA2 encodes the movement and capsid functions. Because the essential viral functionalities are distributed between the two RNAs, they are both necessary for systemic infection. RNA2 is about 2.3 Kbp shorter than RNA1, and therefore, it is RNA2 that has been engineered to accept foreign inserts (Zhang and Ghabrial 2006). The original cloning sites were placed within the open reading frame of RNA2 between the movement protein and the large subunit of the capsid protein (Zhang and Ghabrial 2006; Zhang et al. 2009, 2010). This position requires that any inserted sequence maintains the viral open reading frame, and it has been shown to have great utility for VIGS and for expression of proteins as well. A second position for the cloning site is immediately after the stop codon in the open reading frame of RNA2 (Zhang et al. 2010). This position can only be used for VIGS, and it offers more flexibility in the choice of target sequences and insert orientations because the viral reading frame does not have to be conserved. Using this cloning position, we have found that gene fragments cloned in the anti-sense orientation are most effective at silencing the target genes (Zhang et al. 2010; Juvale et al. 2012). More recently, a third cloning site in the 5′ untranslated of RNA2 was proposed for the insertion of host gene fragments for VIGS (Ali et al. 2014). Similar to the 3′ untranslated region, fragments inserted into the 5′ untranslated region are also not constrained by presence of stop codons.

BPMV VIGS is used in focused, hypothesis-driven studies to investigate functions of one or a few genes and in large-scale screens of candidate genes identified through approaches such as transcriptome profiling or proteomics (Pandey et al. 2011; Zhang et al. 2012; Cooper et al. 2013). The large-scale screening capability is made possible by the transient and rapid nature of VIGS. Recombinant BPMV clones carrying fragments of plant genes approximately 200–300 base pairs in length are simple to assemble (Whitham et al. 2016a). Systemic virus infections occur within two to three weeks after inoculation, and at this time, silencing of the target gene is established in the symptomatic tissues (Zhang et al. 2010). Because the initial inoculation procedures are somewhat laborious and relatively expensive, we typically store the infected tissue from bombarded plants and use the infected leaf sap as inoculum for the actual experiments that involve many plants, which are rub-inoculated (Zhang et al. 2013; Whitham et al. 2016a). Inserts in the 200–300 base pair size range are relatively stable in BPMV, and so, it is possible to passage the virus once without high concern for them being deleted. However, it is prudent to confirm this. Regardless of the inoculation strategy, it is possible to generate information on the functions of genes within about two months using the BPMV VIGS system.

The BPMV system has been mainly applied to studying traits involved in interactions with biotic and abiotic stresses (Atwood et al. 2014; Liu et al. 2015). VIGS candidate genes used in these studies have been identified based on a number sources including transcriptome profiling, proteomics, homologs from model systems like *Arabidopsis thaliana*, and genetic mapping. There has been a lot of success with identifying genes that affect soybean interactions with foliar pathogens such as soybean rust, downy mildew, *Soybean mosaic virus*, and *Pseudomonas syringae* pv. *glycinea* (Liu et al. 2015; Whitham et al. 2016b). In addition, the BPMV system can be used to silence genes in the roots (Juvale et al. 2012), and protocols have been developed that enable identification of genes involved in soybean–SCN interactions (Liu et al. 2012; Kandoth et al. 2013). With regard to abiotic stresses, the BPMV system has been used to assess functions of genes in iron-deficiency chlorosis and salt stress (Rao et al. 2013; Atwood et al. 2014).

The ability of BPMV to induce silencing in a variety of tissues over the course of plant development (Juvale et al. 2012) suggests it will be useful for a variety of traits besides interactions with the environment. The current bias that exists in the literature is likely due to the research interests of the groups who have developed the systems and their collaborators. BPMV has been shown to silence target genes in leaves, roots, petioles, stems, and flowers. However, more work is needed to investigate extent to which VIGS may be transmitted to the seed. Seed transmission of VIGS in soybean has been demonstrated for the *Apple latent spherical virus* (ALSV) vector in the soybean cultivar 'Enrei' (Yamagishi and Yoshikawa 2009). ALSV has been shown to be useful for identifying genes involved in developmental traits in soybean (Takahashi et al. 2013) as well as both defensive and developmental traits in other plant species (Li and Yoshikawa 2015). Detailed methods in the use of ALSV for VIGS in soybean have been published (Yamagishi and Yoshikawa 2013).

10.4.2 IrRNA

DsRNA can be produced in plant cells via transgenes that are designed to express inverted repeats of fragments of target genes. The inverted repeats may be separated by spacers or introns. In the case of spacers, the transcript forms a double-stranded stem with a loop, which is a structure known as a hairpin. A detailed analysis has been conducted of silencing of the soybean *FAD3* homologs, *GmFAD3A*, *GmFAD3B*, and *GmFAD3C*, which share 100, 96.5, and 84.3% identity, respectively, with a 318 base pair hairpin RNA construct (Lu et al. 2015). *GmFAD3A* and *GmFAD3B* mRNAs were silenced very effectively, but *GmFAD3C* mRNA was silenced to a lesser extent. The reduced level of *GmFAD3C* mRNA silencing correlated with the observation that *GmFAD3C* shared perfect complementarity with relatively few of the siRNAs produced by the hairpin transgene, and this in turn led to inefficient cleavage of the mRNA. These data highlight the challenges and limitations of using RNA-silencing approaches to target gene family members in soybean.

In a true functional genomics application of silencing using inverted-repeat transgenes, Danzer et al. (2015) silenced 53 soybean transcription factors that were selected based on detailed transcriptome analyses of soybean seed development. Transgenic soybean lines were made to express an irRNAs targeting a 150–200 base pair fragment in each transcription factor gene. The RNA-silencing lines were screened for defects in seed development and vegetative growth, and mutant phenotypes were observed for silencing of three of the transcription factors. The functions of one of the transcription factors, *SPEECHLESS* (Glyma04g41710), were investigated in depth. The RNA-silencing construct designed against Glyma04g41710 was effective at silencing it and the other three paralogs that shared 98, 85, and 83% nucleotide identity. Unlike the case with *FAD3*, mRNA transcripts of all the *SPEECHLESS* paralogs were silenced to similar levels. The mutant phenotype correlated with decreased mRNA transcript accumulation, and it was similar to Arabidopsis *speechless* mutants based on defects in stomata development. The availability of Arabidopsis *speechless* mutants made it possible to perform a transgene complementation study that confirmed that the soybean *SPEECHLESS* homolog in question was indeed a functional ortholog. This study illustrates the power of utilizing resources, like Arabidopsis mutants, for determining the functions of soybean genes, which can be coupled with gene expression and RNA-silencing studies.

10.4.3 Applications for MiRNA in Soybean Functional Genomics

MiRNAs are small RNAs like the siRNAs, but their biogenesis within the cell occurs via a distinct pathway with processing mediated by DCL1. MiRNAs direct AGO1 to cleave mRNA transcripts containing complementary sequences. MiRNAs can be artificially designed to specifically target plant genes by exchanging the sequence of a natural miRNA with an artificial one (Schwab et al. 2006). Expression of the artificial miRNA precursor can be placed under control of any desired promoter. In soybean, artificial miRNAs have been expressed under the control of the constitutive polyubiquitin promoter (*Gmubi*) and the soybean 7S globulin promoter, which drives seed-specific expression. Melito et al. (2010) used artificial miRNAs in hairy root assays to demonstrate that a leucine-rich repeat receptor kinase was not a candidate for the SCN resistance gene, *Rhg1*. Artificial miRNAs have also been expressed in a seed-specific manner in stably transformed plants to investigate the effects of 7S globulin proteins on seed nitrogen sources and total protein content (Yamada et al. 2014). The 'P-SAMS amiRNA Designer' is a web-based tool that can be used to design artificial miRNAs with novel targets for soybean and other plant species (Carbonell et al. 2014, 2015).

MiRNAs of the 22 nucleotide size class can also be used to trigger the accumulation of siRNAs that are loaded onto AGO proteins and subsequently silence expression of a target gene (s). These siRNAs are known as phasiRNAs, and if they are demonstrated to direct cleavage of a target RNA *in trans*, then they are called trans-acting siRNAs (tasiRNAs). In an approach that has been referred to as miRNA-induced gene silencing (MIGS) (Felippes et al. 2012), the 22 nucleotide miRNA target sequence is placed upstream of a fragment of a target gene, which results in the production of synthetic tasiRNAs (syn-tasiRNAs) that direct silencing of the target gene. In soybean, this approach has been demonstrated to silence genes in both hairy roots and transgenic plants (Jacobs et al. 2016). An advantage of MIGS over inverted repeats is that it is easier to construct the clones, because only a single copy of the target fragment in a single orientation is required. However, it is necessary to first identify a miRNA that initiates production of tasiRNAs in the appropriate tissues or cell types under the desired conditions or co-express a miRNA that is known to induce tasiRNA biogenesis. The P-SAMS syn-tasiRNA Designer Wizard is a web-based tool that can be used to design syn-tasiRNAs for soybean and other plant species (Carbonell et al. 2014).

10.5 Bridging the Genotype-to-Phenotype Gap—Rapidly Developing Areas that Will Accelerate Discovery of Soybean Gene Function

10.5.1 Using Next-Generation Sequencing to Characterize Complex Traits

A high-quality reference genome has enabled researchers to begin to examine complex traits, such as yield, domestication, and drought tolerance. Re-sequencing data from different soybean accessions can be assembled to the existing Williams 82 reference genome. In 2010, Kim et al. (2010b) assembled re-sequencing data from undomesticated soybean (*G. soja*) to the reference. This identified 712 genes that were partially or completely lost in *G. max*, likely during the process of domestication. Lam et al. (2010) re-sequenced 17 wild and 14 cultivated soybean genomes revealing the phylogenetic relationships between lines, identifying linkage disequilibrium blocks and generating over 200,000 SNPs that could be used for mapping and association studies. Li et al. (2013) leveraged the Lam et al. (2010) re-sequencing data with additional re-sequencing data from 25 diverse soybean accessions. The 50 Chinese lines represented wild soybean, land races, and elite cultivars allowing the researchers to differentiate between candidate genes associated with soybean domestication and improvement. Xu et al. (2013) used genome re-sequencing approaches to generate and validate SNPs that were used for QTL mapping of root-knot nematode resistance. Their approach identified two candidate genes responsible for resistance. Chung et al. (2014) used re-sequencing data of ten cultivated and six wild Korean accessions to identify 206 candidate domestication regions with lower diversity than observed within the wild accessions. Some of these regions contained genes with homology to known domestication genes identified in other plant species, while others appear to be novel in soybean. In 2015, Zhou et al. (2015) re-sequenced 302 wild and cultivated accessions allowing the identification of selective sweeps and copy number variants between genomes. In addition, genome-wide association studies were used to identify genomic regions associated with oil content, plant height, and pubescence. Bolon et al. (2014a) used genome re-sequencing to characterize FN radiation mutants.

Rapid decreases in the cost of next-generation sequencing have also facilitated the adoption of de novo sequencing strategies for soybean. In 2014, Li et al. (2014b) used next-generation sequencing of seven phylogenetically and geographically distinct *G. soja* accessions to characterize the *G. soja* pan genome. Eighty percent of the pan genome was present in all seven genomes, representing the core *G. soja* genome. Twenty percent of the pan genome was considered dispensable, with dispensable genes showing greater levels of sequence diversity. Core genes were enriched for biological processes including growth and reproduction. Qi et al. (2014) combined de novo assembly of a wild soybean accession with re-sequencing of a recombinant inbred population to identify the *GmCHX1* salt tolerance gene. Collectively, these studies demonstrate the ability that researchers now have to target complex traits for soybean improvement.

Genes identified using next-generation sequencing base approaches are key targets for characterization using cutting-edge functional genomics tools.

10.5.2 Epigenetics

Soybean epigenetic studies are an exciting area of future research, adding another level to the potential variation resulting in phenotypic changes. It is well recognized that soybean underwent a genetic bottleneck, severely limiting genetic diversity. Methylation studies in soybean, and other species, have shown a large number of allelic variants are silenced by methylation, thus altering methylation patterns may be a mechanism to generate phenotypic diversity (Schmitz 2014). As the majority of epialleles complied with Mendelian inheritance, methylQTLs can be identified for >90% of DMRs (Schmitz et al. 2013). Identifying methylQTLs and the causal epialleles may be a major challenge in soybean genetic studies. Epialleles induced by environmental conditions are a huge area of interest as understanding and identifying these may aide in understanding local adaptation and developing soybeans for extreme climates and climate changes. Finally, generating specific epialleles to test epiQTLs, induce desirable phenotypes, or regulate gene expression in response to stress, all without changing the underlying DNA sequence, has the potential to revolutionize soybean breeding and functional genomics studies.

10.5.3 Genome Editing

Precise genome editing involves the use of technologies that enable researchers to specifically alter the genetic code in a heritable manner without requiring a transgene or even making a transgenic plant (Luo et al. 2015). This sets genome editing apart from the RNA-silencing approaches discussed earlier and from standard methods used to overexpress genes in plants. Genome editing is made possible through the development of site-directed nucleases that can specifically recognize a DNA sequence and cleave it (Baltes and Voytas 2015). Meganucleases, zinc finger nucleases (ZFNs), and transcription activator-like effector nucleases (TALENs) provide site selectivity by protein–DNA interactions. Clustered regularly interspaced short palindromic repeats/Cas9 (CRISPR/Cas9) systems confer site specificity by complementarity between guide RNA and the DNA target site. The ZFNs and TALENs are modular synthetic proteins that fuse the DNA recognition domain with the *Fok*I restriction enzyme, which functions as a dimer. Therefore, ZFNs and TALENs are designed in such a way that it is necessary to express two proteins that have their recognition sites flanking the cleavage site. These nucleases only cleave when both recognition sites are bound and the *Fok*I enzyme is able to dimerize and cleave DNA, which ensures specificity. In the CRISPR/Cas9 system, the guide RNA mediates the target site recognition and DNA cleaved by nuclease activity of the Cas9 protein. Therefore, the CRISPR/Cas9 system is simpler and very easy to program by simply supplying different guide RNAs. Xie et al. (2014) estimated than more than 90% of the soybean transcriptome could be targeted using the CRISPR/Cas9 system.

Cleavage of the DNA target by these nucleases results in a double-strand break that can be repaired by one of two pathways, non-homologous end joining (NHEJ) and homologous recombination (HR) (Baltes and Voytas 2015). NHEJ results in insertions, deletions, and point mutations due to imperfect repair of the double-strand break site. NHEJ, therefore, results in targeted mutagenesis, which is useful for making loss-of-function alleles that are valuable in functional genomics studies. HR involves the double-strand break plus a donor DNA molecule that carries the sequence to insert or modify. The donor DNA is flanked by sequence that is identical to either side of the double-strand break enabling it to be inserted through HR. HR is valuable for crop bioengineering, because it allows for the precise replacement of sequences with the desired ones. However, accomplishing HR is more technically challenging, because both the site-directed nuclease and the donor template must be introduced into the plant to create the specific edit.

The ZFN, TALEN, and CRISPR/Cas9 systems have been used for mutagenesis via NHEJ in soybean (Curtin et al. 2011, 2013; Haun et al. 2014; Jacobs et al. 2015; Sun et al. 2015), and the CRISPR/Cas9 system was used to mediate allele replacement via HR (Li et al. 2015). A general approach is to design the site-directed nuclease and then test its ability to induce mutations at the target locus or loci in hairy roots. Constructs that induce mutations efficiently are then used in soybean transformation. The hairy root step eliminates unproductive site-directed nuclease constructs and saves significant time and resources needed to produce transgenic soybean lines. Mutations in a variety of loci including transgenes and endogenous genes encoding proteins and non-coding RNAs have been reported. The resulting mutations are most frequently small deletions; however, SNPs and insertions occur at significantly lower frequencies. Mutations can occur in a single allele or in both alleles of a gene. Bi-allelic mutations are advantageous because they allow mutant phenotypes to be observed in hairy roots or directly in primary transgenic plants (Jacobs et al. 2015). Even if mutations are heterozygous, it is possible to obtain homozygous, null segregant (lacking the site-directed nuclease transgene) mutants after only a single generation (Haun et al. 2014).

Duplication of the soybean genome complicates functional analysis of genes because a majority of genes have homoeologs that are nearly identical in nucleic acid sequence. The site-directed nucleases offer the ability to create mutations in one or more genes simultaneously (Curtin et al. 2011; Haun et al. 2014; Jacobs et al. 2015; Sun et al. 2015). In regenerated plants, the resulting mutations can be combined or separated via segregation, enabling the effects of the genes to be studied in isolation or combination. Curtin et al. (2011) used a single ZFN pair to create mutations in soybean *DCL4* paralogs (*DCL4a* and *DCL4b*), although only the *dcl4b* mutant produced viable progeny that segregated for the mutation. A single TALEN pair was used by Haun et al. (2014) to target the two fatty acid desaturase 2 genes (*FAD2-1A* and *FAD2-1B*). Four plants in their study carried mutations in both genes, and they were shown to segregate in the progeny of the original transgenic parent. The CRISPR/Cas9 system was also shown to induce mutations in paralogs by using different guide RNAs separately or in combination (Jacobs et al. 2015). Alternatively, the CRISPR/Cas9 system can tolerate mismatches between the guide RNA and its target, so it is also possible to obtain mutations in paralogs through off target activity (Sun et al. 2015). Finally, the site-directed nucleases result in a spectrum of mutations, which opens the possibility for generating allelic series in genes that could be very useful for understanding their functions (Haun et al. 2014).

10.6 Conclusions

Soybean has become an attractive model system for crops in which it is feasible to conduct functional genomics geared toward understanding plant biology, evolution, and domestication. Functional genomics tools and resources are ever expanding, and new technologies are rapidly adopted by the research community to accelerate discoveries related to the function and regulation of soybean genes. A variety of datasets such as mRNA transcript abundance, protein abundance, short RNA populations, and methylation status provide unprecedented insight into gene regulation during developmental programs and responses to the environment. The datasets coupled with sequences of genetically diverse germplasm, expanding mutant collections, RNA-silencing approaches, and precise genome editing provide a fertile ground for generating and testing hypotheses designed to bridge the genotype-to-phenotype gap.

Acknowledgements This work was supported by USDA-ARS Project 3625-21220-005-00D, the Iowa Soybean Association, the United Soybean Board Project Number 1204, USDA NIFA Hatch Project 3808, and State of Iowa funds. The USDA is an equal opportunity provider and employer. Mention of trade names or commercial products in this article is solely for the purpose of providing specific information and does not imply recommendation or endorsement by the US Department of Agriculture.

References

Ali AK, Lin J, Han J, Ibrahim KM, Jarjees MM, Qu F (2014) The 5' untranslated region of *Bean pod mottle virus* RNA2 tolerates unusually large deletions or insertions. Virus Res 179:247–250

Alonso JM, Stepanova AN, Leisse TJ, Kim CJ, Chen H, Shinn P, Stevenson DK, Zimmerman J, Barajas P, Cheuk R, Gadrinab C, Heller C, Jeske A, Koesema E, Meyers CC, Parker H, Prednis L, Ansari Y, Choy N, Deen H, Geralt M, Hazari N, Hom E, Karnes M, Mulholland C, Ndubaku R, Schmidt I, Guzman P, Aguilar-Henonin L, Schmid M, Weigel D, Carter DE, Marchand T, Risseeuw E, Brogden D, Zeko A, Crosby WL, Berry CC, Ecker JR (2003) Genome-wide insertional mutagenesis of *Arabidopsis thaliana*. Science 301:653–657

Anai T, Hoshino T, Imai N, Takagi Y (2012) Molecular characterization of two high-palmitic-acid mutant loci induced by X-ray irradiation in soybean. Breed Sci 61:631–638

Anders S, Huber W (2010) Differential expression analysis for sequence count data. Genome Biol 11: R106

Anderson JE, Kantar MB, Kono TY, Fu F, Stec AO, Song Q, Cregan PB, Specht JE, Diers BW, Cannon SB, McHale LK, Stupar RM (2014) A roadmap for functional structural variants in the soybean genome. Genes Genomes Genet 4:1307–1318

Arikit S, Xia R, Kakrana A, Huang K, Zhai J, Yan Z, Valdes-Lopez O, Prince S, Musket TA, Nguyen HT, Stacey G, Meyers BC (2014) An atlas of soybean small RNAs identifies phased siRNAs from hundreds of coding genes. Plant Cell 26:4584–4601

Atwood SE, O'Rourke JA, Peiffer GA, Yin T, Majumder M, Zhang C, Cianzio SR, Hill JH, Cook D, Whitham SA, Shoemaker RC, Graham MA (2014) Replication protein A subunit 3 and the iron efficiency response in soybean. Plant Cell Environ 37:213–234

Baltes NJ, Voytas DF (2015) Enabling plant synthetic biology through genome engineering. Trends Biotechnol 33:120–131

Becker A (ed) (2013) Virus-induced gene silencing: methods and protocols. Methods in Molecular biology, vol 975. Springer, Berlin

Belfield EJ, Gan X, Mithani A, Brown C, Jiang C, Franklin K, Alvey E, Wibowo A, Jung M, Bailey K, Kalwani S, Ragoussis J, Mott R, Harberd NP (2012) Genome-wide analysis of mutations in mutant lineages selected following fast-neutron irradiation mutagenesis of *Arabidopsis thaliana*. Genome Res 22:1306–1315

Bolon YT, Haun WJ, Xu WW, Grant D, Stacey MG, Nelson RT, Gerhardt DJ, Jeddeloh JA, Stacey G, Muehlbauer GJ, Orf JH, Naeve SL, Stupar RM, Vance CP (2011) Phenotypic and genomic analyses of a fast neutron mutant population resource in soybean. Plant Physiol 156:240–253

Bolon Y-T, Stec AO, Michno J-M, Roessler J, Bhaskar PB, Ries L, Dobbels AA, Campbell BW, Young NP, Anderson JE, Grant DM, Orf JH, Naeve SL, Muehlbauer GJ, Vance CP, Stupar RM (2014a) Genome resilience and prevalence of segmental duplications following fast neutron irradiation of soybean. Genetics 198:967–981

Bolon YT, Stec AO, Michno JM, Roessler J, Bhaskar PB, Ries L, Dobbels AA, Campbell BW, Young NP, Anderson JE, Grant DM, Orf JH, Naeve SL, Muehlbauer GJ, Vance CP, Stupar RM (2014b) Genome resilience and prevalence of segmental duplications following fast neutron irradiation of soybean. Genetics 198:967–981

Cannon SB, Crow JA, Grant D (2012) SoyBase and the legume information system: accessing inormation about soybean and other legume genomes. In: Wilson RF (ed) Designing soybeans for 21st century markets. American Oil Chemists' Society, Urbana, pp 49–62

Carbonell A, Takeda A, Fahlgren N, Johnson SC, Cuperus JT, Carrington JC (2014) New generation of artificial MicroRNA and synthetic trans-acting small interfering RNA vectors for efficient gene silencing in *Arabidopsis*. Plant Physiol 165:15–29

Carbonell A, Fahlgren N, Mitchell S, Cox KL Jr, Reilly KC, Mockler TC, Carrington JC (2015) Highly specific gene silencing in a monocot species by artificial microRNAs derived from chimeric miRNA precursors. Plant J 82:1061–1075

Cho YB, Jones SI, Vodkin LO (2013) The transition from primary siRNAs to amplified secondary siRNAs that regulate chalcone synthase during development of *Glycine max* seed coats. PLoS ONE 8:e76954

Chung W-H, Jeong N, Kim J, Lee WK, Lee Y-G, Lee S-H, Yoon W, Kim J-H, Choi I-Y, Choi H-K, Moon J-K, Kim N, Jeong S-C (2014) Population structure and domestication revealed by high-depth resequencing of Korean cultivated and wild soybean genomes. DNA Res 21:153–167

Cook DE, Bayless AM, Wang K, Guo X, Song Q, Jiang J, Bent AF (2014) Distinct copy number, coding sequence, and locus methylation patterns underlie *Rgh1*-mediated soybean resistance to soybean cyst nematode. Plant Physiol 165:630–647

Cooper JL, Till BJ, Laport RG, Darlow MC, Kleffner JM, Jamai A, El-Mellouki T, Liu S, Ritchie R, Nielsen N, Bilyeu KD, Meksem K, Comai L, Henikoff S (2008) TILLING to detect induced mutations in soybean. BMC Plant Biol 8:9

Cooper B, Campbell KB, McMahon MB, Luster DG (2013) Disruption of *Rpp1*-mediated soybean rust immunity by virus-induced gene silencing. Plant Signal Behav 8:e27543

Cui Y, Barampuram S, Stacey MG, Hancock CN, Findley SS, Mathieu M, Zhang Z, Parrott WA, Stacey G (2013) *Tnt1* retrotransposon mutagenesis: a tool for soybean functional genomics. Plant Physiol 161:36–47

Curtin SJ, Zhang F, Sander JD, Haun WJ, Starker C, Baltes NJ, Reyon D, Dahlborg EJ, Goodwin MJ, Coffman AP, Dobbs D, Joung JK, Voytas DF, Stupar RM (2011) Targeted mutagenesis of duplicated genes in soybean with zinc-finger nucleases. Plant Physiol 156:466–473

Curtin SJ, Anderson JE, Starker CG, Baltes NJ, Mani D, Voytas DF, Stupar RM (2013) Targeted mutagenesis for functional analysis of gene duplication in legumes. Methods Mol Biol 1069:25–42

Danzer J, Mellott E, Bui AQ, Le BH, Martin P, Hashimoto M, Perez-Lesher J, Chen M, Pelletier JM, Somers DA, Goldberg RB, Harada JJ (2015) Down-regulating the expression of 53 soybean transcription factor genes uncovers a role for SPEECHLESS in initiating stomatal cell lineages during embryo development. Plant Physiol 168:1025–1035

Dierking EC, Bilyeu KD (2009) New sources of soybean seed meal and oil composition traits identified through TILLING. BMC Plant Biol 9:89

D'Souza SF, Reddy KS, Badigannavar AM, Manjaya JG, Jambhulkar SJ (2009) Mutation breeding in oilseeds and grain legumes in India: accomplishments and socio-economic impact. In: Shu Q (ed) Induced plant mutations in the genomics era. Food and Agriculture Organization of the United Nations and the International Atomic Energy Agency, Rome, pp 55–57

Dubey A, Farmer A, Schlueter J, Cannon SB, Abernathy B, Tutej R, Woodward J, Shah T, Mulasmanovic B, Kudapa H, Raju NL, Gothalwal R, Pande S, Xiao Y, Town CD, Singh NK, May GD, Jackson S, Varshney RK (2011) Defining the transcriptome assembly and its use for genome dynamics and transcriptome profiling studies in pigeonpea (*Cajanus cajan* L.). DNA Res 18:153–164

Felippes FF, Wang JW, Weigel D (2012) MIGS: miRNA-induced gene silencing. Plant J 70:541–547

Fu L, Niu B, Zhu Z, Wu S, Li W (2012) CD-HIT: accelerated for clustering the next-generation sequencing data. Bioinformatics 28:3150–3152

Gillman JD, Stacey MG, Cui Y, Berg HR, Stacey G (2014) Deletions of the *SACPD-C* locus elevate seed stearic acid levels but also result in fatty acid and morphological alterations in nitrogen fixing nodules. BMC Plant Biol 14:143

Gongora-Castillo E, Buell CR (2013) Bioinformatics challenges in de novo transcriptome assembly using short read sequenes in the absense of a reference genome sequence. Nat Prod Rep 30:490–500

Goodstein DM, Shu S, Howson R, Neupane R, Hayes RD, Fazo J, Mitros T, Dirks W, Hellsten U, Putnam N, Rokhsar DS (2011) Phytozome: a comparative platform for green plant genomics. Nucleic Acids Res 40:D1178–D1186

Grabherr MG, Haas BJ, Yassour M, Levin JZ, Thompson DA, Amit A, Adiconis X, Fan L, Raychowdhury R, Zheng Q, Chen Z, Mauceli E, Hacohen N, Gnirke A, Rhind N, di Palma F, Birren BW, Nusbaum C, Lindblad-Toh K, Friedman N, Regev A (2011) Full-length transcriptome assembly from RNA-Seq data without a reference genome. Nat Biotechnol 15:644–652

Hancock CN, Zhang F, Floyd K, Richardson AO, LaFayette P, Tucker D, Wessler SR, Parrott WA (2011) The rice miniature inverted repeat transposable element *mPing* is an effective insertional mutagen in soybean. Plant Physiol 157:552–562

Haun WJ, Hyten DL, Xu WW, Gerhardt DJ, Albert TJ, Richmond T, Jeddeloh JA, Jia G, Springer NM, Vance CP, Stupar RM (2011) The composition and origins of genomic variation among individuals of the soybean reference cultivar Williams 82. Plant Physiol 155:645–655

Haun W, Coffman A, Clasen BM, Demorest ZL, Lowy A, Ray E, Retterath A, Stoddard T, Juillerat A, Cedrone F, Mathis L, Voytas DF, Zhang F (2014) Improved soybean oil quality by targeted mutagenesis of the fatty acid desaturase 2 gene family. Plant Biotechnol J 12:934–940

Hoshino T, Takagi Y, Anai T (2010) Novel *GmFAD2-1b* mutant alleles created by reverse genetics induce marked elevation of oleic acid content in soybean seeds in combination with *GmFAD2-1a* mutant alleles. Breed Sci 60:419–425

Hossain MS, Joshi T, Stacey G (2015) System approaches to study root hairs as a single cell plant model: current status and future perspectives. Front Plant Sci 6:363

Huang X, Madan A (1999) CAP3: a DNA sequence assembly program. Genome Res 9:868–877

Hwang WJ, Kim MY, Kang YJ, Shim S, Stacey MG, Stacey G, Lee SH (2014) Genome-wide analysis of mutations in a dwarf soybean mutant induced by fast neutron bombardment. Euphytica 203:399–408

Hyten DL, Song Q, Zhu Y, Choi IY, Nelson RL, Costa JM, Specht JE, Shoemaker RC, Cregan PB (2006) Impacts of genetic bottlenecks on soybean genome diversity. Proc Natl Acad Sci USA 103:16666–16671

Jacobs TB, LaFayette PR, Schmitz RJ, Parrott WA (2015) Targeted genome modifications in soybean with CRISPR/Cas9. BMC Biotechnol 15:16

Jacobs TB, Lawler NJ, LaFayette PR, Vodkin LO, Parrott WA (2016) Simple gene silencing using the trans-acting siRNA pathway. Plant Biotechnol J 14:117–127

Jones SI, Vodkin LO (2013) Using RNA-seq to profile soybean seed development from fertilization to maturity. PLoS ONE 8:e59270

Joshi T, Patil K, Fitzpatrick MR, Franklin LD, Yao Q, Cook JR, Wang Z, Libault M, Brechenmacher L, Valliyodan B, Wu X, Cheng J, Stacey G, Nguyen HT, Xu D (2012) Soybean Knowledge Base (SoyKB): a web resource for soybean translational genomics. BMC Genom 13:S15

Juvale PS, Hewezi T, Zhang C, Kandoth PK, Mitchum MG, Hill JH, Whitham SA, Baum TJ (2012) Temporal and spatial *Bean pod mottle virus*-induced gene silencing in soybean. Mol Plant Pathol 13:1140–1148

Kachroo A, Ghabrial S (2012) Virus-induced gene silencing in soybean. Methods Mol Biol 894:287–297

Kandoth PK, Heinz R, Yeckel G, Gross NW, Juvale PS, Hill J, Whitham SA, Baum TJ, Mitchum MG (2013) A virus-induced gene silencing method to study soybean cyst nematode parasitism in *Glycine max*. BMC Res Notes 6:255

Kato KK, Palmer RG (2003) Genetic identification of a female partial-sterile mutant in soybean. Genome 46:128–134

Kaur S, Cogan NO, Pembleton LW, Shinozuka M, Savin KW, Materne M, Forster JW (2011) Transcriptome sequencing of lentil based on second-generation technology permits large-scale unigene assembly and SSR marker discovery. BMC Genom 12:265–276

Kharkwal MC, Shu QY (2009) The role of induced mutations in wold food securities. Paper presented at the induced plant mutations in the genomics era, Vienna, Austria

Kim DS, Lee KJ, Kim JB, Kim SH, Song JY, Seo YW, Lee BM, Kang SY (2010a) Identification of Kunitz trypsin inhibitors mutations using SNAP markers in soybean mutant lines. Theor Appl Genet 121:751–760

Kim MY, Lee S, Van K, Kim T-H, Jeong S-C, Choi I-Y, Kim D-S, Lee Y-S, Park D, Ma J, Kim W-Y, Kim B-C, Park S, Lee K-A, Kim DH, Kim KH, Shin JH, Jang YE, Kim KD, Liu WX, Chaisan T, Kang YJ, Lee Y-H, Kim K-H, Moon J-K, Schmutz J, Jackson SA, Bhak J, Lee S-H (2010b) Whole-genome sequencing and intensive analysis of the undomesticated soybean (*Glycine soja* Sieb. and Zucc.) genome. Proc Natl Acad Sci USA 107:22032–22037

Kim KH, Kang YJ, Kim DH, Yoon MY, Moon J-K, Kim MY, Van K, Lee S-H (2011) RNA-seq analysis of a soybean near-isogenic line carrying bacterial leaf pustule-resistant and -susceptible alleles. DNA Res 18:483–497

Komatsu S, Nanjo Y, Nishimura M (2013) Proteomic analysis of the flooding tolerance mechanism in mutant soybean. J Proteom 79:231–250

Kour A, Boone AM, Vodkin LO (2014) RNA-seq profiling of a defective seed coat mutation in *Glycine max* reveals differential expression of proline-rich and other cell wall protein transcripts. PLoS ONE 9: e96342

Kudapa H, Azam S, Sharpe AG, Taran B, Li R, Deonovic B, Cameron C, Farmer AD, Cannon SB, Varshney RK (2014) Comprehensive transcriptome assembly of chickpea (*Cicer arietinum* L.) using Sanger and next generation sequencing platforms: development and applications. PLoS ONE 9:e86039

Lam H-M, Xu X, Liu X, Chen W, Yang G, Wong F-L, Li M-W, He W, Qin N, Wang B, Li J, Jian M, Wang J, Shao G, Wang J, Sun SS-M, Zhang G (2010) Resequencing of 31 wild and cultivated soybean genomes identifies patterns of genetic diversity and selection. Nat Genet 42:1053–1059

Lee KJ, Kim JB, Kim SH, Ha BK, Lee BM, Kang SY, Kim DS (2011) Alteration of seed storage protein composition in soybean [*Glycine max* (L.) Merrill] mutant lines induced by γ-irradiation mutagenesis. J Agric Food Chem 59:12405–12410

Lee K, Hwant JE, Velusamy V, Ha BK, Kim JB, Kim SH, Ahn JW, Kang SY, Kim DS (2014) Selection and molecular characerization of a lipoxygenase-free soybean mutant line induced by gamma irradiation. Theor Appl Genet 127:2405–2413

Li C, Yoshikawa N (2015) Virus-induced gene silencing of *N* gene in tobacco by *Apple latent spherical virus* vectors. Methods Mol Biol 1236:229–240

Li Y-H, Zhao S-C, Ma J-X, Li D, Yan L, Li J, Qi X-T, Guo X-S, Zhang L, He W-M, Chang R-Z, Liang Q-S, Guo Y, Ye C, Wang X-B, Tao Y, Guan R-X, Wang J-Y, Liu Y-L, Jin L-G, Zhang X-Q, Liu Z-X, Zhang L-J, Chen J, Wang K-J, Nielsen R, Li R-Q, Chen P-Y, Li W-B, Reif JC, Purugganan M, Wang J, Zhang M-C, Wang J, Qiu L-J (2013) Molecular footprints of domestication and improvement in soybean revealed by whole genome re-sequencing. BMC Genom 14:579

Li B, Fillmore N, Bai Y, Collins M, Thomson JA, Stewart R, Dewey CN (2014a) Evaluation of de novo transcriptome assemblies from RNA-Seq data. Genome Biol 15:553

Li Y-H, Zhou G, Ma J, Jiang W, Jin L-G, Zhang Z, Guo Y, Zhang J, Sui Y, Zheng L, Zhang S-S, Zuo Q, Shi X-H, Y-f Li, Zhang W-K, Hu Y, Kong G, Hong H-L, Tan B, Song J, Liu Z-X, Wang Y, Ruan H, Yeung CKL, Liu J, Wang H, Zhang L-J, Guan R-X, Wang K-J, Li W-B, Chen S-Y, Chang R-Z, Jiang Z, Jackson SA, Li R, Qiu L-J (2014b) de novo assembly of soybean wild relatives for pan-genome analysis of diversity and agronomic traits. Nat Biotechnol 32:1045–1052

Li Z, Liu ZB, Xing A, Moon BP, Koellhoffer JP, Huang L, Ward RT, Clifton E, Falco SC, Cigan AM (2015) Cas9-guide RNA directed genome editing in soybean. Plant Physiol 169:960–970

Libault M, Farmer A, Joshi T, Takahashi K, Langley RJ, Franklin LD, He J, Xu D, May G, Stacey G (2010) An integrated transcriptome atlas of the crop model *Glycine max*, and its use in comparative analyses in plants. Plant J 63:86–99

Lin F, Zhao M, Baumann DD, Ping J, Sun L, Liu Y, Zhang B, Tang Z, Hughes E, Doerge RW, Hughes TJ, Ma J (2014) Molecular response to the pathogen *Phytophthora sojae* among ten soybean near isogenic lines revealed by comparative transcriptomics. BMC Genom 15:18

Liu X, Liu S, Jamai A, Bendahmane A, Lightfoot DA, Mitchum MG, Meksem K (2011) Soybean cyst nematode resistance in soybean is independent of the *Rgh4* locus *LRR-RLK* gene. Funct Integr Genom 11:539–549

Liu S, Kandoth PK, Warren SD, Yeckel G, Heinz R, Alden J, Yang C, Jamai A, El-Mellouki T, Juvale PS,

Hill J, Baum TJ, Cianzio S, Whitham SA, Korkin D, Mitchum MG, Meksem K (2012) A soybean cyst nematode resistance gene points to a new mechanism of plant resistance to pathogens. Nature 492:256–260

Liu JZ, Graham MA, Pedley KF, Whitham SA (2015) Gaining insight into soybean defense responses using functional genomics approaches. Brief Funct Genom 14:283–290

Lu S, Yin X, Spollen W, Zhang N, Xu D, Schoelz J, Bilyeu K, Zhang ZJ (2015) Analysis of the siRNA-mediated gene silencing process targeting three homologous genes controlling soybean seed oil quality. PLoS ONE 10:e0129010

Luo R, Liu B, Xie Y, Li Z, Huang W, Yuan J, He G, Chen Y, Pan Q, Liu Y, Tang J, Wu G, Zhang H, Shi Y, Liu Y, Yu C, Wang B, Lu Y, Han C, Cheung DW, Yiu SM, Peng S, Xiaoqian Z, Liu G, Liao X, Li Y, Yang H, Wang J, Lam TW, Wang J (2012) SOAPdenovo2: an empirically improved memory-efficient short-read de novo assembler. GigaScience 1:18

Luo S, Li J, Stoddard TJ, Baltes NJ, Demorest ZL, Clasen BM, Coffman A, Retterath A, Mathis L, Voytas DF, Zhang F (2015) Non-transgenic plant genome editing using purified sequence-specific nucleases. Mol Plant 8:1425–1427

Mathieu M, Winters EK, Kong F, Wan J, Wang S, Eckert H, Luth D, Paz M, Donovan C, Zhang Z, Somers D, Wang K, Nguyen H, Shoemaker RC, Stacey G, Clemente T (2009) Establishment of a soybean (*Glycine max* Merr. L) transposon-based mutagenesis repository. Planta 229:279–289

McCarthy DJ, Chen Y, Smyth GK (2012) Differential expression analysis of multifactor RNA-seq experiments with respect to biological variation. Nucleic Acids Res 40:4288–4297

McHale LK, Haun WJ, Xu WW, Bhaskar PB, Anderson JE, Hyten DL, Gerhardt DJ, Jeddeloh JA, Stupar RM (2012) Structural variants in the soybean genome localize to clusters of biotic stress-response genes. Plant Physiol 159:1295–1308

Melito S, Heuberger AL, Cook D, Diers BW, MacGuidwin AE, Bent AF (2010) A nematode demographics assay in transgenic roots reveals no significant impacts of the *Rhg1* locus LRR-Kinase on soybean cyst nematode resistance. BMC Plant Biol 10:104

Men AE, Laniya TS, Searle IR, Iturbe-Ormaetxe I, Gresshoff I, Jiang Q, Carroll BJ, Gresshoff PM (2002) Fast neutron mutagenesis of soybean (*Glycine soja* L.) produces a supernodulating mutant containing a large deletion in linkage group H. Genome Lett 3:147–155

Meyer JDF, Silva DCG, Yang C, Pedley KF, Zhang C, van de Mortel M, Hill JH, Shoemaker RC, Abdelnoor RV, Whitham SA, Graham MA (2009) Identification and analyses of candidate genes for *Rpp4*-mediated resistance to Asian soybean rust in soybean. Plant Physiol 150:295–307

Moran Lauter AN, Peiffer GA, Yin T, Whitham SA, Cook D, Shoemaker RC, Graham MA (2014) Identification of candidate genes involved in early iron deficiency chlorosis signaling in soybean (*Glycine max*) roots and leaves. BMC Genom 15:702

Nakagawa H (2009) Induced mutations in plant breeding and biological researches in Japan. Paper presented at the induced plant mutations in the genomics era, Vienna, Austria

O'Rourke J, Yang SS, Miller SS, Bucciarelli B, Liu J, Rydeen A, Bozsoki Z, Uhde-Stone C, Tu ZJ, Allan D, Gronwald JW, Vance CP (2013a) An RNA-seq transcriptome analysis of orthophosphate deficient white lupin reveals novel insights into phosphorus acclimation in plants. Plant Physiol 161:705–724

O'Rourke JA, Iniguez LP, Bucciarelli B, Roessler J, Schmutz J, McClean PE, Jackson SA, Hernandez G, Graham MA, Stupar RM, Vance CP (2013b) A re-sequencing based assessment of genomic heterogeneity and fast neutron-induced deletions in a common bean cultivar. Front Plant Sci 4:210

O'Rourke JA, Fu F, Bucciarelli B, Yang SS, Samac DA, Lamb JF, Monteros MJ, Graham MA, Gronwald JW, Krom N, Li J, Dai X, Zhao PX, Vance CP (2015) The *Medicago sativa* gene index 1.2: a web-accessible gene expression atlas for investigating expression differences between *Medicago sativa* subspecies. BMC Genom 16:502

Pandey AK, Yang C, Zhang C, Graham MA, Horstman HD, Lee Y, Zabotina OA, Hill JH, Pedley K, Whitham SA (2011) Functional analysis of the Asian soybean rust resistance pathway mediated by *Rpp2*. Mol Plant Microbe Interact 24:194–206

Peiffer GA, King KE, Severin AJ, May GD, Cianzio SR, Lin SF, Lauter NC, Shoemaker RC (2012) Identification of candidate genes underlying an iron efficiency quantitative trait locus in soybean. Plant Physiol 158:1745–1754

Qi X, Li M-W, Xie M, Liu X, Ni M, Shao G, Song C, Kay-Yuen Yim A, Tao Y, Wong F-L, Isobe S, Wong C-F, Wong K-S, Xu C, Li C, Wang Y, Guan R, Sun F, Fan G, Xiao Z, Zhou F, Phang T-H, Liu X, Tong S-W, Chan T-F, Yiu S-M, Tabata S, Wang J, Xu X, Lam H-M (2014) Identification of a novel salt tolerance gene in wild soybean by whole-genome sequencing. Nat Commun 5:4340

Rahman SM, Takagi Y, Kubota K, Miyamoto K, Kawakita T (1994) High oleic acid mutant in soybean induced by X-ray irradiation. Biosci Biotechnol Biochem 58:1070–1072

Rahman SM, Takagi Y, Miyamoto K, Kawakita T (1995) High stearic acid soybean mutant induced by X-ray irradiation. Biosci Biotechnol Biochem 59:922–923

Rao SS, El-Habbak MH, Havens WM, Singh A, Zheng D, Vaughn L, Haudenshield JS, Hartman GL, Korban SS, Ghabrial SA (2013) Overexpression of *GmCaM4* in soybean enhances resistance to pathogens and tolerance to salt stress. Mol Plant Pathol 15:145–160

Robinson MD, McCarthy DJ, Smyth GK (2010) edgeR: a Bioconductor package for differential expression analysis of digital gene expression data. Bioinformatics 26:1

Rode NW, Bernard RL (1975) Inheritance of a tan saddle mutant. Soybean Genet Newsl 2:39–42

Schmitz RJ (2014) The secret garden—epigenetic alleles underlie complex traits. Science 343:1082–1083

Schmitz RJ, He Y, Valdez-Lopez O, Khan SM, Joshi T, Urich MA, Nery JR, Diers B, Xu D, Stacey G, Ecker JR (2013) Epigenome-wide inheritance of cytosine methylation variants in a recombinant inbred population. Genome Res 23:1633–1674

Schmutz J, Cannon SB, Schlueter J, Ma J, Mitros T, Nelson W, Hyten DL, Song Q, Thelen JJ, Cheng J, Xu D, Hellsten U, May GD, Yu Y, Sakurai T, Umezawa T, Bhattacharyya MK, Sandhu D, Valliyodan B, Lindquist E, Peto M, Grant D, Shu S, Goodstein D, Barry K, Futrell-Griggs M, Abernathy B, Du J, Tian Z, Zhu L, Gill N, Joshi T, Libault M, Sethuraman A, Zhang XC, Shinozaki K, Nguyen HT, Wing RA, Cregan P, Specht J, Grimwood J, Rokhsar D, Stacey G, Shoemaker RC, Jackson SA (2010) Genome sequence of the palaeopolyploid soybean. Nature 463:178–183

Schulz MH, Zerbino DR, Vingron M, Birney E (2012) Oases: robust de novo RNA-seq assembly across the dynamic range of expression levels. Bioinformatics 28:1086–1092

Schwab R, Ossowski S, Riester M, Warthmann N, Weigel D (2006) Highly specific gene silencing by artificial microRNAs in *Arabidopsis*. Plant Cell 18:1121–1133

Severin AJ, Woody JL, Bolon YT, Joseph B, Diers BW, Farmer AD, Muehlbauer GJ, Nelson RT, Grant D, Specht JE, Graham MA, Cannon SB, May GD, Vance CP, Shoemaker RC (2010) RNA-seq atlas of *Glycine max*: a guide to the soybean transcriptome. BMC Plant Biol 10:160

Song Y, Ji D, Li S, Weng P, Li Q, Xiang F (2012) The dynamic changes of DNA methylation and histone modifications of salt responsive transcription factor genes in soybean. PLoS ONE 7:e41274

Song QX, Lu X, Li QT, Chen H, Hu XY, Ma B, Zhang WK, Chen SY, Zhang JS (2013) Genome-wide analysis of DNA methylation in soybean. Mol Plant 6:1961–1974

Srisombun S, Srinives P, Yathaputanon C, Dangpradub S, Malipan A, Srithongchai W, Kumsueb B, Tepjun V, Ngampongsai S, Kornthong A, Impithuksa S (2009) Achievements of grain legume variety improvement using induced mutation of the IAEA/RAS/5/040 project in Thailand. Paper presented at the induced plant mutations in the genomics era, Vienna, Austria

Sun X, Hu Z, Chen R, Jiang Q, Song G, Zhang H, Xi Y (2015) Targeted mutagenesis in soybean using the CRISPR-Cas9 system. Sci Rep 5:10342

Takagi Y, Hossain ABMM, Yanagita T, Kusaba S (1989) High linolenic acid mutant in soybean induced by X-ray irradiation. Jpn J Breed 39:403–409

Takahashi R, Yamagishi N, Yoshikawa N (2013) A MYB transcription factor controls flower color in soybean. J Hered 104:149–153

Tran LS, Mochida K (2010) Functional genomics of soybean for improvement of productivity in adverse conditions. Funct Integr Genom 10:447–462

Trapnell C, Pachter L, Salzberg SL (2009) TopHat: discovering splice junctions with RNA-seq. Bioinformatics 25:1105–1111

Trapnell C, Roberts A, Goff L, Pertea G, Kim D, Kelley DR, Pimentel H, Salzberg SL, Rinn JL, Pachter L (2012) Differential gene and transcript expression analysis of RNA-seq experiments with TopHat and Cufflinks. Nat Protoc 7:562–578

Tremblay A, Hosseini P, Li S, Alkharouf NW, Matthews BF (2013) Analysis of *Phakopsora pachyrhizi* transcript abundance in critical pathways at four time-points during infection of a susceptible soybean cultivar using deep sequencing. BMC Genom 14:614

Whitham SA, Eggenberger AL, Zhang C, Chowda-Reddy RV, Martin KM, Hill JH (2015) Recent advances in *in planta* transient expression and silencing systems for soybean using viral vectors. In: Azhakanandam K, Silverstone A, Daniell H, Davey MR (eds) Recent advancements in gene expression and enabling technologies in crop plants. Springer, New York, pp 423–451

Whitham SA, Lincoln LM, Chowda-Reddy RV, Dittman JD, O'Rourke JA, Graham MA (2016a) Virus-induced gene silencing and transient gene expression in soybean (*Glycine max*) using *Bean pod mottle virus* infectious clones. In: Stacey G, Birchler JA, Ecker J, Martin C, Stitt M, Zhou J-M (eds) Current protocols in plant biology. Wiley, Hoboken

Whitham SA, Qi M, Innes RW, Ma W, Lopes-Caitar V, Hewezi T (2016b) Molecular soybean-pathogen interactions. Annu Rev Phytopathol 54:443–468

Wong J, Gao L, Yang Y, Zhai J, Arikit S, Yu Y, Duan S, Chan V, Xiong Q, Yan J, Li S, Liu R, Wang Y, Tang G, Meyers BC, Chen X, Ma W (2014) Roles of small RNAs in soybean defense against *Phytophthora sojae* infection. Plant J 79:928–940

Xie K, Zhang J, Yang Y (2014) Genome-wide prediction of highly specific guide RNA spacers for CRISPR–Cas9-mediated genome editing in model plants and major crops. Mol Plant 7:923–926

Xu X, Zeng L, Tao Y, Vuong T, Wan J, Boerma R, Noe J, Li Z, Finnerty S, Pathan SM, Shannon JG, Nguyen HT (2013) Pinpointing genes underlying the quantitative trait loci for root-knot nematode resistance in palaeopolyploid soybean by whole genome resequencing. Proc Natl Acad Sci USA 110:13469–13474

Yamada T, Mori Y, Yasue K, Maruyama N, Kitamura K, Abe J (2014) Knockdown of the 7S globulin subunits shifts distribution of nitrogen sources to the residual protein fraction in transgenic soybean seeds. Plant Cell Rep 33:1963–1976

Yamagishi N, Yoshikawa N (2009) Virus-induced gene silencing in soybean seeds and the emergence stage of soybean plants with *Apple latent spherical virus* vectors. Plant Mol Biol 71:15–24

Yamagishi N, Yoshikawa N (2013) Highly efficient virus-induced gene silencing in apple and soybean by *Apple latent spherical virus* vector and biolistic inoculation. Methods Mol Biol 975:167–181

Yang SS, Tu ZJ, Cheung F, Xu WW, Lamb JFS, Jung HJG, Vance CP, Gronwald JW (2011) Using RNA-seq for gene identification, polymorphism detection and transcript profiling in two alfalfa genotypes with divergene cell wall composition in stems. BMC Genom 12:199

Young WP, Schupp JM, Keim P (1999) DNA methylation and AFLP marker distribution in the soybean genome. Theor Appl Genet 99:785–790

Yuan FJ, Zhu DH, Tan YY, Dong DK, Fu XJ, Zhu SL, Li BQ, Shu QY (2012) Identification and characterization of the soybean *IPK1* ortholog of a low phytic acid mutant reveals an exon-excluding splice-site mutation. Theor Appl Genet 125:1413–1423

Zabala G, Vodkin LO (2014) Methylation affects transposition and splicing of a large *CACTA* transposon from a MYB transcription factor regulating anthocyanin synthase genes in soybean seed coats. PLoS ONE 9:e111959

Zabala G, Campos E, Varala KK, Bloomfield S, Jones SI, Win H, Tuteja JH, Calla B, Clough SJ, Hudson M, Vodkin LO (2012) Divergent patterns of endogenous small RNA populations from seed and vegetative tissues of *Glycine max*. BMC Plant Biol 12:177

Zerbino DR, Birney E (2008) Velvet: algorithms for de novo short read assembly using de Bruijn graphs. Genome Res 18:821–829

Zhang C, Ghabrial SA (2006) Development of *Bean pod mottle virus*-based vectors for stable protein expression and sequence-specific virus-induced gene silencing in soybean. Virology 344:401–411

Zhang C, Yang C, Whitham SA, Hill JH (2009) Development and use of an efficient DNA-based viral gene silencing vector for soybean. Mol Plant Microbe Interact 22:123–131

Zhang C, Bradshaw JD, Whitham SA, Hill JH (2010) The development of an efficient multipurpose *Bean pod mottle virus* viral vector set for foreign gene expression and RNA silencing. Plant Physiol 153:52–65

Zhang C, Grosic S, Whitham SA, Hill JH (2012) The requirement of multiple defense genes in soybean *Rsv1*-mediated extreme resistance to *Soybean mosaic virus*. Mol Plant Microbe Interact 25:1307–1313

Zhang C, Whitham SA, Hill JH (2013) Virus-induced gene silencing in soybean and common bean. Methods Mol Biol 975:149–156

Zhong X, Wang Y, Liu X, Gong L, Ma Y, Qi B, Dong Y, Liu B (2009) DNA methylation polymorphism in annual wild soybean (*Glycine soja* Sieb. et Zucc) and cultivated soybean (*G. max* L. Merr.). Can J Plant Sci 89:851–863

Zhou Z, Jiang Y, Wang Z, Gou Z, Lyu J, Li W, Yu Y, Shu L, Zhao Y, Ma Y, Fang C, Shen Y, Liu T, Li C, Li Q, Wu M, Wang M, Wu Y, Dong Y, Wan W, Wang X, Ding Z, Gao Y, Xiang H, Zhu B, Lee S-H, Wang W, Tian Z (2015) Resequencing 302 wild and cultivated accessions identifies genes related to domestication and improvement in soybean. Nat Biotechnol 33:408–414

Zhu T, Schupp JM, Oliphant A, Keim P (1994) Hypomethylated sequences: characterization of the duplicate soybean genome. Mol Genet Genom 244:638–645

11 Transposable Elements

Meixia Zhao and Jianxin Ma

Abstract

Transposable elements (TEs) including retrotransposons and DNA transposons are the major DNA components in soybean (*Glycine max* L. Merr.), accounting for approximately 60% of the soybean reference genome. The majority of soybean TEs are long terminal repeat retrotransposons that were amplified in the past a few million years (myr). Overall, the TEs were preferentially accumulated in the pericentromeric regions of all chromosomes, but different classes/superfamilies/families of TEs generally exhibited different patterns of distribution along chromosomes. Such a distribution pattern appears to be the outcome of TE insertion bias as well as natural selection purging deleterious genic TE insertions. Despite their periodic proliferation, many TEs have accumulated various deletions through unequal homologous recombination and illegitimate recombination to become "dead" copies and to counteract genome expansion caused by periodic TE amplification. The majority of intact TEs appear to be inactivated, either transcriptionally or transpositionally, by epigenetic mechanisms such as DNA methylation, histone modification, and small RNA-mediated silencing. Nevertheless, active endogenous TEs have been found and are being used to develop TE-based soybean mutants for discovery of genes underlying traits of agronomic importance.

11.1 Introduction

Transposable elements (TEs) were first discovered in maize by the cytogeneticist Barbara McClintock in the mid-1950s (McClintock 1956a, b) and are now recognized to be ubiquitous in eukaryotes with few exceptions (Feschotte et al. 2002; Bennetzen 2005). In all plant

M. Zhao · J. Ma (✉)
Department of Agronomy, Purdue University, West Lafayette, IN 47907, USA
e-mail: maj@purdue.edu

M. Zhao
e-mail: Zhao185@purdue.edu

H.T. Nguyen and M.K. Bhattacharyya (eds.), *The Soybean Genome*, Compendium of Plant Genomes, DOI 10.1007/978-3-319-64198-0_11

genomes that have been sequenced and annotated to date, TEs are found to be the most abundant DNA components, making up ~15% (e.g., in Arabidopsis) up to >80% (e.g., in wheat) of those genomes (The International Arabidopsis Initiative 2000; The International Wheat Genome Sequencing Consortium (IWGSC) 2014). Despite their relative abundance, TEs in the host genomes are generally inactivated genetically and epigenetically, preventing or restricting them from generating additional copies, but some TEs can be reactivated periodically to further proliferate or burst within very short time frames (Hirochika et al. 1996; Jiang et al. 2003; Piegu et al. 2006; Schnable et al. 2009). Such dynamic processes, in addition to the polyploidization events that also commonly occurred in plants, are largely responsible for size variation of the plant genomes (Bennetzen et al. 2005). Besides their roles in mediating structural genomic variation, accumulating evidence indicates that TEs are also key players in regulating gene expression, altering gene functions, and creating new genes (Kashkush et al. 2002, 2003; Jiang et al. 2004; Lai et al. 2005; Zabala and Vodkin 2005, 2008; Du et al. 2009; Lisch 2009; Yang and Bennetzen 2009; Zhao et al. 2013). As such, TEs have been drawing more and more attentions from broad research communities.

11.2 Classification

Based on the transposition mechanisms, TEs are generally categorized into two classes: retrotransposons and DNA transposons (Finnegan 1989; Wicker et al. 2007; Zhao and Ma 2013). The former transpose through an RNA intermediate and involve reverse transcription of the RNA template to produce new copies that subsequently integrate into the host genome, while the latter transpose through a DNA intermediate, with the majority excising from its original site using transposase and integrating elsewhere in the host genome. Retrotransposons, mainly based on their structures, are further divided into two subfamilies, the long terminal repeat (LTR) retrotransposons (LTR-RTs) and non-LTR retrotransposons including long interspersed repetitive elements (LINEs) and short interspersed repetitive elements (SINEs) (Grandbastien 1992; Kumar and Bennetzen 1999; Lander et al. 2001). Multi-step process of reverse transcription of an LTR-RT results in the placement of two identical LTRs, each consisting of a U3, R, and U5 regions that contain signals for initiation and termination of transcription, at either end of the element, and subsequently participate in integration of the element into the host genome (Kumar and Bennetzen 1999). The majority of DNA transposons, according to their structural features and transposase similarities, are categorized into seven superfamilies, including Tc1/Mariner, hAT, Mutator, PIF/Harbinger, Pong, CACTA, and miniature inverted-repeat TEs (MITEs) (Wicker et al. 2007; Fedoroff 2012). In the past several years, a new category of DNA transposons, designated Helitrons, have been reported in a number of plants (Lai et al. 2005; Hollister and Gaut 2007; Du et al. 2009; Yang and Bennetzen 2009). It is proposed that this category of DNA transposons amplify through a rolling-circle transposition process, which does not involve the excision of TE sequences (Pritham and Feschotte 2007).

Upon the generation of the soybean Williams 82 reference genome sequence, a complete or nearly complete set of TEs in the genome were identified using a combination of structure-based and homology-based approaches (Du et al. 2010a; Schmutz et al. 2010). In the 975-Mb genomic DNA assembled and mapped to the 20 chromosomes, a total of 32,552 retrotransposons and 6029 DNA transposons with clearly defined boundaries were annotated (Table 11.1). Of these 32,552 retrotransposons, 32,370 are LTR-RTs belonging to 510 distinct families including 353 Ty3/Gypsy-like families and 157 Ty1/copia-like families, and 182 are LINEs. The 6,029 DNA transposons are classified into Tc1/Mariner, hAT, Mutator, PIF/Harbinger, Pong, CACTA, MITEs, and Helitrons. These elements together with numerous truncated TE fragments or remnants account for approximately 59% of the soybean genome, among which, 42% are LTR-RTs and 17% are DNA transposons (Du et al. 2010a, b).

Table 11.1 Transposable elements with clear boundaries and signatures of insertion sites identified in the soybean reference genome

Classification	Copy numbers
LTR-Retrotransposon	32,370
Ty1/copia	13,318
Intact elements	4,913
Solo LTRs	8,405
Ty3/gypsy	19,052
Intact elements	9,193
Solo LTRs	9,859
Non-LTR retrotransposon	182
LINE	182
Class II: DNA transposon	6,029
Subclass I	5,947
Tc1/Mariner	9
hAT	65
Mutator	2,373
PIF/Harbinger	90
Pong	12
CACTA	65
MITE	3,333
Tourist	1,575
Stowaway	1,758
Subclass II	82
Helitron	82
Total	38,581

Du et al. (2010a)

11.3 Amplification

As observed in many other plants, TE copy numbers in soybean vary dramatically among different superfamilies/families (Du et al. 2010a, b). More than 77% of the LTR-RT families have fewer than 10 complete copies, whereas 36 families each have more than 100 complete copies. The three largest families are Gmr9 (i.e., SNARE), Gmr4, and Gmr5, which have 4724, 3370, and 2925 copies with clearly defined boundaries, accounting for more than 1/3 of all LTR-RT copies identified in the reference genome. When all fragments were included, the SNARE family alone makes up ~12% of the reference genome. Based on the divergence levels between two LTRs of individual LTR-RT elements, approximately 91% of the 14,006 intact elements were estimated to have been amplified within the last 3 million years (myr), with 3248 elements generated within the last 0.5 myr (Du et al. 2010b). It appears that many LTR-RT families were primarily amplified within distinct evolutionary time frames, demonstrating the variability of the spectrum of transpositional activities among different families. However, how a particular family was periodically activated or reactivated to proliferate or burst remains unclear.

Compared with LTR-RTs, which had two identical LTR sequences upon their birth, DNA TEs lacked such intra-element identical sequences upon their transposition, and thus, the transposition events are not datable or difficult to be dated. Nevertheless, a few hundreds of nonreference DNA TEs were identified in a soybean population (Tian et al. 2012), suggesting their recent transpositional activities. In addition, the observations of highly identical elements within a same superfamily such as Mutator and CACTA were also an indication of their recent amplification.

11.4 Biased Insertion

One of the striking features of the soybean genome is the large fraction of the recombination-suppressed heterochromatic regions surrounding centromeres (Schmutz et al. 2010), which are referred to as pericentromeric regions, contrasting to the rest of the genomes that are referred to as chromosomal arms. The pericentromeric regions account for ~57% (i.e., 556 Mb) of the soybean genome (Du et al. 2010a; Schmutz et al. 2010). By contrast, such regions make up only ~12% (i.e., 46 Mb) of the rice genome, which is composed of ~400 Mb of DNA (International Rice Genome Sequencing Project 2005; Tian et al. 2009). As generally observed in many other plants (SanMiguel et al. 1996, 1998; Wang and Dooner 2006; Dooner and He 2008; Tian et al. 2009), TEs, particularly LTR-RTs in soybean, tend to be organized in nested patterns (Du et al.

2010a, b) and are preferentially accumulated in the pericentromeric regions (Du et al. 2012). It was estimated that LTR-RTs and DNA TEs in soybean make up approximately 47.2 and 21.5% of its pericentromeric regions versus 8.8 and 8.9% of its chromosomal arms (Du et al. 2012). In rice, LTR-RTs account for 38.8% of the pericentromeric regions. Intriguingly, the proportion of LRT-RTs in the chromosomal arms of rice ($\sim 17\%$) is substantially higher than that in the chromosomal arms of soybean ($\sim 9\%$) (Tian et al. 2009), demonstrating the different intensities of preferential accumulation of TEs in pericentromeric regions relative to chromosomal arms between the two genomes.

Such a bias of TE accumulation for or against distinct types of genomic regions can be interpreted as the outcome of insertion bias. Actually, many families of TEs exhibit strong insertion bias in the plant genomes. For example, de novo insertions of a LTR-RT, termed Tal1, in *Arabidopsis lyrata* preferentially target centromeric regions (Tsukahara et al. 2012). By contrast, de novo insertions of an active rice DNA TE, designated mPing, preferentially occur within a few kb upstream or downstream regions of genes in soybean (Hancock et al. 2011). In maize, the *Mutator* elements have a tendency to target the 5′ ends of genes and the regions with high frequency of recombination rates (Liu et al. 2009). Recent characterization and profiling of nonreference TE insertions, which are absent in the soybean reference genome and often detected in only a single accession of a re-sequenced soybean population (Lam et al. 2010) and thought to have recently occurred during varietal diversification, revealed consistent distribution patterns of the nonreference LTR-RT insertions and all LTR-RT insertions accumulated in the reference genome, many of which occurred a few million years ago (mya) (Tian et al. 2012). These observations suggest that the distribution pattern of LTR-RTs in the soybean genome is largely determined by their insertion bias. Indeed, different types/superfamilies/families of TEs in the soybean genome often exhibit different distribution patterns (Du et al. 2010b), and TE families enriched in chromosomal arms and pericentromeric regions were both observed (Du et al. 2010b), reflecting their distinct insertion biases. Insertional preferences of TEs were also observed in many other plants (Jiang and Wessler 2001; Dietrich et al. 2002; Miyao et al. 2003; Baucom et al. 2009; Liu et al. 2009; Naito et al. 2009). It remains unclear how the specificity of TE insertions are determined, although the integration specificity of Ty3/gypsy LTR-RTs in yeast was found to be associated with specific interactions between the integrases/transposases and unique chromatin proteins (Kirchner et al. 1995; Zou and Voytas 1997).

Nevertheless, the distribution patterns of nonreference DNA TEs and the accumulated ones in the reference genome were inconsistent (Tian et al. 2012). If these nonreference DNA TEs and accumulated ones are indeed representative of those proliferated within two distinct evolutionary time frames, such a difference in pattern of TE distribution would be the outcome of natural selection such as purifying selection, which is believed to be the major force purging deleterious TE insertions, particularly the ones within genes, from the host genome (Hanada et al. 2009). Because pericentromeric regions are generally gene-poor regions, where TE insertions would cause less frequent deleterious mutations than those occur in chromosomal arms, biased accumulation of both LTR-RTs and DNA TEs in the pericentromeric regions is expected, regardless of their distinct insertion preferences for the two contrasting chromatin environments. Purifying selection against TE insertions in pericentromeric regions is thought to be inefficient or less efficient than in chromosomal arms due to the suppressed or severely reduced rates of genetic recombination in the former regions than in the latter regions (Gaut et al. 2007); thus, it would be logical to deduce that the intensities and effectiveness of purifying selection against different TE classes or different TE families that exhibit distinct insertion preferences in the soybean genome are likely to vary. It appears that the distinct TE insertion preferences are primary factors that resulted in different levels of effectiveness of purifying selection against deleterious TE insertions, reshaping the distribution patterns of TEs after their insertion into the host genome (Tian et al. 2009, 2012).

11.5 Elimination

In flowering plants, unequal homologous recombination (UR) and illegitimate recombination (IR) have been proposed to be two major mechanisms responsible for removal of TE DNA, counteracting host genome expansion primarily caused by periodic TE proliferation (Devos et al. 2002; Ma et al. 2004; Bennetzen et al. 2005). It is obvious that TEs in the soybean genome have undergone rapid elimination by these two mechanisms. Among the 32,370 LTR-RTs identified in the soybean reference genome, 18,264 were found to be solo LTRs (Du et al. 2010a). Each solo LTR is thought to be formed by intra-element UR between the two highly identical LTRs, leading to a net loss of one LTR and the internal portion from the original intact element. Overall, the ratio (1.29:1) of solo LTRs to intact elements in soybean (Du et al. 2010b) is significantly lower than observed in rice (1.62:1) (Ma et al. 2004; Tian et al. 2009) but significantly highly than reported in Arabidopsis (0.50:1) (Devos et al. 2002). Why the activeness of UR underlying solo LTR formation differs among these plant species remains to be investigated. Nevertheless, the ratios of solo LTRs to intact elements in recombination-suppressed/reduced pericentromeric regions were found to be significantly lower than in chromosomal arms in both soybean and rice (Tian et al. 2009; Du et al. 2010b). In addition, a positive correlation between the ratios of solo LTRs to intact elements and the rates of genetic recombination along the rice chromosomal arms was observed in rice (Tian et al. 2009). Therefore, genetic or epigenetic factors that affect the rates of local genetic recombination appear to play a primary role in regulating solo LTR formation. It was found that the ratios of solo LTRs to intact elements vary among LTR-RT families, and such variation appears to be associated with the overall ages of intact elements and the length of LTRs (Tian et al. 2009; Du et al. 2010b).

In contrast to solo LTRs, partially deleted or truncated elements including LTR-RTs and DNA TEs are thought to be the outcome of IR, which does not appear to involve crossing-over between homologous sequences (Devos et al. 2002; Ma et al. 2004; Vitte and Bennetzen 2006). How many truncated elements in the soybean reference genome have not been determined due to the difficulty in defining the boundaries of highly truncated, deleted, or fragmented elements, in which multiple overlapping deletion events may have been involved, and due to the numerous sequence gaps remained in the reference genome, many of which are adjacent to TE sequences (Du et al. 2010b; Schmutz et al. 2010). Therefore, the effectiveness of IR for removal of TE DNA has not been precisely assessed. However, tens and thousands of TE fragments, besides the 38,581 soybean TEs with clearly defined boundaries, were detected by homology-based searches against soybean TEs deposited in the soybean TE database (Du et al. 2010a), suggesting that the process that accumulated small TE deletions by IR in soybean, as observed in other plants (Devos et al. 2002; Ma et al. 2004; Vitte and Bennetzen 2006), were also considerably active.

11.6 RNA-Mediated Recombination

Chimeric TEs are generally considered to be the products of genomic UR between different TEs sharing sequence similarity (Devos et al. 2002; Ma et al. 2004; Sharma et al. 2008). Theoretically, such interelement recombination, if occurred, could produce chimeric solo LTRs, intact elements, or LTR-internal-LTR-internal-LTR complexes, and none of these chimeric structures, would be flanked by the short target site duplication (TSD) sequences (Devos et al. 2002; Ma et al. 2004). Indeed, a few dozens or hundreds of solo LTRs and intact LTR-RTs without TSDs were observed in Arabidopsis (Devos et al. 2002), rice (Vitte and Panaud 2003; Ma et al. 2004; Ma and

Bennetzen 2006; Tian et al. 2009), and soybean (Du et al. 2010b, c), respectively. Particularly, a total of 13 LTR-internal-LTR-internal-LTR complexes were found in the rice reference genome (Tian et al. 2009). Phylogenetic and structural analyses of the intact LTR-RTs belonging to the SNARE family in soybean revealed that a few hundreds of intact elements possess chimeric LTR sequences derived from two clearly defined, highly diverged subfamilies: the autonomous SARE subfamily and the nonautonomous SNRE subfamilies, but nearly all these elements with chimeric LTRs were flanked by TSDs (Du et al. 2010c). This unexpected observation led to the discovery of RNA-based interelement recombination that was proposed to have occurred during reverse transcription through template switches within a same virus-like particles (VLPs) (Feng et al. 2000; Wicker et al. 2007), to produce unique chimeric LTR sequences (Du et al. 2010c; Fig. 11.1). Such a recombination process seems to have led to homogenization of LTR sequences between the autonomous SARE elements and the nonautonomous SNRE elements. This process appears to be essential for enhancement and maintenance of their autonomous–nonautonomous partnership (Du et al. 2010c).

The reverse transcription process by which retrotransposons are amplified has not been experimentally validated in most eukaryotes, but copackaging of multiple RNA templates from LTR-RTs during VLP formation was observed in yeast (*Saccharomyces cerevisiae*) (Feng et al. 2000). In addition, switches between different RNA templates that led to retroviral recombination of genomic RNA were experimentally detected (Hu and Temin 1990; Luo and Taylor 1990). If the process for LTR-RT amplification is, indeed, similar to that of retroviruses, frequent interelement recombination of LTR-RTs via template switching would be expected, although such recombination if occurs between highly similar templates is unlikely to be detected (Du et al. 2010c).

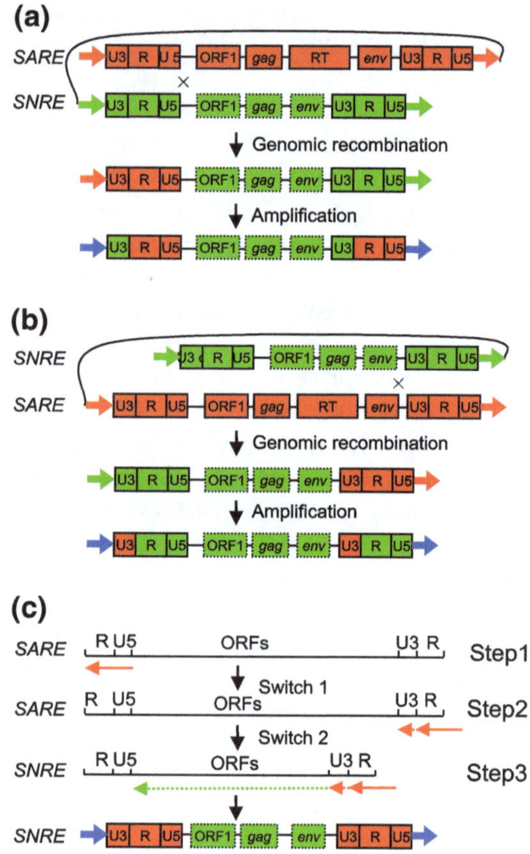

Fig. 11.1 Models for recombination between autonomous (SARE) and nonautonomous (SNRE) elements. **a**, **b** Intra-strand unequal recombination between two SARE and SNRE elements to form chimeric structures of nonautonomous recombinants. "×" represents proposed crossover. **c** Initiation of reverse transcription (step 1) from an autonomous element, intra-element template switch (step 2), followed by an interelement template switch (step 3) to form a nonautonomous recombinant with LTRs from the autonomous partner. U5, R, and U3 are three typical components of an LTR that contain signals for initiation and termination of transcription. The internal region in a SARE element generally contains genes encoding gag, reverse transcriptase (RT) envelope (env)-like protein, and an additional open reading frame 1 (ORF1) that is unique to the SARE elements. Other sequences featured are DR (flanking target direct repeat) as indicated by thick arrows. Dotted frames indicated degraded sequences in SNRE elements that share similarities with genes encoding ORF1, gag, RT, and env-like proteins in SARE elements. The figure was previously presented by Du et al. (2012)

11.7 Epigenetic Regulation

In addition to structural variation, such as the formation of solo LTRs through UR and accumulation of small deletions through IR, that have produced numerous incomplete TE copies permanently lacking transcriptional activity and transpositional capacity in their host genomes, epigenetic mechanisms including DNA methylation, histone modification, and small RNA interfering have been suggested to play important roles in TE silencing or inactivation (Kasschau et al. 2007; Nobuta et al. 2007; Zilberman et al. 2007). For example, cytosine methylation of TE sequences was found be associated with histone methylation and heterochromatinization, giving rise to transcriptional repression and TE inactivation (Lippman and Martienssen 2004). Such TE methylation requires small interfering RNAs generally derived from TEs to act via the RNA-directed DNA methylation (RdDM) pathway (Aufsatz et al. 2002; Cao and Jacobsen 2002; Hamilton et al. 2002; Huettel et al. 2007; Matzke et al. 2007). The epigenetic features of the soybean genome have not been comprehensively profiled. Nevertheless, heavy cytosine methylation of TEs, as observed in many other plants, was also seen in soybean (Kim et al. 2015). In addition, the transcriptional levels of the majority of TE families are extremely low relative to those of protein-encoding genes per the transcriptomic data available from Soybase (www.soybase.org). Furthermore, the majority of TEs in Arabidopsis, rice, and soybean were found to match 24 nt-siRNAs derived from their respective hosts (El Baidouri et al. 2015). These observations suggest that the RdDM pathways for regulating TE methylation between soybean and other plants are conserved. The epigenetic features are highly heritable during DNA replication (McNairn and Gilbert 2003; Kinoshita et al. 2004; Slotkin et al. 2005; Dowen et al. 2012), and such heritability appears to be essential for maintenance of the status of silenced TEs across generations.

While the majority of TEs are silenced epigenetically, some elements, such as the CACTA-like Tgm families, that were often found within gene bodies may be active, or can be reactivated under specific conditions such as tissue culture and tissue wounding by insects (Grandbastien, 2015). These conditions appear to be able to trigger demethylation of these elements and also their transcription by inducing their own stress-related promoters (Hirochika 1993, 1996, 2000; Takeda et al. 1999; Hashida et al. 2003), thereby restoring their transpositional activities, yet the actual mechanisms by which these elements are activated remain to be investigated.

11.8 Effects on Host Genes

Although the majority of genes, particularly, the protein-encoding regions annotated in all sequenced plant genomes do not harbor TEs, a number of null mutations of genes naturally formed by TE insertions have been identified and functionally characterized. Such TE insertions can either block host gene expression (Salvi et al. 2007) or produce altered transcripts or proteins that differ from the wild-type forms. In soybean, nine potential active CACTA-type DNA transposons, named as Tgm1, Tgm2, Tgm3, Tgm4, Tgm5, Tgm6, Tgm7, Tgm-Express1, Tgm9 and Tgmt*, have been reported to present in gene bodies in soybean (Rhodes and Vodkin 1988; Zabala and Vodkin 2005, 2008; Xu et al. 2010; Takahashi et al. 2011; Tsukahara et al. 2012; Yan et al. 2015). For instance, the insertion of Tgm1 within the soybean seed lectin gene resulted in a loss-of-function mutation that produced seed lectinless phenotype (Rhodes and Vodkin 1985). The insertion of Tgm-Express1 in the *Wp* locus and the insertion of Tgm9 resided in the *W4* locus that encodes dihydroflavonol-4-reductase 2 (DFR2) changed flower color (Zabala and Vodkin 2005, Xu et al. 2010). Actually, excision of Tgm9 from the *W4* locus can occur naturally, resulting in variegated flowers as a phenotypic marker (Xu et al. 2010), which is being used to develop Tgm9-based soybean mutant population (Chap. 12). Because most TEs are heavily methylated and often associated with inactive chromatin (Slotkin and Martienssen 2007), genes adjacent to methylated TEs can be influenced to become silenced (Hollister and Gaut 2009). A recent survey of RNA-seq reads from two soybean callus tissues identified ~1000 putative TE readouts (Zhao and Ma, unpublished obs.), although whether or how these TE readouts may influence the expression of their flanking genes is unknown. In spite of the frequently seen gene interruption by TEs within or adjacent to genes, TE insertions that induce novel transcriptional regulation to host genes were also observed. For instance, an insertion of a TE ~50 kb upstream of the *teosinte branched 1* (*tb1*) gene resulted in its overexpression that represses branching in domesticated maize (Studer et al. 2011). Indeed, many TEs, particularly, LTR-RTs, exhibit read-out transcription toward their flanking regions including gene spaces in plants (Kashkush et al. 2003; Kashkush and Khasdan 2007). Because TEs carry regulatory sequences such as promoters and enhancers for their own transcription, the TE readouts into flanking genes would alter their expression (Lisch 2013).

In addition to the effects of TEs on cis-regulation of gene expression, some TEs were found to be able to produce siRNAs that target host genes (Hanada et al. 2009). By a combination of computational and experimental approaches, 12 TEs in soybean including nine Mutators and three LTR-RTs were identified to have been triggered by miRNAs to produce secondary siRNAs, termed "transacting siRNAs" (tasiRNAs) or "phased siRNAs" (phasiRNAs), which were predicted to be able to target TEs as well as a number of protein-encoding genes (Zhao and Ma unpublished obs.).

11.9 Conclusions and Perspectives

With the recent availability of the soybean reference genome sequence, a nearly complete set of soybean TEs have been identified and characterized, and a soybean TE database has been constructed (www.SoyTEbase.org), providing the foundation for the study of TE biology, particularly TE function. As observed in many other plants, the majority of TEs in the soybean genome are heavily methylated and have undergone various structural variations to become defective copies such as solo LTRs and truncated elements or remnants and to counteract expansion of the soybean genome caused by periodic amplification of numerous TEs at various scales. Many TEs exhibited haplotype variation at population levels, as an indication of their recent transpositional activities, but few are found to be active. To date, several hundreds of soybean varieties have been re-sequenced, and re-sequencing thousands of varieties are underway. Such re-sequencing data from a large population, or eventually the entire soybean germplasm collection worldwide will be extremely valuable toward the discovery of a full set of endogenous active TEs, which may have substantially shaped the genetic or epigenetic diversity (e.g., natural de novo mutations) underlying various traits, including the ones of agronomic importance, in soybean. Characterization of such a set of TEs will also enhance our understanding of TE properties with regard to their insertional specificity, distribution patterns, evolutionary fates, and interaction with the host genes and genome.

References

Aufsatz W, Mette MF, van der Winden J, Matzke AJM, Matzke M (2002) RNA-directed DNA methylation in *Arabidopsis*. Proc Natl Acad Sci USA 99:16499–16506

Baucom RS, Estill JC, Chaparro C, Upshaw N, Jogi A, Deragon JM, Westerman RP, Sanmiguel PJ, Bennetzen JL (2009) Exceptional diversity, non-random distribution, and rapid evolution of retroelements in the B73 maize genome. PLoS Genet 5:e1000732

Bennetzen J (2005) Transposable elements, gene creation and genome rearrangement in flowering plants. Curr Opin Genet 15:621–627

Bennetzen JL, Ma J, Devos K (2005) Mechanisms of recent genome size variation in flowering plants. Ann Bot 95:127–132

Cao X, Jacobsen SE (2002) Role of the *Arabidopsis* DRM methyltransferases in de novo DNA methylation and gene silencing. Curr Biol 12:1138–1144

Devos KM, Brown JKM, Bennetzen JL (2002) Genome size reduction through illegitimate recombination counteracts genome expansion in *Arabidopsis*. Genome Res 12:1075–1079

Dietrich CR, Cui F, Packila ML, Li J, Ashlock DA, Nikolau BJ, Schnable PS (2002) Maize Mu transposons are targeted to the 59 untranslated region of the *gl8* gene and sequences flanking Mu target-site duplications exhibit nonrandom nucleotide composition throughout the genome. Genetics 160:697–716

Dooner HK, He L (2008) Maize genome structure variation: interplay between retrotransposon polymorphisms and genic recombination. Plant Cell 20: 249–258

Dowen RH, Pelizzola M, Schmitz RJ, Lister R, Dowen JM, Nery JR, Dixon JE, Ecker JR (2012) Widespread dynamic DNA methylation in response to biotic stress. Proc Natl Acad Sci USA 109:E2183–E2191

Du C, Fefelova N, Caronna J, He L, Dooner HK (2009) The polychromatic Helitron landscape of the maize genome. Proc Natl Acad Sci USA 106:19916–19921

Du J, Grant D, Tian Z, Nelson RT, Zhu L, Shoemaker RC, Ma J (2010a) SoyTEdb: a comprehensive database of transposable elements in the soybean genome. BMC Genom 11:113

Du J, Tian Z, Hans CS, Laten HM, Cannon SB, Jackson SA, Shoemaker RC, Ma J (2010b) Evolutionary conservation, diversity and specificity of LTR-retrotransposons in flowering plants: insights from genome-wide analysis and multi-specific comparison. Plant J 63:584–598

Du J, Tian Z, Bowen NJ, Schmutz J, Shoemaker RC, Ma J (2010c) Bifurcation and enhancement of autonomous-nonautonomous retrotransposon partnership through LTR swapping in soybean. Plant Cell 22:48–61

Du J, Tian Z, Sui Y, Zhao M, Song Q, Cannon SB, Cregan P, Ma J (2012) Pericentromeric effects shape the patterns of divergence, retention, and expression of duplicated genes in the paleopolyploid soybean. Plant Cell 24:21–32

El Baidouri M, Kim KD, Abernathy B, Arikit S, Maumus F, Panaud O, Meyers BC, Jackson SA (2015) A new approach for annotation of transposable elements using small RNA mapping. Nucleic Acids Res 43:e84

Fedoroff NV (2012) Transposable elements, epigenetics, and genome evolution. Science 338:758–767

Feng YX, Moore SP, Garfinkel DJ, Rein A (2000) The genomic RNA in Ty1 virus-like particles is dimeric. J Virol 74:10819–10821

Feschotte C, Jiang N, Wessler S (2002) Plant transposable elements: where genetics meets genomics. Nat Rev Genet 3:329–341

Finnegan DJ (1989) Eukaryotic transposable elements and genome evolution. Trends Genet 5:103–107

Gaut BS, Wright SI, Rizzon C, Dvorak J, Anderson LK (2007) Recombination: an underappreciated factor in the evolution of plant genomes. Nat Rev Genet 8:77–84

Grandbastien MA (1992) Retroelements in higher-plants. Trends Gene 8:103–108

Grandbastien MA (2015) LTR retrotransposons, handy hitchhikers of plant regulation and stress response. Biochim Biophys Acta 1849:403–416

Hamilton A, Voinnet O, Chappell L, Baulcombe D (2002) Two classes of short interfering RNA in RNA silencing. EMBO J 21:4671–4679

Hanada K, Vallejo V, Nobuta K, Slotkin RK, Lisch D, Meyers BC, Shiu SH, Jiang N (2009) The functional role of pack-MULEs in rice inferred from purifying selection and expression profile. Plant Cell 21:25–38

Hancock CN, Zhang F, Floyd K, Richardson AO, Lafayette P, Tucker D, Wessler SR, Parrott WA (2011) The rice miniature inverted repeat transposable element mPing is an effective insertional mutagen in soybean. Plant Physiol 157:552–562

Hashida S, Kitamura K, Mikami T, Kishima Y (2003) Temperature shift coordinately changes the activity and the methylation state of transposon Tam3 in *Antirrhinum majus*. Plant Physiol 132:1207–1216

Hirochika H (1993) Activation of tobacco retrotransposons during tissue-culture. EMBO J 12:2521–2528

Hirochika H, Sugimoto K, Otsuki Y, Tsugawa H, Kanda M (1996) Retrotransposons of rice involved in mutations induced by tissue culture. Proc Natl Acad Sci USA 93:7783–7788

Hirochika H, Okamoto H, Kakutani T (2000) Silencing of retrotransposons in *Arabidopsis* and reactivation by the *ddm1* mutation. Plant Cell 12:357–368

Hollister JD, Gaut BS (2007) Population and evolutionary dynamics of Helitron transposable elements in *Arabidopsis thaliana*. Mol Biol Evol 11:2515–2524

Hollister JD, Gaut BS (2009) Epigenetic silencing of transposable elements: a trade-off between reduced transposition and deleterious effects on neighboring gene expression. Genome Res 19:1419–1428

Hu WS, Temin HM (1990) Retroviral recombination and reverse transcription. Science 250:1227–1233

Huettel B, Kanno T, Daxinger L, Bucher E, van der Winden J, Matzke AJ, Matzke M (2007) RNA-directed DNA methylation mediated by *DRD1* and *Pol IVb*: a versatile pathway for transcriptional gene silencing in plants. Biochim Biophys Acta 1769:358–374

International Rice Genome Sequencing Project (2005) The map-based sequence of the rice genome. Nature 436:793–800

Jiang N, Wessler SR (2001) Insertion preference of maize and rice miniature inverted repeat transposable elements as revealed by the analysis of nested elements. Plant Cell 13:2553–2564

Jiang N, Bao Z, Zhang X, Hirochika H, Eddy SR, McCouch SR, Wessler SR (2003) An active DNA transposon family in rice. Nature 421:163–167

Jiang N, Bao Z, Zhang X, Eddy SR, Wessler SR (2004) Pack-MULE transposable elements mediate gene evolution in plants. Nature 431:569–573

Kashkush K, Khasdan V (2007) Large-scale survey of cytosine methylation of retrotransposons and the impact of readout transcription from long terminal repeats on expression of adjacent rice genes. Genetics 177:1975–1985

Kashkush K, Feldman M, Levy AA (2002) Gene loss, silencing and activation in a newly synthesized wheat allotetraploid. Genetics 160:1651–1659

Kashkush K, Feldman M, Levy AA (2003) Transcriptional activation of retrotransposons alters the expression of adjacent genes in wheat. Nat Genet 33:102–106

Kasschau KD, Fahlgren N, Chapman EJ, Sullivan CM, Cumbie JS, Givan SA, Carrington JC (2007) Genome-wide profiling and analysis of *Arabidopsis* siRNAs. PLoS Biol 5:479–493

Kim KD, El Baidouri M, Abernathy B, Iwata-Otsubo A, Chavarro C, Gonzales M, Libault M, Grimwood J, Jackson SA (2015) A comparative epigenomic analysis of polyploidy-derived genes in soybean and commonbean. Plant Physiol 168:1433–1447

Kinoshita T, Miura A, Choi YH, Kinoshita Y, Cao X, Jacobsen SE, Fischer RL, Kakutani T (2004) One-way control of *FWA* imprinting in *Arabidopsis* endosperm by DNA methylation. Science 303:521–523

Kirchner J, Connolly CM, Sandmeyer SB (1995) Requirement of RNA polymerase III transcription factors for in vitro position-specific integration of a retroviruslike element. Science 267:1488–1491

Kumar A, Bennetzen JL (1999) Plant retrotransposons. Annu Rev Genet 33:479–532

Lai J, Li Y, Messing J, Dooner HK (2005) Gene movement by Helitron transposons contributes to the haplotype variability of maize. Proc Natl Acad Sci USA 102:9068–9073

Lam HM, Xu X, Liu X, Chen W, Yang G, Wong FL, Li MW, He W, Qin N, Wang B, Li J, Jian M, Wang J, Shao G, Wang J, Sun SS, Zhang G (2010) Resequencing of 31 wild and cultivated soybean genomes identifies patterns of genetic diversity and selection. Nat Genet 42:1053–1059

Lander ES, Linton LM, Birren B et al (2001) Initial sequencing and analysis of the human genome. Nature 409:860–921

Lippman Z, Martienssen R (2004) The role of RNA interference in heterochromatic silencing. Nature 431:364–370

Lisch D (2009) Epigenetic regulation of transposable elements in plants. Annu Rev Plant Biol 60:43–66

Lisch D (2013) How important are transposons for plant evolution? Nat Rev Genet 14:49–61

Liu S, Yeh CT, Ji T, Ying K, Wu H, Tang HM, Fu Y, Nettleton D, Schnable PS (2009) Mu transposon insertion sites and meiotic recombination events co-localize with epigenetic marks for open chromatin across the maize genome. PLoS Genet 5:e1000733

Luo GX, Taylor J (1990) Template switching by reverse transcriptase during DNA synthesis. J Virol 64:4321–4328

Ma J, Bennetzen JL (2006) Recombination, rearrangement, reshuffling, and divergence in a centromeric region of rice. Proc Natl Acad Sci 103:383–388

Ma J, Devos KM, Bennetzen JL (2004) Analyses of LTR-retrotransposon structures reveal recent and rapid genomic DNA loss in rice. Genome Res 14:860–869

Matzke MA, Kanno T, Huettel B, Daxinger L, Matzke AJM (2007) Targets of RNA-directed DNA methylation. Curr Opin Plant Biol 10:512–519

McClintock B (1956a) Intranuclear systems controlling gene action and mutation. Brookhaven Symp Biol 8:58–74

McClintock B (1956b) Controlling elements and the gene. Brookhaven Symp Biol 21:197–216

McNairn AJ, Gilbert DM (2003) Epigenomic replication: linking epigenetics to DNA replication. BioEssays 25:647–656

Miyao A, Tanaka K, Murata K, Sawaki H, Takeda S, Abe K, Shinozuka Y, Onosato K, Hirochika H (2003) Target site specificity of the Tos17 retrotransposon shows a preference for insertion within genes and against insertion in retrotransposon-rich regions of the genome. Plant Cell 15:1771–1780

Naito K, Zhang F, Tsukiyama T, Saito H, Hancock CN, Richardson AO, Okumoto Y, Tanisaka T, Wessler SR (2009) Unexpected consequences of a sudden and massive transposon amplification on rice gene expression. Nature 461:1130–1134

Nobuta K, Venu RC, Lu C, Beló A, Vemaraju K, Kulkarni K, Wang W, Pillay M, Green PJ, Wang GL, Meyers BC (2007) An expression atlas of rice mRNAs and small RNAs. Nat Biotechnol 25:473–477

Piegu B, Guyot R, Picault N, Roulin A, Sanyal A, Kim H, Collura K, Brar DS, Jackson S, Wing RA, Panaud O (2006) Doubling genome size without polyploidization: dynamics of retrotransposition-driven genomic expansions in *Oryza australiensis*, a wild relative of rice. Genome Res 16:1262–1269

Pritham EJ, Feschotte C (2007) Massive amplification of rolling-circle transposons in the lineage of the bat *Myotis lucifugus*. Proc Natl Acad Sci USA 6:895–1900

Rhodes PR, Vodkin LO (1985) Highly structured sequence homology between an insertion element and the gene in which it resides. Proc Natl Acad Sci USA 82:493–497

Rhodes PR, Vodkin LO (1988) Organization of the Tgm family of transposable elements in soybean. Genetics 120:597–604

Salvi S, Sponza G, Morgante M, Tomes D, Niu X, Fengler KA, Meeley R, Ananiev EV, Svitashev S, Bruggemann E, Li B, Hainey CF, Radovic S, Zaina G, Rafalski JA, Tingey SV, Miao GH, Phillips RL, Tuberosa R (2007) Conserved noncoding genomic sequences associated with a flowering-time quantitative trait locus in maize. Proc Natl Acad Sci USA 104:11376–11381

SanMiguel P, Tikhonov A, Jin YK, Motchoulskaia N, Zakharov D, Melake-Berhan A, Springer PS, Edwards KJ, Lee M, Avramova Z, Bennetzen JL (1996) Nested retrotransposons in the intergenic regions of the maize genome. Science 274:765–768

SanMiguel P, Gaut BS, Tikhonov A, Nakajima Y, Bennetzen JL (1998) The paleontology of intergene retrotransposons of maize. Nat Genet 20:43–45

Schmutz J, Cannon SB, Schlueter J et al (2010) Genome sequence of the palaeopolyploid soybean. Nature 463:178–183

Schnable PS, Ware D, Fulton RS et al (2009) The B73 maize genome: complexity, diversity, and dynamics. Science 326:1112–1115

Sharma A, Schneider KL, Presting GG (2008) Sustained retrotransposition is mediated by nucleotide deletions and interelement recombinations. Proc Natl Acad Sci USA 105:15470–15474

Slotkin RK, Martienssen R (2007) Transposable elements and the epigenetic regulation of the genome. Nat Rev Genet 8:272–285

Slotkin RK, Freeling M, Lisch D (2005) Heritable transposon silencing initiated by a naturally occurring transposon inverted duplication. Nat Genet 37:641–644

Studer A, Zhao Q, Ross-Ibarra J, Doebley J (2011) Identification of a functional transposon insertion in the maize domestication gene *tb1*. Nat Genet 43: 1160–1163

Takahashi R, Morita Y, Nakayama M, Kanazawac A, Abe J (2011) An active CACTA-family transposable element is responsible for flower variegation in wild soybean *Glycine soja*. Plant Genome 11:0028

Takeda S, Sugimoto K, Otsuki H, Hirochika H (1999) A 13-bp cis-regulatory element in the LTR promoter of the tobacco retrotransposon Tto1 is involved in responsiveness to tissue culture, wounding, methyl jasmonate and fungal elicitors. Plant J 18:383–393

The Arabidopsis Genome Initiative (2000) Analysis of the genome sequence of the flowering plant *Arabidopsis thaliana*. Nature 408:796–815

The International Wheat Genome Sequencing Consortium (IWGSC) (2014) A chromosome-based draft sequence of the hexaploid bread wheat (*Triticum aestivum*) genome. Science 345:1251788

Tian Z, Rizzon C, Du J, Zhu L, Bennetze JL, Jackson SA, Gaut BS, Ma J (2009) Do genetic recombination and gene density shape the pattern of DNA elimination in rice long terminal repeat retrotransposons? Genome Res 19:2221–2230

Tian Z, Zhao M, She M, Du J, Cannon SB, Liu X, Xu X, Qi X, Li MW, Lam HM, Ma J et al (2012) Genome-wide characterization of nonreference transposons reveals evolutionary propensities of transposons in soybean. Plant Cell 24:4422–4436

Tsukahara S, Kawabe A, Kobayashi A, Ito T, Aizu T, Shin-I T, Toyoda A, Fujiyama A, Tarutani Y, Kakutani T (2012) Centromere-targeted de novo integrations of an LTR retrotransposon of *Arabidopsis lyrata*. Genes Dev 26:705–713

Vitte C, Bennetzen JL (2006) Analysis of retrotransposon structural diversity uncovers properties and propensities in angiosperm genome evolution. Proc Natl Acad Sci USA 103:17638–17643

Vitte C, Panaud O (2003) Formation of solo-LTRs through unequal homologous recombination counterbalances amplifications of LTR-retrotransposons in rice *Oryza sativa* L. Mol Biol Evol 20:528–540

Wang Q, Dooner HK (2006) Remarkable variation in maize genome structure inferred from haplotype diversity at the *bz* locus. Proc Natl Acad Sci USA 103:17644–17649

Wicker T, Sabot F, Hua-Van A, Bennetzen JL, Capy P, Chalhoub B, Flavell A, Leroy P, Morgante M, Panaud O, Paux E, SanMiguel P, Schulman AH (2007) A unified classification system for eukaryotic transposable elements. Nat Rev Genet 8:973–982

Xu M, Brar HK, Grosic S, Palmer RG, Bhattacharyya MK (2010) Excision of an active CACTA-like transposable element from *DFR2* causes variegated flowers in soybean [*Glycine max* (L.) Merr.]. Genetics 184:53–63

Yan F, Di S, Takahashi R (2015) CACTA-superfamily transposable element is inserted in *MYB* transcription factor gene of soybean line producing variegated seeds. Genome 58:365–374

Yang L, Bennetzen JL (2009) Distribution, diversity, evolution, and survival of Helitrons in the maize genome. Proc Natl Acad Sci USA 106:19922–19927

Zabala G, Vodkin LO (2005) The wp mutation of *Glycine max* carries a gene-fragment-rich transposon of the CACTA superfamily. Plant Cell 17:2619–2632

Zabala G, Vodkin L (2008) A putative autonomous 20.5 kb-CACTA transposon insertion in an *F3'H* allele identifies a new CACTA transposon subfamily in Glycine max. BMC Plant Biol 8:124

Zhao M, Ma J (2013) Co-evolution of plant LTR-retrotransposons and their host genomes. Protein Cell 4:493–501

Zhao M, Du J, Lin F, Tong C, Yu J, Huang S, Wang X, Liu S, Ma J (2013) Shifts in evolutionary rate and intensity of purifying selection between two *Brassica* genomes revealed by analyses of orthologous transposons and relics of a whole genome triplication. Plant J 76:211–220

Zilberman D, Gehring M, Tran RK, Ballinger T, Henikoff S (2007) Genome-wide analysis of *Arabidopsis thaliana* DNA methylation uncovers an interdependence between methylation and transcription. Nat Genet 39:61–69

Zou S, Voytas DF (1997) Silent chromatin determines target preference of the *Saccharomyces* retrotransposon Ty5. Proc Natl Acad Sci USA 94:7412–7416

Transposon-Based Functional Characterization of Soybean Genes

12

Devinder Sandhu and Madan K. Bhattacharyya

Abstract

Type II transposable elements that use a cut-and-paste mechanism for jumping from one genomic region to another are ideal for use in tagging and cloning genes. Precise excision from an insertion site in a mutant gene leads to regaining the wild-type function. Thus, the function of a gene can be established based on the mutant phenotype and the regaining of the wild-type phenotype following precise excision of the element. Heterologous type II transposable elements including the *Ac/Ds* system from maize, the miniature inverted repeat system, *mPing* from rice, and the *Tnt1* retrotransposon from tobacco have been successfully applied in functional analyses of soybean genes. Although several endogenous transposable elements have been identified in soybean, evidence of an active type II transposable element in soybean was largely lacking. We have previously reported the isolation of the type II soybean transposon *Tgm9* from intron II of the *dihyroflavonol-4-reductase 2* (*DFR2*) gene of the *W4* locus. *Tgm9* is an active element and produces variegated flowers through somatic excision. Excision of the *Tgm9* element from the progenitor cells of flower buds results in genotypes with purple flowers that are known as germinal revertants. The element was discovered from a commercial soybean cultivar, and the line carrying the element was termed T322. The T322 genome contains only one active *Tgm9* copy in the *W4* locus. In a recent study, the utility of *Tgm9* was assessed by studying a set of random germinal revertants. The new mutations created following excision of *Tgm9* from *DFR2* were evaluated using a transposon display assay. This

D. Sandhu (✉)
US Salinity Laboratory, USDA-ARS, Riverside, CA 92507, USA
e-mail: devinder.sandhu@usda.ars.gov

M.K. Bhattacharyya
Department of Agronomy, Iowa State University, Ames, IA 50011, USA
e-mail: mbhattac@iastate.edu

H.T. Nguyen and M.K. Bhattacharyya (eds.), *The Soybean Genome*, Compendium of Plant Genomes, DOI 10.1007/978-3-319-64198-0_12

study revealed that *Tgm9* transposes to all 20 soybean chromosomes from its original site in the *DFR2* gene. Although *Tgm9* exhibited preferential transposition to a few genomic regions, across the entire genome 25.7% of the new *Tgm9* mutants were detected in exon or intron sequences. Thus, *Tgm9* is a suitable endogenous type II transposon to generate an indexed insertional mutant collection for functional characterization of most of the soybean genes.

12.1 Introduction

Soybean is the world's most valuable crop with high levels of protein (~40%) and oil (~20%), hence it is an important source of human nutrition and livestock and aquaculture feed (Masudaa and Goldsmith 2009). It has also become an important source of biodiesel in recent years. In 2014, the USA produced ~35% of the world's soybean crop with a value approaching $40 billion, and the US export of soybean and soya products including biodiesel was over $30 billion (http://www.soystats.com; http://unitedsoybean.org/article/u-s-soy-exports-hit-new-milestones/). Despite the global economic importance of soybean, the molecular and genetic bases of the physiological processes controlling agronomically important traits are largely unknown. The soybean genome has been sequenced, and expression patterns of soybean genes are mostly known (http://www.soybase.org/SequenceIntro.php; http://soykb.org). However, functional analyses of soybean genes are still very arduous because of the lack of an efficient and rapid transformation procedure.

In the last decade, transposable elements have been successfully utilized in functional characterization of soybean genes. Transposon tagging has been turned out to be an attractive approach in soybean. In this chapter, we focus on the history of transposon tagging approaches in soybean and recent progress in isolating soybean genes using an endogenous transposable element.

12.2 Transposable Elements and Gene Tagging

Transposable elements, originally identified and studied by Barbara McClintock (1956), have become an ideal means of gene isolation. Transposable elements provide unique tools for functional characterization of eukaryotic genomes and have played a major role in understanding gene function and genome organization (Walbot 1992; Martienssen 1998; Peterson 2013; Naito et al. 2014).

After the initial isolation of the bronze locus in maize using the *Activator/Dissociation (Ac/Ds)* system, many plant genes have been isolated using transposon tagging (Fedoroff et al. 1983; Dooner and Belachew 1989; Jones et al. 1994). The *Ac/Ds* transposon system was extensively used for gene tagging and functional genomics in maize (Lazarow et al. 2013). However, due to the lack of active and well-characterized transposable elements in many crops, transposon tagging has been limited to a few plant species. Heterologous transposable elements have been exploited in plant species that do not carry any characterized active endogenous transposable element (Bancroft et al. 1993; Chuck et al. 1993; Martienssen 1998). For example, heterologous transposon tagging has been utilized in crop species including tomatoes, tobacco, flax, Arabidopsis, and barley (Jones et al. 1994; Whitham et al. 1994; James et al. 1995; Lawrence et al. 1995; Singh et al. 2012). The most commonly used transposable element system is *Ac/Ds,* which transposes to closely linked loci and thus is most suitable for tagging genes in a specific genomic region of a chromosome (Dooner and Belachew 1989).

12.3 Transposition of *Ac/Ds* in Soybean

The feasibility of using the *Ds* transposon as an insertional mutagenesis tool was tested in soybean by building a collection of *Ds*-insertion mutants for different genomic regions through transformation of soybean (Mathieu et al. 2009). As the transformation in soybean is slow, laborious, and expensive, tactical targeting of the transposon is important for maximizing gene-tagging efficiency. To improve the efficiency of functional characterization of soybean genes, an enhancer trap element is added to the T-DNA molecule carrying the *Ac/Ds* system so that promoter activity of the gene at the insertion site can be monitored (Mathieu et al. 2009). Combining T-DNA with a transposon-based system allowed strategic placement of launch sites in the gene-rich regions. To avoid somatic transposition that can complicate gene cloning, expression of *Ac* was controlled by a meiosis specific promoter (Mathieu et al. 2009). A total of 900 soybean events were generated, and insertion sites were determined for 200 of them. This revealed that this construct showed a strong bias for insertion into gene-rich regions (Mathieu et al. 2009). This system has been successfully utilized in isolating and characterizing a gene involved in male fertility (Mathieu et al. 2009).

12.4 *mPing*-Based Mutagenesis in Soybean

mPing is a nonautonomous miniature inverted repeat transposable element (MITE) from rice (Jiang et al. 2003; Kikuchi et al. 2003; Nakazaki et al. 2003). It is a deletion derivative of the *Ping* element lacking two open reading frames required for transposition (Naito et al. 2009). *mPing* is an active element that has reached to large copy numbers in some rice lines. For instance, the copy number of *mPing* is several times higher in Gimbozu compared to its progenitor line Aikoku and related landraces (Naito et al. 2014). In addition to knocking out gene functions through interruption of coding regions, stress-responsive *cis*-elements of *mPing* alter expression of adjacent genes (Naito et al. 2009). To study the efficiency of *mPing* in insertional mutagenesis, *mPing* and other genes involved in its transposition were transformed into soybean (Hancock et al. 2011). This study showed that *mPing* produces heritable insertions in soybean over multiple generations. Transposition is developmentally regulated with escalated transposition events observed during late development stages (Hancock et al. 2011). Analysis of the 72 *mPing* transposition sites in soybean revealed that the insertion sites were distributed on 19 of the 20 soybean chromosomes suggesting that it transposes to unlinked genomic regions (Hancock et al. 2011). Only four insertion sites were located in the annotated pericentromeric region indicating its preference for the euchromatic regions (Hancock et al. 2011). Eighty-five percent of *mPing* insertions were located within 5 kb, and 51% were within 2.5 kb of an annotated gene (Hancock et al. 2011). In rice, although *mPing* showed a bias toward gene-rich regions, it was underrepresented in coding sequences, perhaps due to its target site preference for AT-rich regions (Naito et al. 2014). In soybean, *mPing* did not depict any exon avoidance like in rice, maybe due to lower G/C content in soybean exons as compared to that in rice.

12.5 *Tnt1* Retrotransposon Mutagenesis in Soybean

Retrotransposons provide certain advantages over type II transposons and have been used for insertional mutagenesis in several plants (Kumar and Hirochika 2001). The *Tnt1* retrotransposon identified from tobacco has been successfully used in plants including *Medicago truncatula* (d'Erfurth et al. 2003), Arabidopsis (Courtial et al. 2001), lettuce (Mazier et al. 2007), and soybean (Cui et al. 2013) for tagging genes. In soybean, DNA gel blot analysis of *Tnt1* in 27 independent transformants revealed that the number of insertions ranged from four to nineteen and were detected in all 20 chromosomes (Cui et al. 2013). Fiber-FISH analysis showed that some of the insertion sites contained single copy *Tnt1* insertions, while others contained multiple copy tandem inserts. Tissue culture treatments induced transposition of the element increasing the number of insertions (Cui et al. 2013).

For a successful mutagenesis experiment in a large genome-like soybean, where genes constitute only a small proportion of the total DNA, it is imperative to have transposons that have a preference for gene-rich regions. Analysis of the *Tnt1* insertion sites in soybean revealed that it preferentially transposes to gene-containing regions (Cui et al. 2013). Of the 99 *Tnt1* sites analyzed, 62% were localized to annotated genes of all 20 chromosomes. As *Tnt1* has a preference for gene-containing regions, it is a promising transposon for the functional characterization of the protein-coding genes in soybean (Cui et al. 2013).

Stability and heritability of *Tnt1* was tested by growing progenies from plants harboring *Tnt1*. The locations of *Tnt1* among the progenies were stable, and the progenies showed Mendelian inheritance for *Tnt1* (Cui et al. 2013). Although *Tnt1* transcripts were detected in the vegetative tissues, there was no indication of transposition during development (Cui et al. 2013). Both cotyledonary node and somatic embryogenesis approaches induced activation of *Tnt1* among the stable transgenic soybean lines carrying the element (Cui et al. 2013). Thus, through tissue culture of a few transformation events, we can generate a large population of *Tnt1*-induced mutants.

12.6 *Tgm9*, An Endogenous Active Transposable Element in Soybean

Application of a well-characterized heterologous transposon in inducing mutation in a target species requires development of a large number of independent transgenic lines with the element. In *Ac/Ds* system, transposition is mainly to linked regions. This system therefore requires a large collection of transgenic lines with transposons distributed uniformly throughout the genome. *Tnt1* requires tissue culture, which may activate endogenous retrotransposons and/or cause epigenomic changes resulting in heritable somaclonal variation. Due to reduced efficiency of transposable elements in the heterologous systems and expenses associated with this approach, the search to identify endogenous active transposons in various plant species continues.

Several endogenous transposable elements have been identified and characterized in soybean (Rhodes and Vodkin 1988; Zabala and Vodkin 2008, 2014; Xu et al. 2010). Of these, *Tgm9* is the only known active endogenous transposable element that has been used in gene-tagging experiments. It is a 20,548 bp CACTA-type element that was isolated from the second intron of the *dihyroflavonol-4-reductase 2* (*DFR2*) gene of the *W4* locus (Xu et al. 2010). The *W4* locus controls pigment formation in flowers and hypocotyls, and the *w4-m* allele shows variegated flowers due to altered pigment accumulation (Groose et al. 1988; Xu et al. 2010). When the element is excised from *DFR2*, the petals regain the wild-type purple color. Somatic transposition from petals results in variegated flowers. The *w4-m* plants produce germinal revertants that carry only purple flowers. In those lines, the excised element transposes into new genetic loci causing heritable mutations; and therefore, the line is called mutable (*w4-m*). The *w4-m* line registered as T322 is unstable and undergoes germinal reversion at a high frequency of ~6% per generation (Groose et al. 1990).

12.7 The Mutable Line T322 Contains a Single Active Copy of *Tgm9*

For a successful transposon tagging experiment, it is important to know the number of active transposon copies in the genome. Two probes, designed from the 5′ end and 3′ end of *Tgm9,* were compared with the T322 genome sequence to determine the *Tgm9* insertion sites in the T322 genome (Sandhu et al. 2017). A total of 6 and 18 insertion sites were detected by the 5′-end and the 3′-end probes, respectively. Two copies of the element including one in the *DFR2* gene on chromosome 17 were detected by both probes. The second copy was found in heterozygous condition and was not detected among mutants previously generated from T322. Therefore, it represents a recent transposition event. Elements detected only by one probe presumably represent truncated copies of *Tgm9*.

12.8 Forward Genetics: Identification of Genes by Studying *Tgm9*-Induced Soybean Mutants

Tgm9 tagging can be used as a tool to conduct forward genetics for isolating soybean genes by using characterized mutant phenotypes generated from insertion of *Tgm9*. Several mutants such as male-sterile, female-fertile (MSFF) mutant T359 (*ms9*) (Palmer and Horner 2000) and male-sterile, female-sterile (MSFS) mutants (Kato and Palmer 2003; Palmer et al. 2008a; Raval et al. 2013; Baumbach et al. 2016), partial female-sterile mutants {T364 (*Fsp2*), T365 (*Fsp3*), and T367 (*Fsp5*)} (Kato and Palmer 2004), necrotic root mutants (Palmer et al. 2008b), and chlorophyll-deficient mutants {T323 (*y20, Ames 2*), T325 (*y20, Ames 4*), and T346 (*y20, Ames 17*)} (Palmer et al. 1989) were identified by studying germinal revertants that carry only purple flowers.

The insertion sites of *Tgm9* among some of the mutants were determined by applying a genome walking approach (Sandhu et al. 2017). For *st8*, a MSFS mutant, *Tgm9* insertion site was located in the 10th exon of *Glyma.16G072300* that codes for a MER3 DNA helicase involved in crossing over (Baumbach et al. 2016). The *Tgm9* insertion site in the *ms9* mutant was located in the first intron of *Glyma.03G152300* that has no functional annotation (Table 12.1). The *Tgm9* insertion sites in *Fsp2*, *Fsp3*, and *Fsp5* were found in Intron 7 of *Glyma.06G174200*, Exon 2 of *Glyma.08G359000,* and the 272 bp upstream of *Glyma.18G169500, respectively.* Glyma.06G174200 *encodes a haloacid dehalogenase-like hydrolase, and* Glyma.08G 359000 *and* Glyma.18G169500 *both encode embryo-specific protein 3, (ATS3) (Table 12.1). All three chlorophyll-deficient mutants were due to insertion of* Tgm9 *in the first intron of* Glyma.12G19520 *that is predicted to encode a lactate/malate dehydrogenase (Table 12.1).*

The genetic linkage map positions of the mutant genes were used to confirm *Tgm9* insertion sites in these mutants. The location of *Tgm9* insertion sites among the studied mutants agreed with the genetic map positions of the mutant genes; for example, the *st8* gene was flanked by two microsatellite markers in a ~62 kb region on Chromosome *16* that contains the *MER3* gene tagged by *Tgm9* (Raval et al. 2013; Baumbach et al. 2016). The genetic location of *ms9* and the physical location of *Tgm9* insertion site in the *ms9* mutant are on Chromosome *3* (Cervantes-Martinez et al. 2007). *Tgm9* in the *Fsp2* mutant was found in a gene mapped to a region to which the *Fsp2* locus was mapped (Kato and Palmer 2004) (Table 12.1). Similarly, *Tgm9* insertion sites in *Fsp3* and *Fsp5* mutants were localized to Chromosomes *8* and *18*, to which the respective mutant genes were previously mapped (Kato and Palmer 2004). The *Tgm9* insertion sites in the three chlorophyll-deficient mutants (T323, T325, and T346) were localized to Chromosome *12*, ~1.9 Mb from Satt253, to which the three *y20 Mdh1-n* mutations were mapped (Kato and Palmer 2004).

Table 12.1 Forward genetics approach showing identification of *Tgm9* insertion sites in known soybean mutants identified from the progenies of mutable line T322 (*w4-m*)

T #	Mutant	Mutant phenotype	Gene involved	Upstream/exon/intron/downstream	Annotation
T323	*y20 (Ames 2) Mdh1-n (Ames 2)*	Yellow-green leaves, malate dehydrogenase 1 null	*Glyma.12G159300*	1st Intron	Lactate/malate dehydrogenase, NAD binding domain
T325	*y20 (Ames 4) Mdh 1-n (Ames 4)*	Yellow-green leaves, malate dehydrogenase 1 null	*Glyma.12G159300*	1st Intron	Lactate/malate dehydrogenase, NAD binding domain
T346	*y20 (Ames 17) Mdh1-n (Ames 19)*	Yellow-green leaves, malate dehydrogenase null	*Glyma.12G159300*	1st Intron	Lactate/malate dehydrogenase, NAD binding domain
T359	*ms9*	Male sterile, female fertile	*Glyma.03G152300*	1st intron	No Functional annotation
T364	*Fsp2*	Female partial sterile	*Glyma.06G174200*	7th intron	Haloacid dehalogenase-like hydrolase
T365	*Fsp3*	Female partial sterile	*Glyma.08G359000*	2nd exon	Embryo-specific protein 3, (ATS3)
T367	*Fsp5*	Female partial sterile	*Glyma.18G169500*	272 bp upstream	Embryo-specific protein 3, (ATS3)
ASR-10-181	*St8*	Male sterile, female sterile	*Glyma.16G072300*	10th intron	MER3 DNA helicase

12.9 Reverse Genetics: Identification of *Tgm9* Insertion Sites Among Germinal Revertants

Following excision of *Tgm9* from *DFR2,* it inserts in new genetic locations resulting in insertion mutations (Xu et al. 2010). We observed that sometimes both copies of *Tgm9* excised from *DFR2*. Therefore, at most two mutations in a single mutant plant can be expected. The element can be used to generate hundreds of thousands mutations. A high-throughput sequencing of the *Tgm9* insertion sites of the population can then identify mutations in a target gene for its functional analyses. This approach is known as reverse genetics.

Seeds of individual T322 plants are harvested, and progenies of each plant are grown in a row. Seeds from a large number of mutable plants with variegated flowers are harvested to identify additional germinal revertants. Progeny rows are grown from individual mutable plants. With the germinal reversion frequency of 6%, we should get about six plants with purple flowers in a progeny row of 100 plants. Plants with only purple flowers are tagged, and leaf tissues are harvested for determining *Tgm9* insertion sites (Fig. 12.1).

Recently, we studied *Tgm9* insertion sites among 124 germinal revertant plants with purple flowers (Sandhu et al. 2017). *Tgm9* transposed multiple times to six genomic regions suggesting its preference for transposition into some genomic regions. About 16.2% of the insertion sites were present in exons, 9.5% in introns, and 2.9%

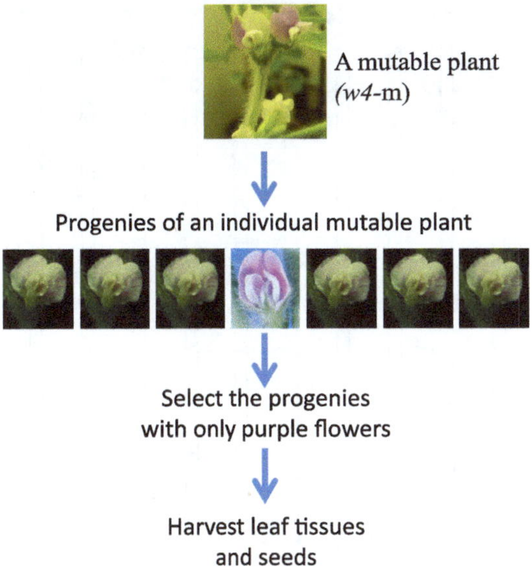

Fig. 12.1 Experimental plan for obtaining germinal revertants with mutations in novel genetic loci. Approximately 150 progenies of an individual mutable plant are grown in 15-ft-long plots, and a single germinal revertant with only purple flowers tagged for collecting leaf tissues for determining *Tgm9* insertion sites and harvesting seeds at maturity

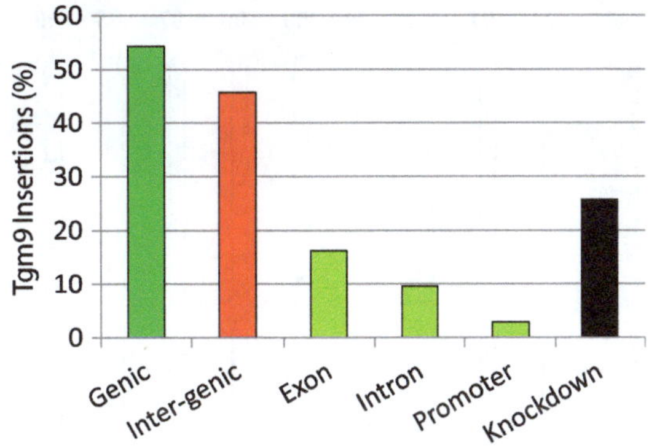

Fig. 12.2 Percent distribution of *Tgm9* insertion sites across the genomes of 105 unique novel soybean mutants. Note that ~28.6% contain insertions in exons, introns or the promoter region and are expected to exhibit complete loss of function (knockout mutations). Other mutations in the genic regions may result in reduced function and are classified as knockdown mutants. Here we consider a genic region to contain a predicted transcript and 2-kb sequences at both 5′- and 3′-ends of the transcript. An inter-genic region spans in between two genic regions

within promoter (300 bp upstream of the transcription initiation site) sequences (Fig. 12.2) (Sandhu et al. 2017). These mutations are most likely knockout mutants suitable for studying gene function.

Tgm9 moves from its original location in the *DFR2* locus on Chromosome 17 (Xu et al. 2010) to all 20 soybean chromosomes (Fig. 12.3). The number of insertion sites varied from one (Chromosomes *5*, *16*, and *19*) to sixteen (Chromosome *9*) per chromosome (Sandhu et al. 2017). Insertion frequency was higher toward the telomeres. Of the 105 unique insertion sites, 85 (81%) were located in distal 50% of the chromosome arms, and only 20 (19%) were present in proximal half of chromosome arms (Sandhu et al. 2017). *Tgm9* showed preference toward euchromatic regions as only 22.9% of the insertion sites were present in the heterochromatic pericentromeric repeat regions, which constitute 52.9% of the soybean genome (Fig. 12.3). Preferential transposition of *Tgm9* to the gene-rich regions is particularly important in tagging genes in a large genome-like soybean, where only 16% of the genome contains genes (Sandhu et al. 2017). From the above, it is apparent that the *Tgm9* system is suitable for gene identification through forward and reverse genetics studies.

12.10 Transposon Tagging and the Future of Soybean Research

As the soybean genome has been sequenced, the research priority for soybean now is to determine functions of these genes, especially the ones that control the agronomic traits such as stress tolerance, yield, and quality traits. In the last decade, due to addition of a variety of genomic resources in soybean, transposon tagging has emerged as an attractive alternative in determining soybean gene functions. The availability of molecularly characterized transposon-induced soybean mutants will expedite both basic and applied research in this economically important crop. A high-throughput sequencing approach can be used to determine the *Tgm9* insertion sites among hundreds of thousands mutants. The

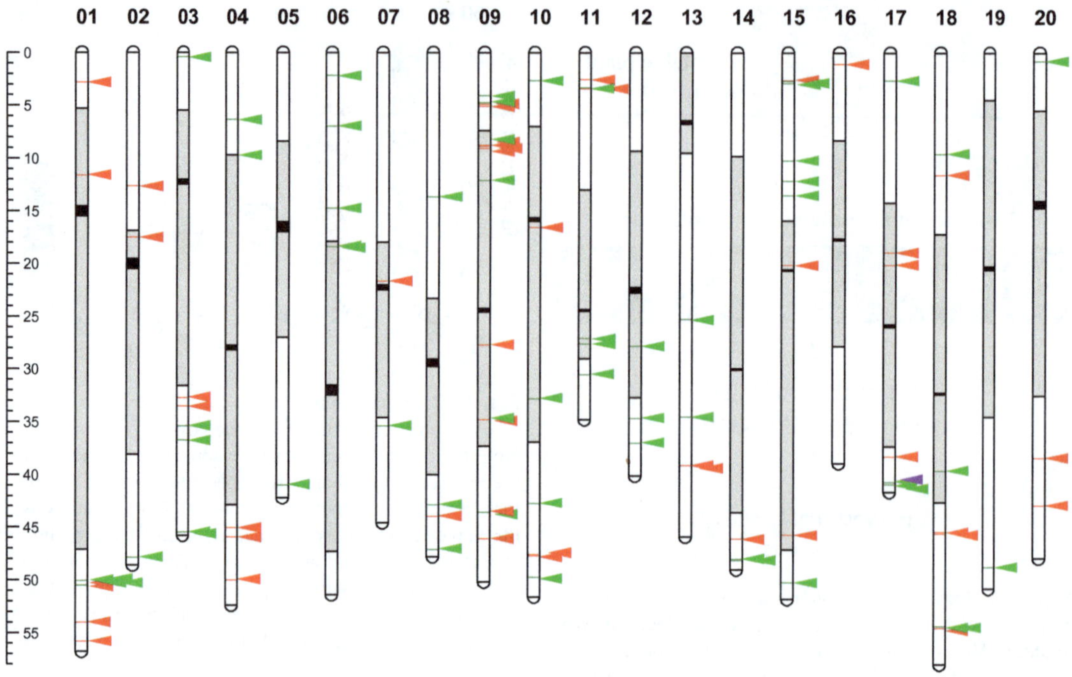

Fig. 12.3 Chromosomal distribution of *Tgm9* insertion sites among mutants. *Green arrows* represent locations of *Tgm9* in genic regions, and *red arrows* represent locations of *Tgm9* in inter-genic regions. A purple arrow represents the original location of *Tgm9* in *DFR2* in the *w4-m* (T322) line. Centromeres (*black rectangles*) and heterochromatic regions (*gray areas*) are shown on individual chromosomes. *Scale* is represented in million base pairs (Mb) of DNA

display of those *Tgm9* insertion sites in the SoyBase genome browser will facilitate identification of mutants for most single copy soybean genes. Soybean researchers will be able to use this resource to identify suitable mutants for genetically improved soybean for disease and pest resistance, quality and quantity of oil, proteins and seed compositions, and also for adaptation to adverse climatic conditions. Soybean biologists will be able to investigate novel traits unique to soybean as well as conduct translational genomics based on knowledge gained in other model plant species such as Arabidopsis, rice, and maize. The soybean research community will use this resource to gain a better understanding of the molecular basis of physiological processes in soybean. The fundamental knowledge and genetic variation created using *Tgm9* will provide means to improve yield and quality in soybean for securing a sustainable supply of this nutritionally important crop in the twenty-first century.

Acknowledgements We are thankful to United Soybean Board and University of Wisconsin at Stevens Point for grant support to conduct the *Tgm9* research. We are thankful to David Grant for kindly reviewing the chapter and for his suggestions.

References

Bancroft I, Jones JD, Dean C (1993) Heterologous transposon tagging of the *DRL1* locus in *Arabidopsis*. Plant Cell 5:631–638

Baumbach J, Pudake RN, Johnson C, Kleinhans K, Ollhoff A et al (2016) Transposon tagging of a male-sterility, female-sterility gene, *St8*, revealed that the meiotic MER3 DNA helicase activity is essential for fertility in soybean. PLoS ONE 11:e0150482. doi:10.1371/journal.pone.0150482

Cervantes-Martinez I, Xu M, Zhang L, Huang Z, Kato KK et al (2007) Molecular mapping of male-sterility loci *ms2* and *ms9* in soybean. Crop Sci 47:374–379

Chuck G, Robbins T, Nijjar C, Ralston E, Courtney-Gutterson N et al (1993) Tagging and cloning of a petunia flower color gene with the maize transposable element activator. Plant Cell 5:371–378. doi:10.1105/tpc.5.4.371

Courtial B, Feuerbach F, Eberhard S, Rohmer L, Chiapello H et al (2001) *Tnt*1 transposition events are induced by in vitro transformation of *Arabidopsis thaliana*, and transposed copies integrate into genes. Mol Genet Genom 265:32–42

Cui Y, Barampuram S, Stacey MG, Hancock CN, Findley S et al (2013) *Tnt1* retrotransposon mutagenesis: a tool for soybean functional genomics. Plant Physiol 161:36–47. doi:10.1104/pp.112.205369

d'Erfurth I, Cosson V, Eschstruth A, Lucas H, Kondorosi A et al (2003) Efficient transposition of the *Tnt1* tobacco retrotransposon in the model legume *Medicago truncatula*. Plant J 34:95–106

Dooner HK, Belachew A (1989) Transposition pattern of the maize element *Ac* from the *Bz-M2(Ac)* allele. Genetics 122:447–457

Fedoroff N, Wessler S, Shure M (1983) Isolation of the transposable maize controlling elements *Ac* and *Ds*. Cell 35:235–242

Groose RW, Weigelt HD, Palmer RG (1988) Somatic analysis of an unstable mutation for anthocyanin pigmentation in soybean. J Hered 79:263–267

Groose RW, Schulte SM, Palmer RG (1990) Germinal reversion of an unstable mutation for anthocyanin pigmentation in soybean. Theor Appl Genet 79:161–167

Hancock CN, Zhang F, Floyd K, Richardson AO, Lafayette P et al (2011) The rice miniature inverted repeat transposable element *mPing* is an effective insertional mutagen in soybean. Plant Physiol 157:552–562. doi:10.1104/pp.111.181206

James DW Jr, Lim E, Keller J, Plooy I, Ralston E et al (1995) Directed tagging of the *Arabidopsis FATTY ACID ELONGATION1* (*FAE1*) gene with the maize transposon activator. Plant Cell 7:309–319. doi:10.1105/tpc.7.3.309

Jiang N, Bao Z, Zhang X, Hirochika H, Eddy SR et al (2003) An active DNA transposon family in rice. Nature 421:163–167. doi:10.1038/nature01214

Jones DA, Thomas CM, Hammond-Kosack KE, Balint-Kurti PJ, Jones JD (1994) Isolation of the tomato *Cf-9* gene for resistance to *Cladosporium fulvum* by transposon tagging. Science 266:789–793

Kato KK, Palmer RG (2003) Molecular mapping of the male-sterile, female-sterile mutant gene (*st8*) in soybean. J Hered 94:425–428

Kato KK, Palmer RG (2004) Molecular mapping of four ovule lethal mutants in soybean. Theor Appl Genet 108:577–585

Kikuchi K, Terauchi K, Wada M, Hirano HY (2003) The plant MITE *mPing* is mobilized in anther culture. Nature 421:167–170. doi:10.1038/nature01218

Kumar A, Hirochika H (2001) Applications of retrotransposons as genetic tools in plant biology. Trends Plant Sci 6:127–134

Lawrence GJ, Finnegan EJ, Ayliffe MA, Ellis JG (1995) The *L6* gene for flax rust resistance is related to the Arabidopsis bacterial resistance gene *RPS2* and the tobacco viral resistance gene *N*. Plant Cell 7:1195–1206. doi:10.1105/tpc.7.8.1195

Lazarow K, Doll ML, Kunze R (2013) Molecular biology of maize *Ac/Ds* elements: an overview. Methods Mol Biol 1057:59–82. doi:10.1007/978-1-62703-568-2_5

Martienssen RA (1998) Functional genomics: probing plant gene function and expression with transposons. Proc Natl Acad Sci USA 95:2021–2026

Masudaa T, Goldsmith PD (2009) World soybean production: area harvested, yield, and long term projections. Int Food Agribus Manag Rev 12:143–162

Mathieu M, Winters EK, Kong F, Wan J, Wang S et al (2009) Establishment of a soybean (*Glycine max* Merr. L) transposon-based mutagenesis repository. Planta 229:279–289. doi:10.1007/s00425-008-0827-9

Mazier M, Botton E, Flamain F, Bouchet JP, Courtial B et al (2007) Successful gene tagging in lettuce using the *Tnt1* retrotransposon from tobacco. Plant Physiol 144:18–31. doi:10.1104/pp.106.090365

McClintock B (1956) Controlling elements and the gene. Cold Spring Harb Symp Quant Biol 221:197–216

Naito K, Zhang F, Tsukiyama T, Saito H, Hancock CN et al (2009) Unexpected consequences of a sudden and massive transposon amplification on rice gene expression. Nature 461:1130–1134. doi:10.1038/nature08479

Naito K, Monden Y, Yasuda K, Saito H, Okumoto Y (2014) *mPing*: the bursting transposon. Breed Sci 64:109–114. doi:10.1270/jsbbs.64.109

Nakazaki T, Okumoto Y, Horibata A, Yamahira S, Teraishi M et al (2003) Mobilization of a transposon in the rice genome. Nature 421:170–172. doi:10.1038/nature01219

Palmer RG, Horner HT (2000) Genetics and cytology of a genic male-sterile, female-sterile mutant from a transposon-containing soybean population. J Hered 91:378–383. doi:10.1093/jhered/91.5.378

Palmer RG, Hedges BR, Benavente RS, Groose RW (1989) *w4*-mutable line in soybean. Dev Genet 10:542–551. doi:10.1002/dvg.1020100613

Palmer RG, Sandhu D, Curran K, Bhattacharyya MK (2008a) Molecular mapping of 36 soybean male-sterile, female-sterile mutants. Theor Appl Genet 117:711–719. doi:10.1007/s00122-008-0812-5

Palmer RG, Zhang L, Huang Z, Xu M (2008b) Allelism and molecular mapping of soybean necrotic root mutants. Genome 51:243–250. doi:10.1139/g08-001

Peterson T (ed) (2013) Plant transposable elements: methods and protocols. Methods in molecular biology. Springer, New York

Raval J, Baumbach J, Ollhoff AR, Pudake RN, Palmer RG et al (2013) A candidate male-fertility female-fertility gene tagged by the soybean endogenous transposon, *Tgm9*. Funct Integr Genom 13:67–73. doi:10.1007/s10142-012-0304-1

Rhodes PR, Vodkin LO (1988) Organization of the *Tgm* family of transposable elements in soybean. Genetics 120:597–604

Sandhu D, Baumbach J, Ghosh J, Johnson C, Cina T et al (2017) The endogenous transposable element *Tgm9* is suitable for generating knockout mutants for functional analyses of soybean genes and genetic improvement in soybean. PLOS ONE (under review)

Singh S, Tan HQ, Singh J (2012) Mutagenesis of barley malting quality QTLs with *Ds* transposons. Funct Integr Genom 12:131–141. doi:10.1007/s10142-011-0258-8

Walbot V (1992) Strategies for mutagenesis and gene cloning using transposon tagging and T-DNA insertional mutagenesis. Annu Rev Plant Physiol Plant Mol Biol 43:49–82

Whitham S, Dinesh-Kumar SP, Choi D, Hehl R, Corr C et al (1994) The product of the tobacco mosaic virus resistance gene *N*: similarity to toll and the interleukin-1 receptor. Cell 78:1101–1115

Xu M, Brar HK, Grosic S, Palmer RG, Bhattacharyya MK (2010) Excision of an active CACTA-Like transposable element from *DFR2* causes variegated flowers in soybean *Glycine max* (L.) Merr. Genetics 184:53–63. doi:10.1534/genetics.109.107904

Zabala G, Vodkin LO (2008) A putative autonomous 20.5 kb-CACTA transposon insertion in an F3'H allele identifies a new CACTA transposon subfamily in *Glycine max*. BMC Plant Biol 8:124. doi:10.1186/1471-2229-8-124

Zabala G, Vodkin LO (2014) Methylation affects transposition and splicing of a large CACTA transposon from a MYB transcription factor regulating anthocyanin synthase genes in soybean seed coats. PLoS ONE 9:e111959. doi:10.1371/journal.pone.0111959

SoyBase: A Comprehensive Database for Soybean Genetic and Genomic Data

13

David Grant and Rex T. Nelson

Abstract

SoyBase, the USDA-ARS soybean genetics and genomics database, provides a comprehensive collection of data, analysis tools, and links to external resources of interest to soybean researchers. The SoyBase home page (https://soybase.org) contains the SoyBase Toolbox which provides quick access to a search of the SoyBase database or the SoyCyc metabolic pathways database, access to the data download page, a genome sequence BLAST tool, and direct links to the genetic and sequence maps. An extensive navigation menu and site description provides facile access to all sections of SoyBase. Comprehensive information for a number of data types is available at SoyBase including the current genetic maps, the soybean reference genome sequence with tracks covering genetic markers, genome organization, gene annotation and expression, and gene knockout mutants. SoyBase includes an extensive RNA-Seq gene atlas and innovative tools for identifying fast neutron-induced mutants. SoyBase is an actively curated database, with new data regularly being incorporated, including additions to the controlled vocabularies (ontologies) for soybean growth, development and phenotypic traits, soybean genes, and quantitative trait loci. New "omics" tools enable sophisticated queries on lists of genes. These features are all accessed using intuitive interfaces and are linked together wherever possible.

D. Grant (✉) · R.T. Nelson
USDA-ARS, Corn Insects and Crop Genetics Research Unit, Iowa State University, Ames, IA 50011, USA
e-mail: david.grant@ars.usda.gov

R.T. Nelson
e-mail: rex.nelson@ars.usda.gov

D. Grant
Department of Agronomy, Iowa State University, Ames, IA 50011, USA

13.1 Introduction

SoyBase is the soybean breeder's and molecular biologist's "toolbox" for discovering genetic traits useful in basic research and elite cultivar development. SoyBase has existed in this role for more than 20 years (Grant et al. 1996, 2010) and is continually updated with significant genetic

H.T. Nguyen and M.K. Bhattacharyya (eds.), *The Soybean Genome*, Compendium of Plant Genomes, DOI 10.1007/978-3-319-64198-0_13

and genomic data. The pace of change has increased rapidly over the past few years due to publication of the Williams 82 reference genome sequence in 2010 which, along with rapidly declining sequencing costs, has allowed many more sequence data sets to be generated. Here, we review the major features of SoyBase, focusing particularly on recently added data and capabilities.

The SoyBase home page allows users to quickly access several commonly used tools and data sets using the SoyBase Toolbox (Fig. 13.1). The home page also announces a few of the most recent updates to SoyBase along with a list of upcoming conferences and other items of general community interest.

Searching at SoyBase is done using a single free-form text entry box in the SoyBase Toolbox. The search term is used to identify all of the data and types of data in SoyBase that are relevant to the search. These results are presented to the user in a comprehensive summary page (Fig. 13.2). Each cell in the summary page table reports the number of items of that type in SoyBase that relate to the search term, and provides a link to those data.

Providing breeders and researchers with an intuitive "toolbox" to make use of complex data requires SoyBase to interact with users in two complementary ways: first, to provide answers to a specific question (e.g., returning the DNA sequence of the *Adh1* gene or the map location of a QTL); and second, to allow facile and intuitive browsing of the database contents through integrated genetic and genomic map viewers or metabolic pathway diagrams. A user can enter SoyBase with as little as a DNA sequence or gene function and after a single search be presented with a concise summary table containing links to everything in SoyBase related to the query. Navigation from this summary and between the various data pages is quick and

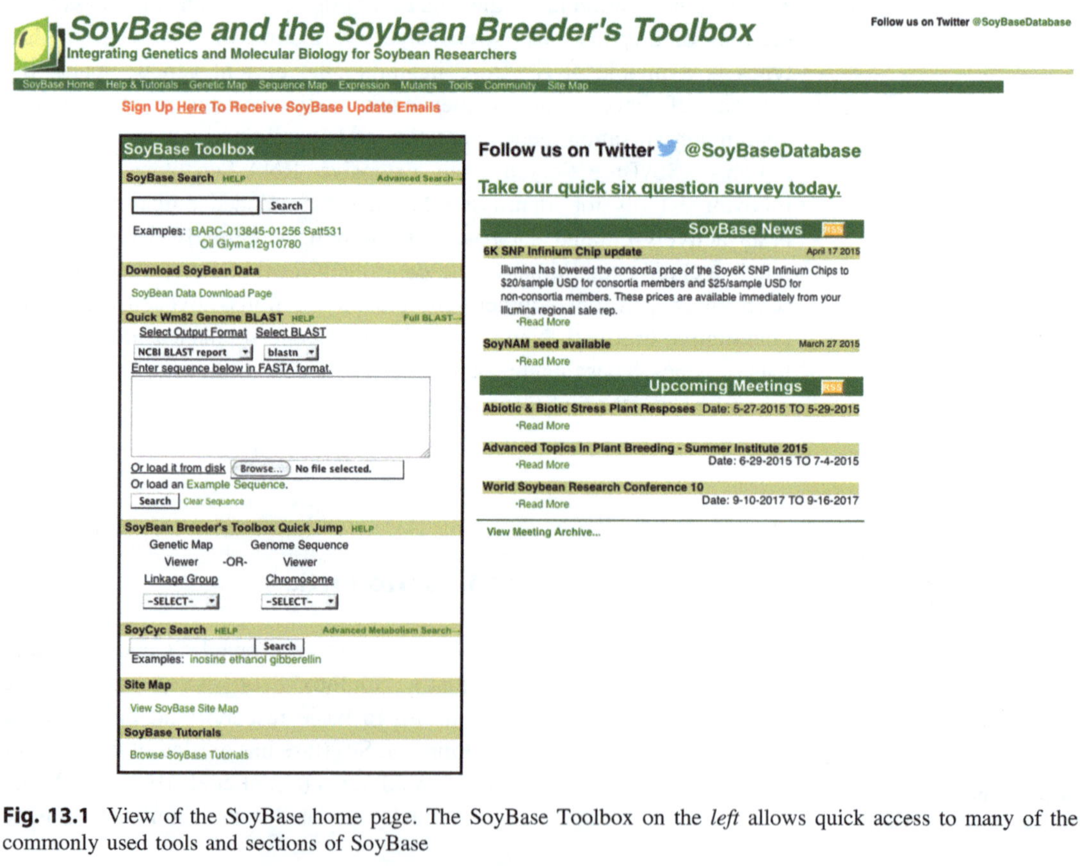

Fig. 13.1 View of the SoyBase home page. The SoyBase Toolbox on the *left* allows quick access to many of the commonly used tools and sections of SoyBase

Search Domain	Query Term	Genetic Maps	Genomic Maps	Details Pages	Annotation	Expression Data
Marker	*Oil*					
QTL	*Oil*	186		188	5	
Gene Call	*Oil*		81	81	5	71
Trait	*Oil*			5	11	
Gene	*Oil*					

Fig. 13.2 Results of a search of SoyBase. *Rows in the table* represent the major data types, while *columns* describe various kinds of data related to those data types. *A number in a cell* refers to the number of records in SoyBase. *Each number* is a hyperlink to a page containing a summary of these data

Sequence Map Resources

- SoyBase Wm82 Genome Browser
- All Potential SSRs Indentified in Wm82
- SoySNP50K Haplotypes and Data
- NSF SoyMap II Project: Comparative sequence maps for G. max vs. 7 wild Glycine species
- Soybean Transposable Elements
  - View TE family relationships and download TE data by structural classifications
  - Visualize TEs in the context of the soybean genetic map
  - Visualize TEs in the context of the soybean genomic sequence
  - Retrieve TE information based on a TE name
  - Download TE information based on proximity to a gene or genomic sequence coordinate
  - Download all TE sequences in FASTA format
  - Download a summary of the TE information as tab delimited text
- Download Sequence Data from SoyBase
  - Convert Wm82.a1.v1.1 Gene Model Names to Wm82.a2.v1 Names
  - Download genome sequence coordinates for selected features
  - Download genome sequence coordinates for selected features by chromosome
  - Download genome or predicted protein sequence for gene calls
  - Download annotations for selected gene calls
  - Download gene model flanking sequence
  - Download gene model 3' and 5' UTR sequences
  - Download SoySNP50K Data
- External Soybean Sequence Resources
  - Phytozome Soybean Sequence Data
  - PlantGDB Soybean Sequence Data
- How to Cite SoyBase
- Submit Your Data to SoyBase
- Contact SoyBase

Fig. 13.3 SoyBase site map. Part of the site map is shown. *Each line* is a hyperlink to the appropriate section of SoyBase

efficient due to the extensive links provided on each page.

Navigation within SoyBase is typically done via the Menu bar, which is available at the top of every page. In addition, a comprehensive site map is provided (Fig. 13.3) which allows a user to explore all of the parts of SoyBase that are relevant to a high-level concept (e.g., sequence map or analysis tools) without requiring prior knowledge as to how this information is organized at SoyBase.

The sections below describe some of the major data types contained in SoyBase and give examples of how they comprise an integrated whole.

13.2 SoyBase Organization and Content

SoyBase is organized into seven sections:

- Genetic maps—the genetic, QTL and GWAS QTL, and BAC-based physical maps;
- Sequence map—the Williams 82 reference genome sequence and extensive annotation including soybean and other legume gene expression, soybean transposable elements, Glycineae diversity data from the SoyMapII project, and SoySNP50K and other SNP data sets;
- Expression—search and visualize soybean gene expression atlases and GEO data sets;
- Mutants—searchable collections of fast neutron and transposable element mutants;
- Tools—numerous tools for searching SoyBase and other data sets, including sequence similarity searches using BLAST against various soybean target data sets, several tools for use with the Affymetrix SoyChip1 expression microarray, tools to analyze and retrieve gene sets based on gene annotations, and SoyCyc, a tool for visualizing soybean genes identified as coding for enzymatic steps in metabolic pathways;
- Community—links to community resources at other sites as well as communications from the community-led Soybean Genetics Executive committee, copies of presentations from the annual Soybean Breeder's Workshop and the Biennial Molecular and Cellular Biology of the Soybean Meeting, links to soybean producer sites and other legume research programs, soybean research reports, and data set from various academic and government sources;
- SoyBase *Tutorials*—an extensive collection of video tutorials devoted to the effective use and searching of the database and using SoyBase analysis tools as well as video content created by the community regarding soybean growth, development, and abiotic and biotic stresses.

13.2.1 Genetic Map

The current soybean composite genetic map contains more than 4000 molecular and gene markers and over 3100 QTLs describing more than 90 traits. Two much denser single nucleotide polymorphism (SNP) maps for crosses used in the reference genome assembly are also available. Bacterial artificial chromosome (BAC)-based physical maps are available for use in moving between the genetic and sequence maps (described in Shoemaker et al. 2008 and Grant et al. 2010).

The genetic and physical maps in SoyBase are shown using CMap, the comparative map viewer component of the Generic Model Organism Project (GMOD, Stein et al. 2002). Individual linkage groups are shown in two panes: a composite genetic map, constructed by the SoyBase staff using genetic markers and over 3100 QTLs from more than 90 published trait association studies, and a consensus genetic map which contains several thousand additional sequence-based genetic markers identified from the genomic sequence and mapped in three reference populations (Hyten et al. 2010) (Fig. 13.4). Placing the cursor over a marker or QTL generates highlighting that shows all positionally related markers and QTLs. Zooming within a linkage group is accomplished using the up- and down-pointing triangles on the map backbones.

Many of the genetic markers have been placed on the genome sequence. Contextual menus are provided in both the genetic and genome viewers which allow easy movement between them, as well as providing links to the extensive summary pages of associated data about the markers and QTLs from the database.

In addition to the more common biparental QTLs, SoyBase also includes QTLs identified in a genome-wide association study (GWAS QTL). These population-level marker-trait associations

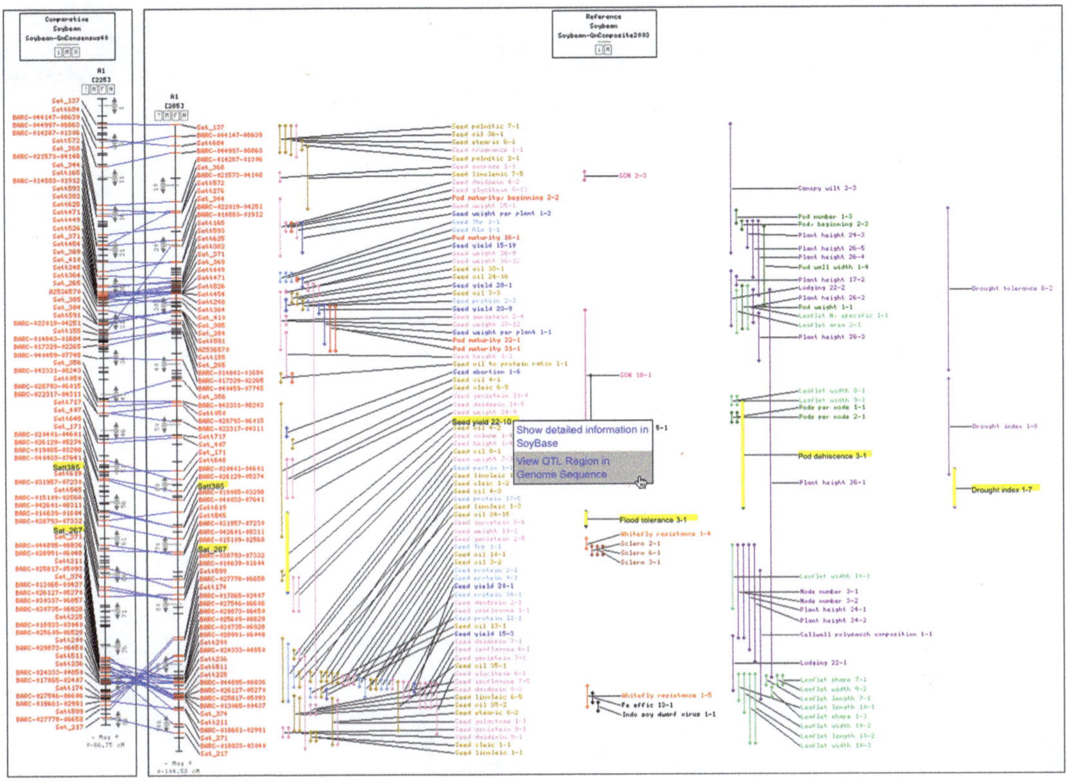

Fig. 13.4 SoyBase genetic map. Default view of Linkage Group A1/Chromosome 5 genetic map. Genetic loci are shown on the map backbone with QTL to the *right*. *Yellow regions* demonstrate how moving the cursor over any map object highlights all related objects. Clicking on a map object brings up a contextual menu of navigation choices. Maps can be zoomed using the *up/down arrows* at each cM indicator

can be shown relative to the biparental QTL in the genetic map viewer (Fig. 13.5a). Since GWAS and biparental QTL are identified using different methods, increased confidence can be attached to locations where the marker-trait association regions are in common. GWAS QTL can also be shown along with the gene calls and other annotation tracks in the sequence map viewer (Fig. 13.5b) to aid in identifying candidate genes for the trait. For ease of comparison, the same color coding and column/track categories are used in the genetic and sequence viewers.

13.2.2 Sequence Map

The soybean cultivar Williams 82 reference genomic sequence (Schmutz et al. 2010) and the associated annotations are contained in SoyBase. The SoyBase sequence viewer uses the GMOD genome browser GBrowse (Stein et al. 2002). Because the reference genome assembly has been improved over time, SoyBase makes available sequence viewers for each of the releases: the initial release Glyma.Wm82.a1 (Gmax1.01), the re-annotation of the Glyma. Wm82.a1 (Gmax1.1), and Glyma.Wm82.a2 (Gmax2.0). Any new releases of the Wm82 or other reference genomes will be incorporated into SoyBase as needed. The most recent reference genome assembly is always the default shown at SoyBase.

The Wm82.a1 and Wm82.a2 genome assemblies differ due to many small inversions and a few large translocations. These differences preclude developing an algorithm to show correspondences between the two coordinate systems.

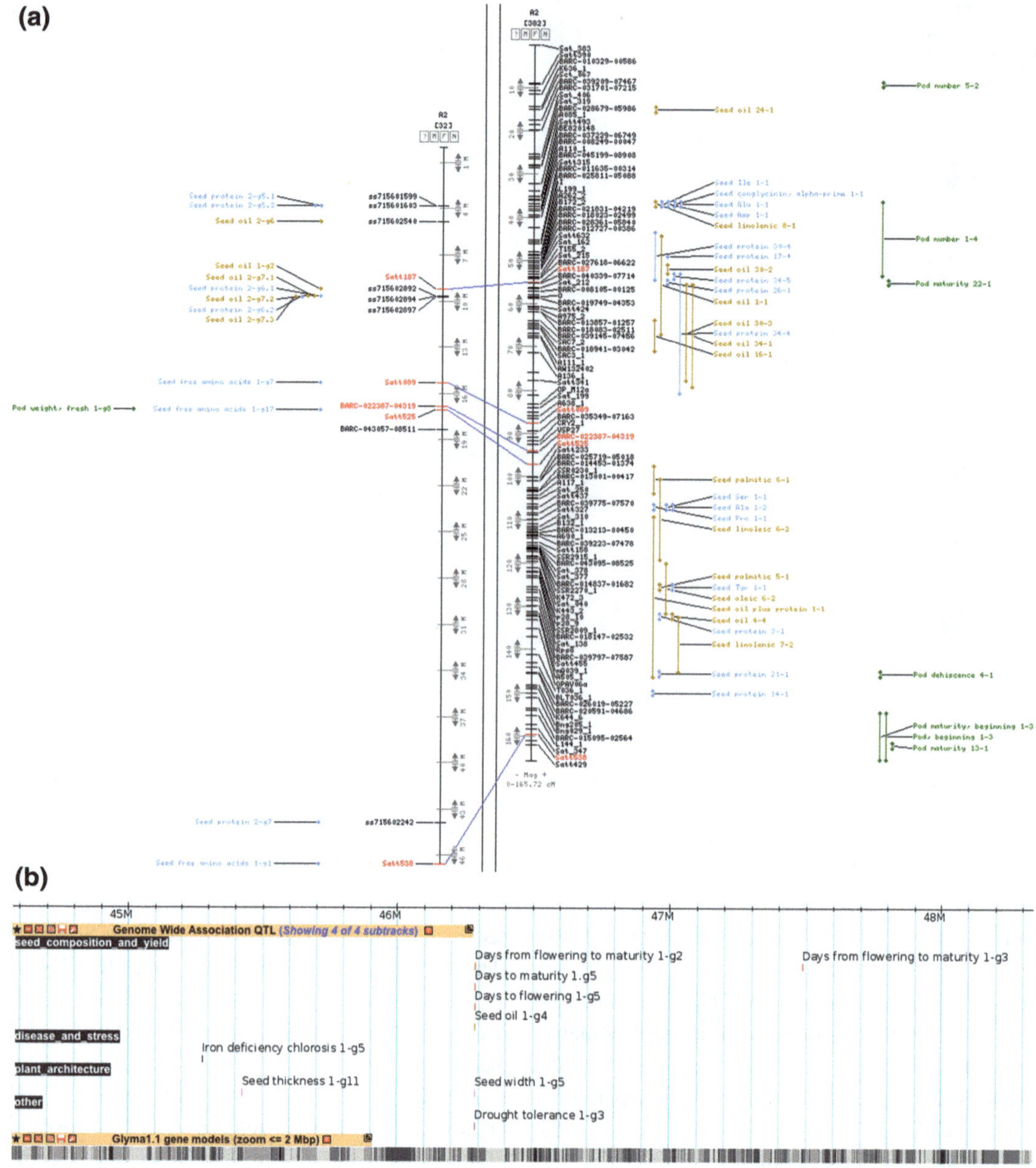

Fig. 13.5 GWAS QTL in SoyBase. **a** Selected biparental QTL and GWAS QTL shown on the SoyBase genetic map. The same color scheme and column positions are used for both displays. **b** Selected GWAS QTL shown in the SoyBase genome browser. Subtrack names correspond to the QTL and GWAS QTL columns in the genetic map display

For this reason, although many analyses have been done using both assemblies, some annotation tracks may appear in only one of the sequence viewers.

Over 50 annotation tracks can be displayed, including the BACs used to construct the physical map, sequence-based genetic markers with contextual menu links to the genetic map, the computationally computed gene calls and associated annotations, RNA-seq expression profiles, and the insertions and/or deletions (indels) associated with the fast neutron mutant

Table 13.1 Annotation tracks in the SoyBase genome browser

Annotation Class	Examples of available tracks
General soybean features	Centromere, pericentromere
Genes	JGI gene models, NCBI RefSeq gene models, published genes
Genome structure	Soybean intragenomic synteny blocks, synteny with other legumes
Mutants	Fast neutron-induced mutants
BAC clones	BACs used in soybean physical map
Expressed sequence tags	Soybean ESTs, ESTs from many other legumes
Expression—microarray	Affymetrix SoyChip1
Markers	Soybean genetic markers, markers from many other legumes
RNA-Seq	Soybean gene atlas, seed development atlas, others
Repetitive sequences	Soybean transposable elements, telomeric repeats
Germplasm SNP haplotypes	USDA soybean germplasm collection analyzed with the SoySNP50 K Illumina Infinium Chip

populations. Table 13.1 shows the annotation tracks available for the Wm82 reference genome.

Figure 13.6a shows the default genome browser view for chromosome Gm01. Of note is the track that shows synteny (corresponding genomic regions) within soybean. Since the soybean genome is a mosaic resulting from at least two whole genome duplications (Schlueter et al. 2004; Pfeil et al. 2005), this track, along with the ones that show synteny between soybean and other genome sequences, is particularly useful when evaluating regions that contain similar QTLs or genes (Fig. 13.6b). At the left of each annotation track, a series of red glyphs can be used to customize both the entire view (i.e., collapse or delete specific tracks) and a single track (change glyph color and shape, etc.). Of particular interest, the rightmost red "question mark" glyph provides important information about the track and its contents. Tracks can be rearranged by a simple click&drag on the track title. Although the annotation tracks can be chosen in any combination, several example displays composed of conceptually related tracks are provided (see Example Views near the top of Fig. 13.6a).

In addition to tracks related to genome structure, a number of gene expression tracks are also available. Figure 13.7 shows an RNA-seq-based gene atlas annotation track. The relative number of RNA-seq reads that mapped to each region is indicated by intensity in the heat map of gene expression. This view of gene expression can be used to identify possible instances of alternate splicing or changes in gene expression during development or externally applied treatments.

13.2.3 Gene Expression

Gene expression data from three multi-tissue and -developmental stage RNA-seq atlases are available at SoyBase, along with approximately 50 gene expression data sets covering different biotic and abiotic treatments. A user can submit a list of gene names and receive a table showing their expression values (Fig. 13.8) based on the results from Gene Atlas data sets.

13.2.4 Mutant Populations

The mutant populations section of SoyBase contains extensive data for the fast neutron-generated (FN) mutant populations (Bolon et al. 2011, 2014). As described below, the primary entry page allows mutants to be selected by mutant name, gene model(s) or loci covered by an indel, indel name, mutant line pedigree, visual or biochemical phenotype,

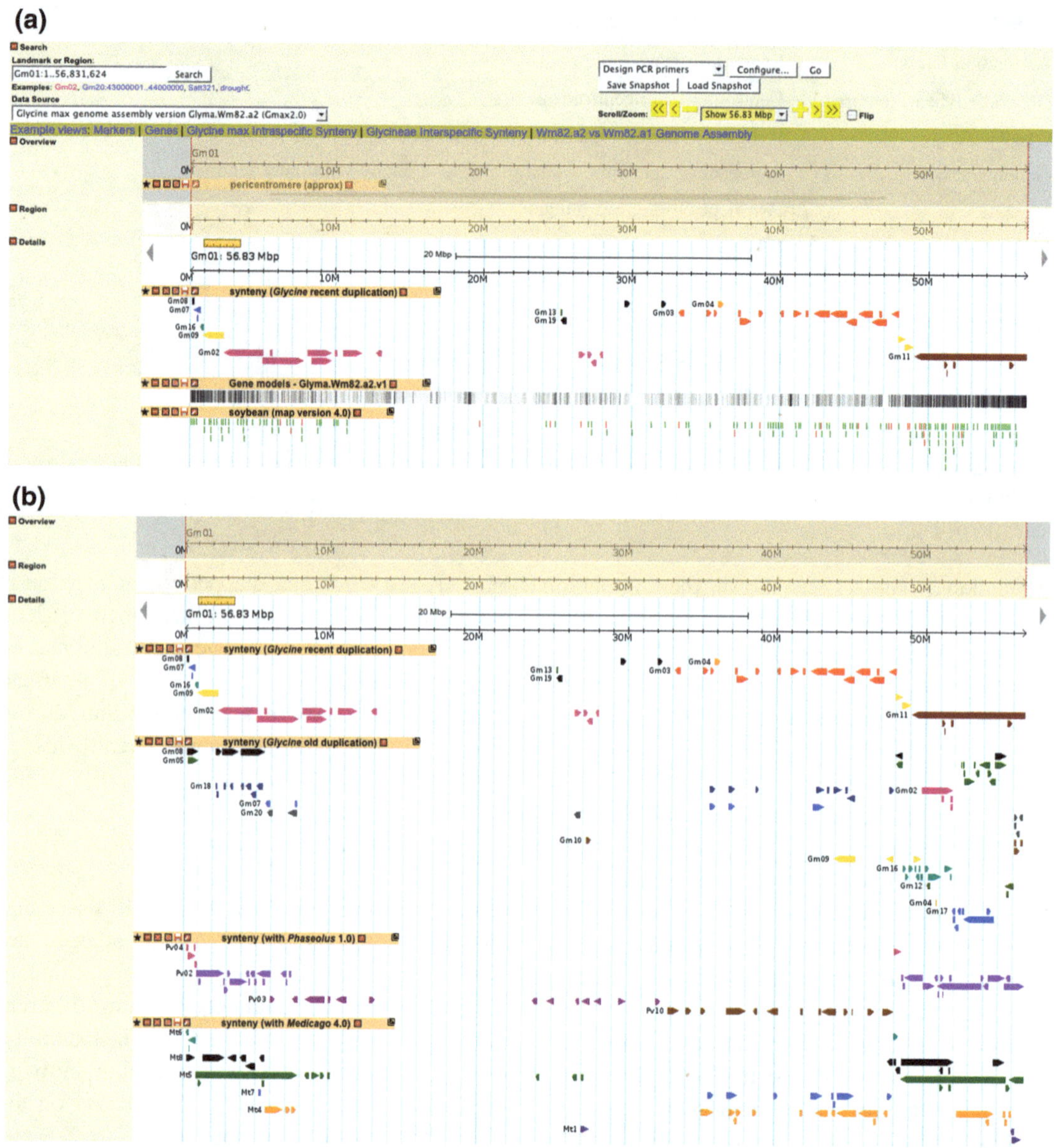

Fig. 13.6 SoyBase genome browser. **a** The default view in the SoyBase genome browser. Because of the large number of gene visible in this whole chromosome view, the gene tracks are shown as a density heat map. Zooming into less than 2 MB shows the individual genes. Additional annotation tracks can be chosen using the "Select Tracks" tab at the *top* of the screen. Examples of some common annotation track sets are provided in the "Example Views" line just above the genome browser. **b** SoyBase genome browser with tracks showing intraspecific synteny for the two most recent whole genome duplication events in *Glycine max*, and interspecific synteny between *G. max*, *Phaseolus vulgaris,* and *Medicago truncatula*

or sequence similarity of an indel to a query sequence (Fig. 13.9).

Biochemical and Phenotypic Data: A subset of the mutants has been characterized for a number of seed and whole plant biochemical and phenotype traits. Several tools are provided that allow the identification of mutants using either phenotypic descriptors or values for biochemical traits.

Genome Changes: For some mutants, comparative genome hybridization (CGH) has been

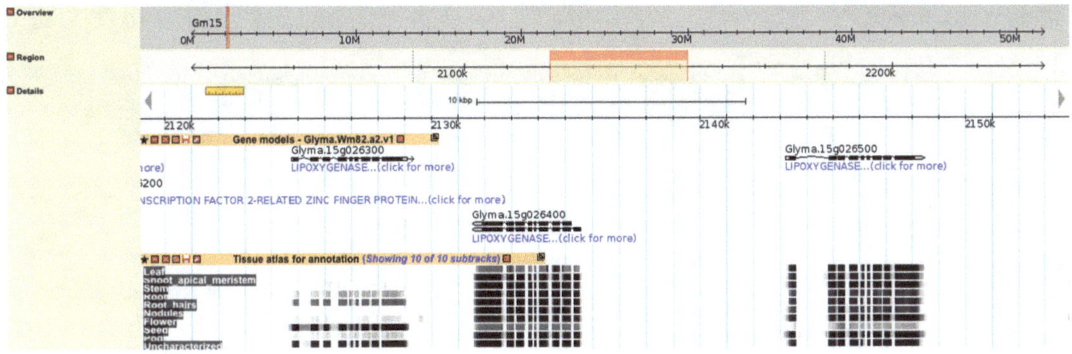

Fig. 13.7 SoyBase genome browser zoomed into the region around 3 segmentally duplicated lipoxygenase genes. Annotation tracks showing the RNA-seq transcript density for 10 tissue and developmental time points are shown

Download List	young_leaf	flower	one cm pod	pod shell 10DAF	pod shell 14DAF	seed 10DAF	seed 14DAF	seed 21DAF	seed 25DAF	seed 28DAF	seed 35DAF	seed 42DAF	root	nodule
Glyma09g41910	13	20	14	14	18	6	18	6	17	7	23	11	26	13
Glyma09g41920	3	1	2	0	1	4	6	3	10	3	12	4	17	0
Glyma09g41930	0	0	0	0	1	0	0	0	1	0	0	0	86	25
Glyma09g41940	0	10	78	13	0	288	235	84	168	104	84	12	0	278
Glyma09g41950	24	33	23	14	14	11	13	3	10	12	21	5	21	10
Glyma09g41960	64	2	5	2	11	4	0	0	2	0	2	10	2	0

Fig. 13.8 Gene expression values for genes. A table showing gene expression for a user-submitted set of genes. The RNA-seq read count values are from the Soybean Gene Atlas described by Severin et al (2010)

used to determine the number and end points of the indels induced by fast neutron treatment. These data can be displayed as a track in the SoyBase Genome Browser to visualize the genomic region and genes involved in the deletion event. A searchable database of all fast neutron deleted genes is also available to allow users to quickly identify mutants that span a gene of interest.

Figure 13.10 gives an example of how the genome sequence can be used to determine a short list of candidate genes for a trait. The fast neutron mutant FN0172306.01.M3 has an early maturing phenotype. One of the genes covered by a deletion in this mutant, Glyma09g38241, is annotated as being related to circadian rhythm and thus might be of interest in this context. This deletion (FN0172306.01.M3_1) and the duplication FN0186954.01.M3_2 in an independent mutant overlap in this chromosomal region and could be used to investigate how a deletion or duplication of the candidate gene could affect the phenotype.

Image Search: Photographs of mutant plants and plant parts are presented in a user-friendly browser. Images chosen for further analysis can be accumulated in a "shopping cart" during browsing (Fig. 13.11). All of the images in this list can be viewed together on a single page to facilitate comparisons between mutants with similar phenotypes.

13.2.5 Tools and Data Resources

SoyBase provides a number of search and analysis tools including:

- Sequence Similarity Searches

User supplied sequences can be compared to the Williams 82 reference genome and gene models,

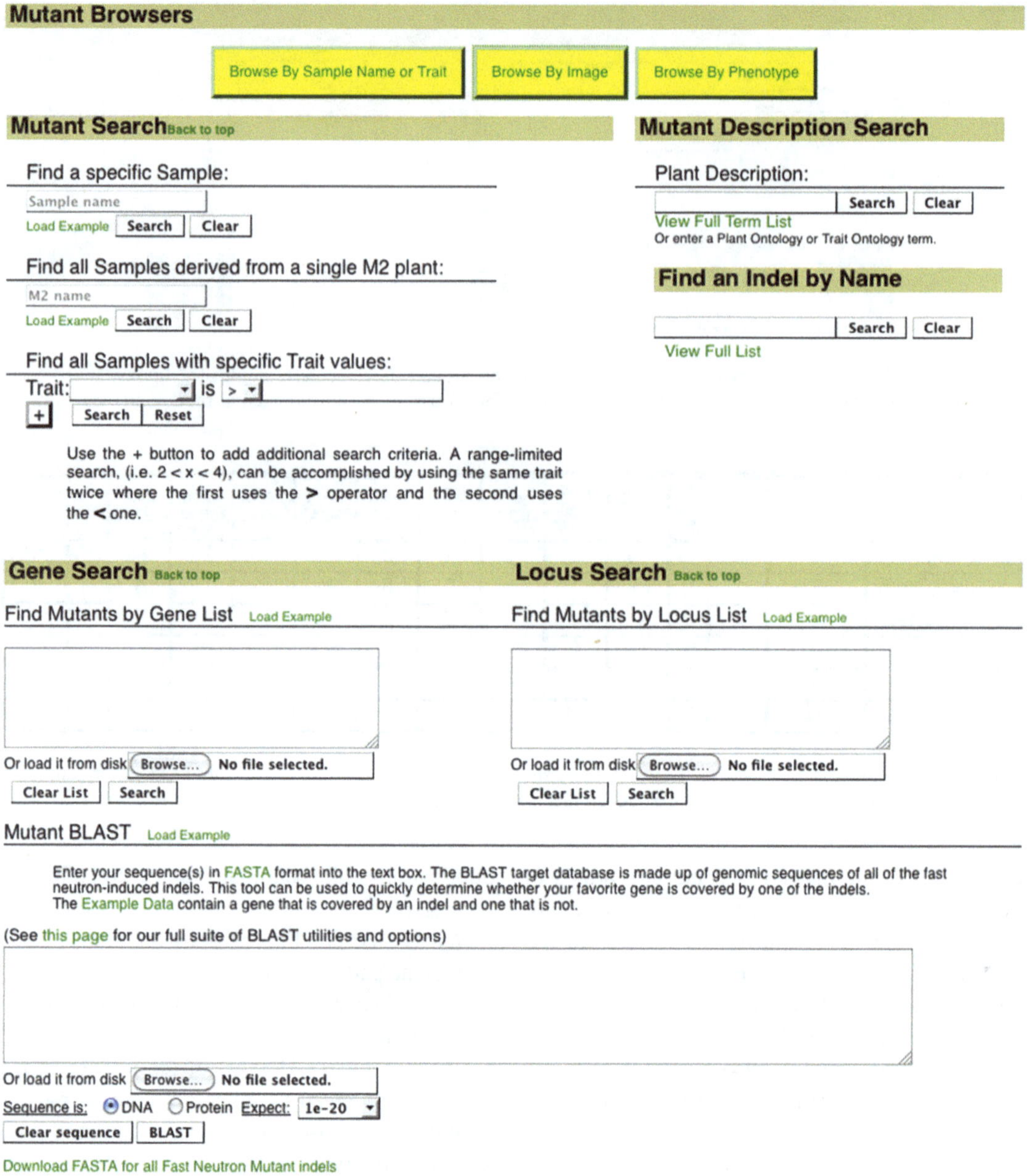

Fig. 13.9 Fast neutron-induced mutants page. All of the searching and reporting tools for the fast neutron-induced mutants can be accessed from this page

the NCBI Ref-Seq gene models, the NCBI soybean expressed sequence tag (EST) collection, the mitochondrial and chloroplast genomes and gene calls, fast neutron and transposon affected genes, soybean transposable elements identified by Du et al. (2010), and legume gene family consensus sequences using BLAST (Altschul et al. 1997) or BLAT (Kent 2002. BLAST results are shown in the familiar WebBLAST text format (Fig. 13.12a) or on a graphical representation of all 20 soybean chromosomes (Fig. 13.12b). BLAT similarity searches against

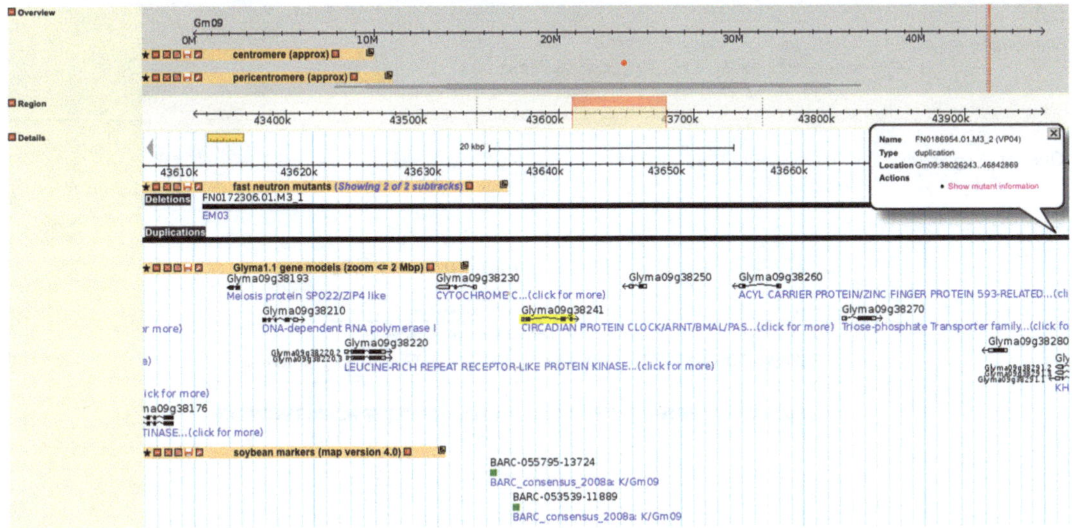

Fig. 13.10 Fast neutron-induced mutants in SoyBase genome browser. A region of chromosome 9 (Gm09) shown in the SoyBase genome browser. Tracks for fast neutron-induced mutants, gene models, and genetic markers are shown

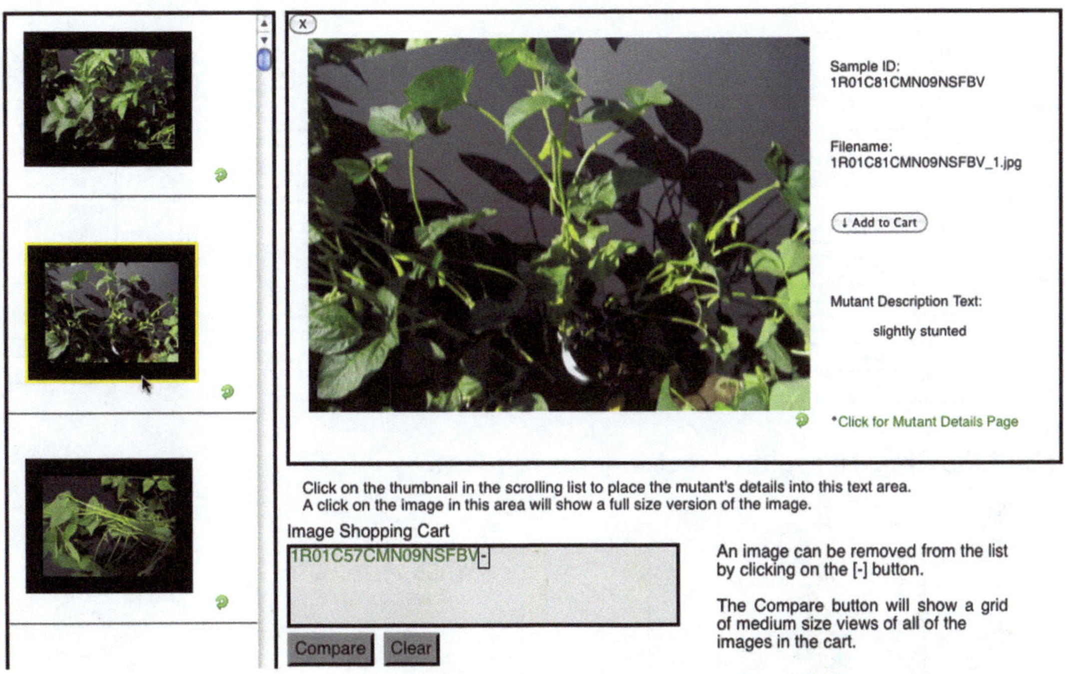

Fig. 13.11 Fast neutron-induced whole plant images. This tool allows image-based browsing of visual mutant phenotypes. Clicking on a thumbnail in the *left pane* shows a larger version along with additional information about the mutant. Mutant images can be saved to a "shopping cart" for further analysis

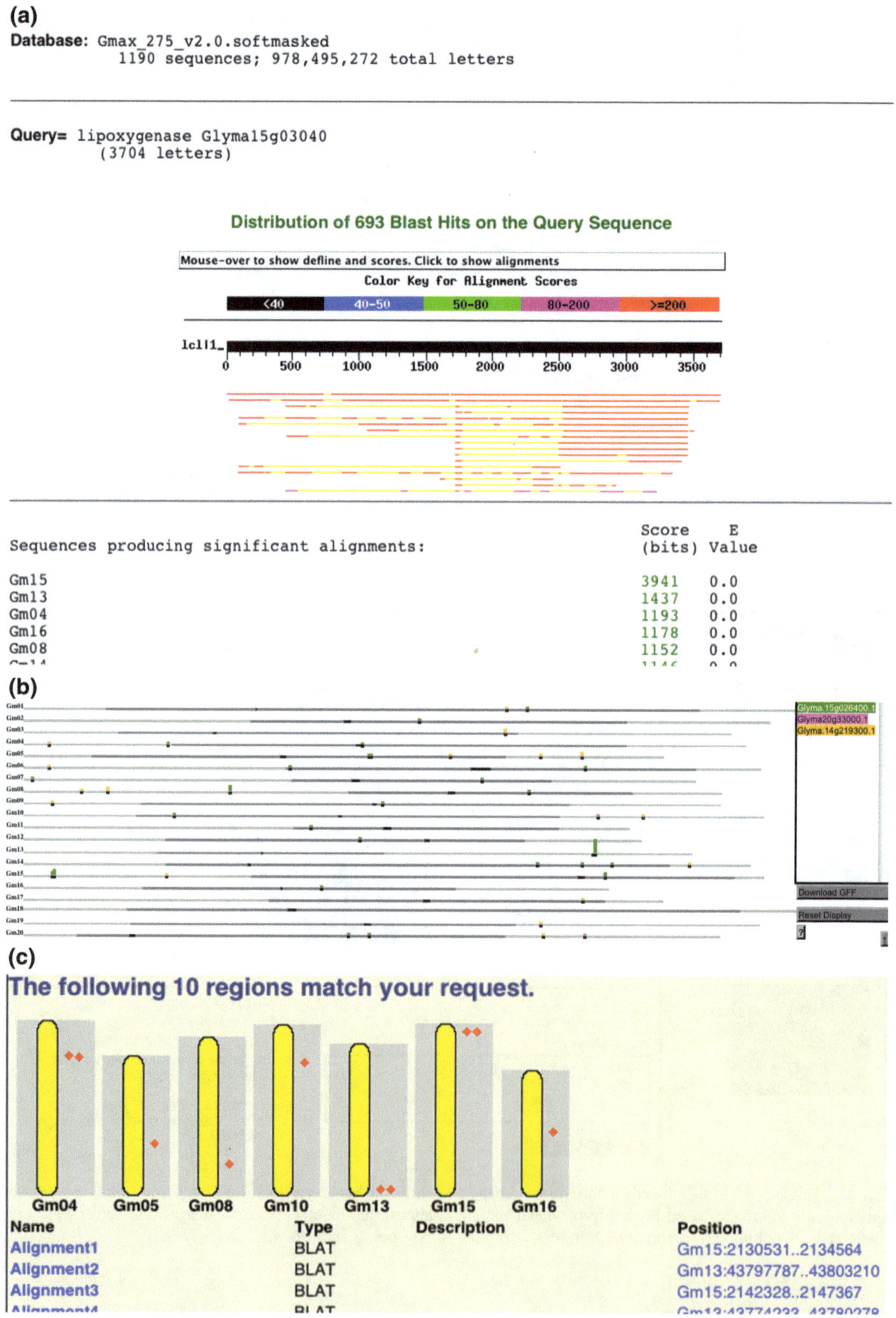
(a)
Database: Gmax_275_v2.0.softmasked
1190 sequences; 978,495,272 total letters
Query= lipoxygenase Glyma15g03040
(3704 letters)
Distribution of 693 Blast Hits on the Query Sequence
Mouse-over to show defline and scores. Click to show alignments
Color Key for Alignment Scores
<40
40-50
50-80
80-200
>=200
lcl|1_
0
500
1000
1500
2000
2500
3000
3500
Sequences producing significant alignments:
Score (bits)
E Value
Gm15 3941 0.0
Gm13 1437 0.0
Gm04 1193 0.0
Gm16 1178 0.0
Gm08 1152 0.0
(b)
Glyma.15g026400.1
Glyma20g33000.1
Glyma.14g219300.1
Download GFF
Reset Display
(c)
The following 10 regions match your request.
Gm04
Gm05
Gm08
Gm10
Gm13
Gm15
Gm16
Name
Type
Description
Position
Alignment1 BLAT Gm15:2130531..2134564
Alignment2 BLAT Gm13:43797787..43803210
Alignment3 BLAT Gm15:2142328..2147367

◀ **Fig. 13.12** Sequence similarity searches. **a** Result of BLAST similarity search using the NCBI WebBLAST. Links are provided from individual HSPs to the SoyBase genome browser. **b** Results of BLAST similarity searches using three sequences in a multiple FASTA file. *Colors* indicate positions of HSPs for the different sequences. **c** Result of BLAT similarity search using the SoyBase genome browser

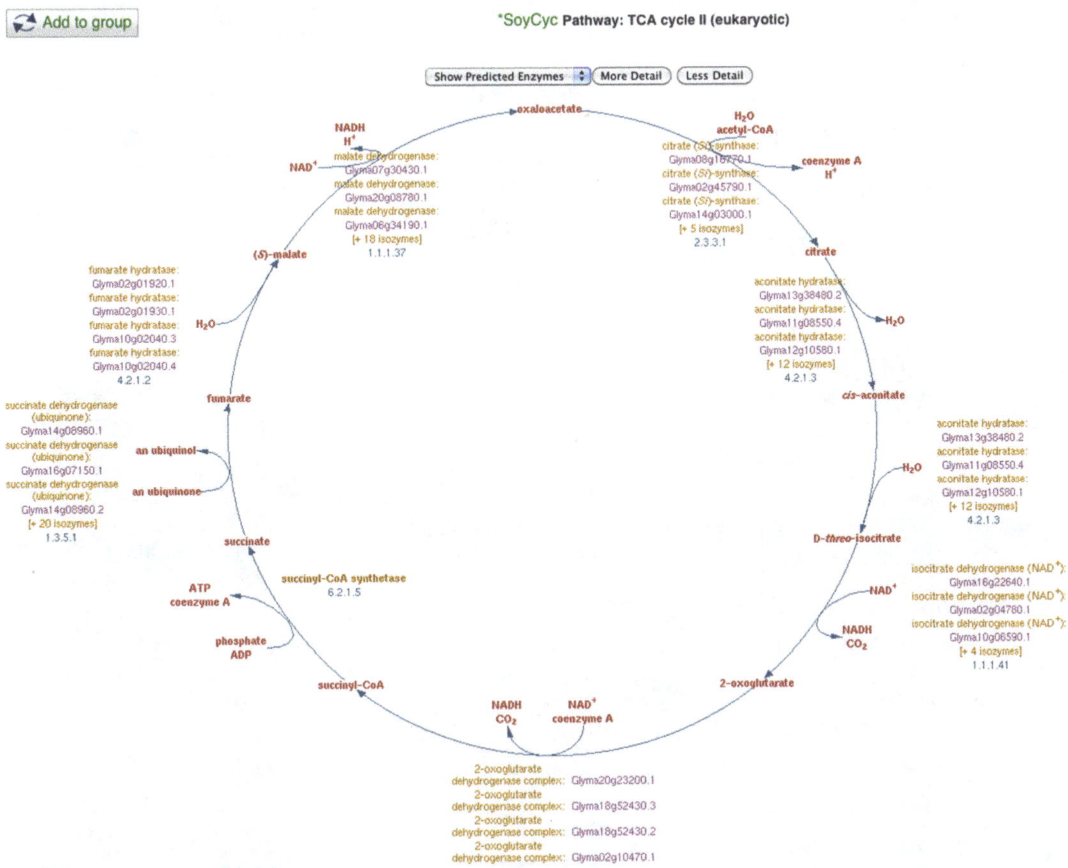

Fig. 13.13 SoyCyc Metabolism Browser. The TCA Cycle II (eukaryotic) pathway is shown. The gene models associated with each enzymatic step are given. Clicking on a name opens a page with additional information about that gene

the whole genome sequence can be performed directly in the genome sequence browser, with the results returned in that viewer (Fig. 13.12c).

- SoyCyc Metabolic Pathways Browser

Genes involved in metabolic reactions of soybean were identified computationally by the Plant Metabolic Network (Zhang et al. 2010). These data have been incorporated into SoyCyc and are displayed at SoyBase using PathwayTools (Karp et al. 2010). SoyCyc allows users to explore the metabolic pathways graphically and to quickly collect details about pathways, reactions, enzymes, and reactants (Fig. 13.13). In addition, results from expression or other experiments can easily be "painted" onto the various reactions in the pathways, thus allowing a global view of soybean metabolism under varying treatments.

- Gene List Analysis Tools

Many "Omics" analyses produce lists of soybean gene models that are putatively associated with a particular treatment or tissue type. SoyBase

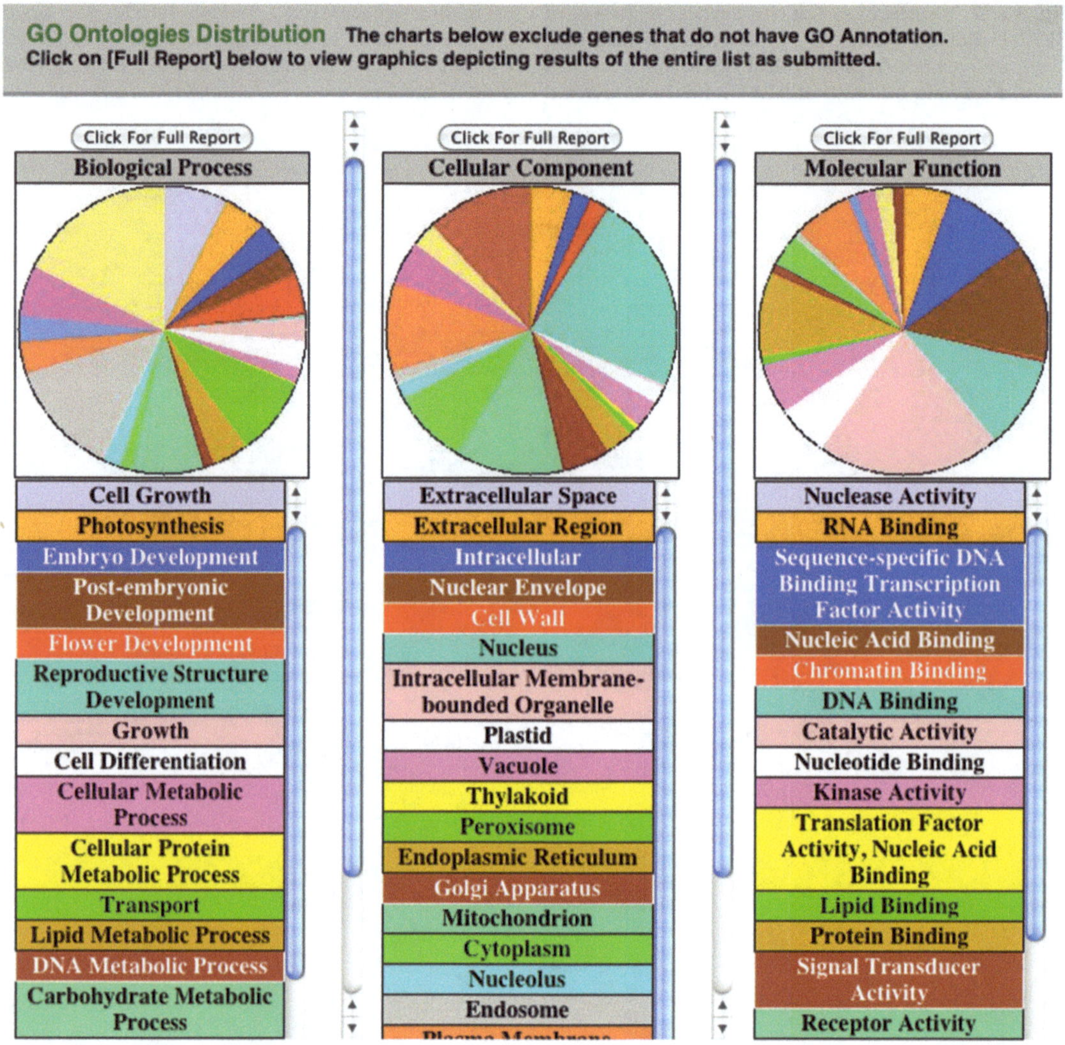

Fig. 13.14 GO Term Enrichment Tool. Pie charts showing the annotation distribution for a user-submitted list of genes for the three GO aspects

provides several tools that allow these lists to be analyzed with the results presented in the context of other data. The entry page for these tools allows users to upload a list of soybean gene models for analysis. Two of the frequently accessed tools are:

- Gene Model Annotation

Annotations for the gene models can be retrieved. Annotations available include GO, Uniref100, TAIR10, PFAM, PANTHER, and KOG.

- GO Term Enrichment Tool

The list of gene models can be analyzed for their GO annotations in all three GO aspects (molecular function, cellular component, and biological process). This tool calculates any enrichment of GO annotations (i.e., over- or underrepresentation) based on the relative frequencies of the GO terms associated with the genes. Results are shown in the familiar pie chart (Fig. 13.14). The approximate location of each of the gene models is also presented in the SoyBase whole genome

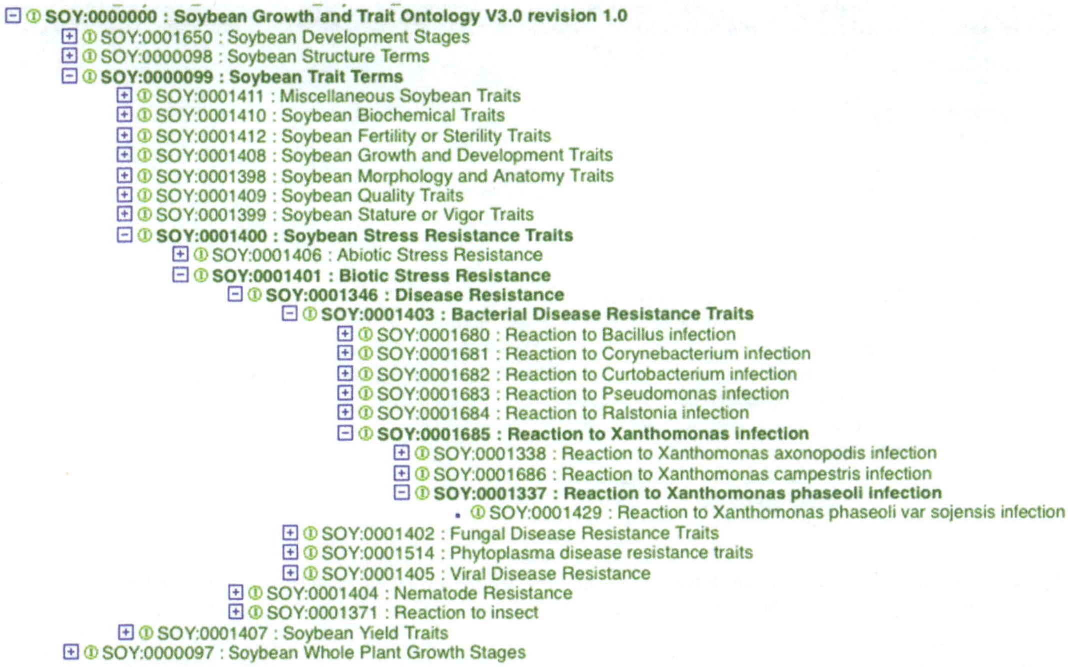

Fig. 13.15 SoyBase ontologies and controlled vocabulary. An expanded view of part of the SoyBase ontology browsing tool. This view allows a user to identify the controlled vocabulary term used at SoyBase for a trait of interest and to quickly find all records related to that trait

viewer, where all 20 soybean chromosomes can be simultaneously examined.

- Ontologies and Controlled Vocabulary Search Terms

SoyBase provides a comprehensive collection of controlled vocabularies to describe soybean whole plant growth, plant structure, phenotypic traits, and development (Fig. 13.15). These ontologies provide a common vocabulary that facilitates interspecific searches. Development of these ontologies is coordinated with Plant Ontology (PO, Avraham et al. 2008), Gramene Plant Trait Ontology (TO, Jaiswal et al. 2002), Gene Ontology (GO, Ashburner et al. 2000), and the BBCH scale (Munger et al. 1997). The soybean trait ontology has also been integrated with the Integrated Breeding Platform's Soybean Trait Dictionary (Ribault 2015).

- Data Download

A number of tools are provided that allow downloading selected subsets of most SoyBase data types including gene calls from the Wm82.a1 and Wm82.a2 genome assemblies, a tool to identify the correspondences between the gene call names from these genome assemblies, flanking sequences of the gene calls, annotation of gene calls, genetic and genomic marker coordinates, QTL and gene lists, and diversity data from analysis of the soybean germplasm (Fig. 13.16).

13.2.6 Community Resources

A large number of data resources are provided as links to external sites. These include other soybean-centric Web sites and laboratories, and

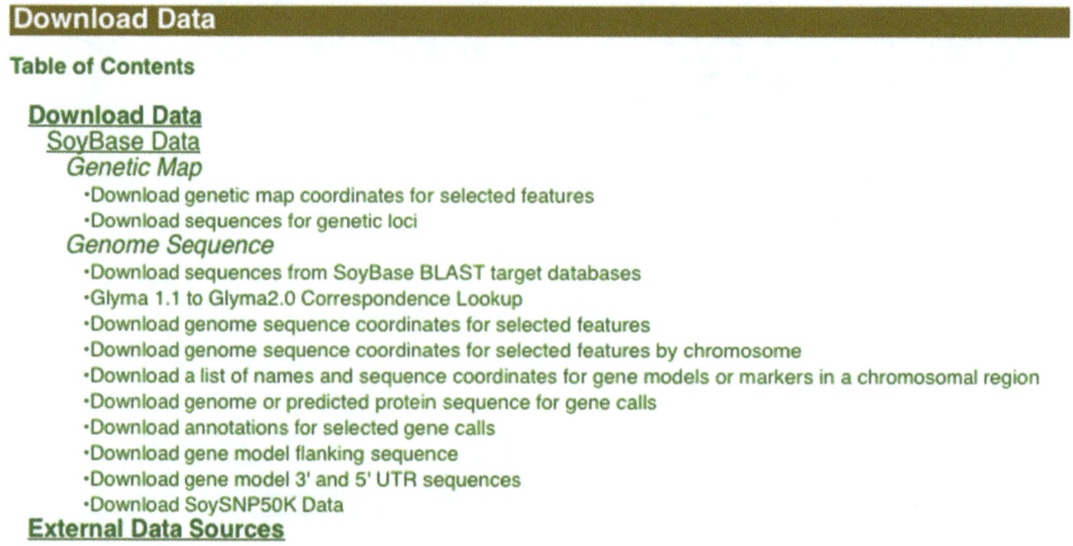

Download Data

Table of Contents

Download Data
SoyBase Data
Genetic Map
•Download genetic map coordinates for selected features
•Download sequences for genetic loci
Genome Sequence
•Download sequences from SoyBase BLAST target databases
•Glyma 1.1 to Glyma2.0 Correspondence Lookup
•Download genome sequence coordinates for selected features
•Download genome sequence coordinates for selected features by chromosome
•Download a list of names and sequence coordinates for gene models or markers in a chromosomal region
•Download genome or predicted protein sequence for gene calls
•Download annotations for selected gene calls
•Download gene model flanking sequence
•Download gene model 3' and 5' UTR sequences
•Download SoySNP50K Data
External Data Sources

Fig. 13.16 Downloading data from SoyBase. A number of preformatted searches are provided for commonly requested subsets of the data in SoyBase. The figure shows some of the genetic and sequence searches available

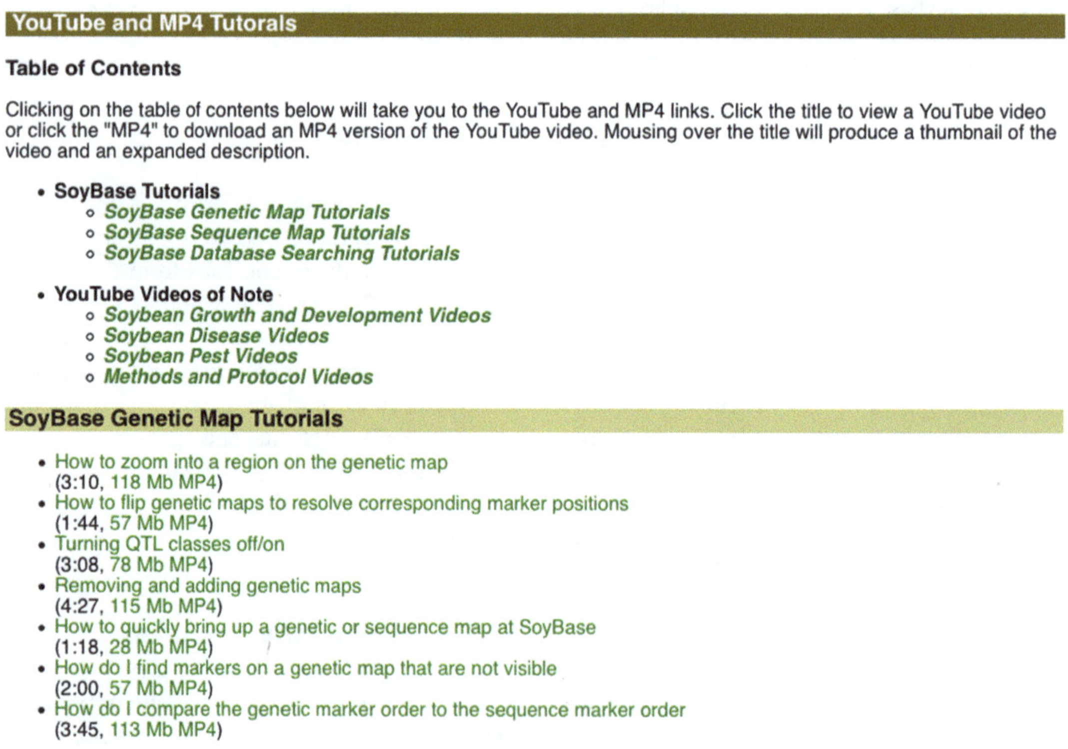

YouTube and MP4 Tutorals

Table of Contents

Clicking on the table of contents below will take you to the YouTube and MP4 links. Click the title to view a YouTube video or click the "MP4" to download an MP4 version of the YouTube video. Mousing over the title will produce a thumbnail of the video and an expanded description.

- SoyBase Tutorials
  - SoyBase Genetic Map Tutorials
  - SoyBase Sequence Map Tutorials
  - SoyBase Database Searching Tutorials
- YouTube Videos of Note
  - Soybean Growth and Development Videos
  - Soybean Disease Videos
  - Soybean Pest Videos
  - Methods and Protocol Videos

SoyBase Genetic Map Tutorials

- How to zoom into a region on the genetic map (3:10, 118 Mb MP4)
- How to flip genetic maps to resolve corresponding marker positions (1:44, 57 Mb MP4)
- Turning QTL classes off/on (3:08, 78 Mb MP4)
- Removing and adding genetic maps (4:27, 115 Mb MP4)
- How to quickly bring up a genetic or sequence map at SoyBase (1:18, 28 Mb MP4)
- How do I find markers on a genetic map that are not visible (2:00, 57 Mb MP4)
- How do I compare the genetic marker order to the sequence marker order (3:45, 113 Mb MP4)

Fig. 13.17 SoyBase video tutorials. A number of short video tutorials covering specific tools and use cases are provided. The figure gives examples of some of the tutorials covering the SoyBase genetic maps

an extensive collection of USDA sites about soybean breeding and production.

13.2.7 Tutorials

SoyBase maintains an expanding collection of short video tutorials describing the use of the various data analysis and retrieval tools available at the database (Fig. 13.17). These tutorials cover aspects of database use such as general searching, BLAST analysis of sequence using over 30 specialized soybean sequence databases, and manipulating and interpreting the genomic map viewer (GBrowse) and genetic map viewer (CMap) interfaces. These tutorials also educate the user in how to customize the browsing experience using both viewers. Other tutorials focus on specific tool use such as the SoyCyc soybean metabolic database and using the Mutant collection search tool. SoyBase also contains links to tutorials and videos created by the community that cover more general soybean-related subjects such as growth and development, identifying insect pests and scouting fields for insect damage and other biotic and abiotic stresses. Protocol videos are also listed that demonstrate laboratory or soybean husbandry techniques.

13.3 Future Plans

SoyBase is continuously updated to incorporate newly published or released data. New data types are added to SoyBase as they become available, and are internally linked to the existing data when possible. Some data types that will soon be added to SoyBase include:

- resequenced genomes for approximately 200 historically important soybean cultivars;
- mutant populations generated by insertional mutagenesis using several transposable elements.

13.4 Availability

All data in SoyBase are freely available without restrictions. As described above, a number of commonly requested data collections are available from dedicated download pages (e.g., genetic markers and QTL in a region or linkage group, or gene models in a region or chromosome, flanking sequences for gene models). Requests for subsets of the data in SoyBase not currently provided are welcomed and should be sent to the SoyBase Curator (soybase@soybase.org).

Direct data submission is actively solicited. Questions, comments, and requests can be sent using the Contact Us links in SoyBase or directly by e-mail to the SoyBase Curator (soybase@soybase.org).

Acknowledgements The SoyBase Development Group includes David Grant, Rex T. Nelson, Kevin Feeley, Nathan Weeks, Jacqueline D. Campbell, Victoria Carollo Blake, Wei Huang, and Steven B. Cannon.Data in SoyBase were generously provided by many cooperators or were collected from the literature by SoyBase staff. Funding: This work was supported by the US Department of Agriculture, Agricultural Research Service (USDA-ARS) CRIS Project 3625-21000-062-00D.Conflict of interest: No conflicts of interest declared.

Disclaimer This digital data set publication was prepared by an agency of the US Government. Neither the US Government nor any agency thereof, nor any of their employees, makes any warranty, expressed or implied, or assumes any legal liability or responsibility for the accuracy, completeness, or misuse of the data, or for damage, transmission of viruses, or computer contamination through the distribution of these data sets or for the usefulness of any information, apparatus, product, or process disclosed in this report, or represents that its use would not infringe privately owned rights. Reference therein to any specific commercial product, process, or service by trade name, trademark, manufacturer, or otherwise does not necessarily constitute or imply its endorsement, recommendation, or favoring by the US Government or any agency thereof. Any views and opinions of authors expressed herein do not necessarily state or reflect those.

The US Department of Agriculture (USDA) prohibits discrimination in all its programs and activities on the basis of race, color, national origin, age, disability, and where applicable, sex, marital status, familial status, parental status, religion, sexual orientation, genetic information, political beliefs, reprisal, or because all or a part of an individual's income is derived from any public assistance program. (Not all prohibited bases apply to all programs.) Persons with disabilities who require alternative means for communication of program information (Braille, large print, audiotape, etc.) should contact USDA's TARGET Center at (202) 720-2600 (voice and TDD). To file a complaint of discrimination, write to USDA, Director, Office of Civil Rights, 1400 Independence Avenue, SW., Washington, DC 20250–9410, or call (800) 795-3272 (voice) or (202) 720-6382 (TDD). USDA is an equal opportunity provider and employer.

References

Altschul SF, Madden TL, Schaffer AA, Zhang J, Zhang Z, Miller W, Lipman DJ (1997) Gapped BLAST and PSI-BLAST: a new generation of protein database search programs. Nucleic Acids Res 25:3389–3402

Ashburner M, Ball CA, Blake JA, Botstein D, Butler H, Cherry JM, David AP, Dolinski K, Dwight SS, Eppig JT, Harris MA, Hill DP, Issel-Tarver L, Kasarskis A, Lewis S, Matese JC, Richardson JE, Ringwald M, Rubin GM, Sherlock G (2000) Gene Ontology: tool for the unification of biology. Nat Genet 25(1):25–29

Avraham S, Tung CW, Ilic K, Jaiswal P, Kellog EA, McCouch S, Pujar A, Reiser L, Rhee SY, Sachs MM, Schaeffer M, Stein L, Stevens P, Vincent L, Zapata F, Ware D (2008) The plant ontology database: a community resource for plant structure and developmental stages controlled vocabulary and annotations. Nucleic Acids Res 36(suppl 1):D449–D454

Bolon Y-T, Haun WJ, Xu WW, Grant D, Stacey MG, Nelson RT, Gerhardt DJ, Jeddeloh JA, Stacey G, Muellbauer GJ, Orf JH, Naeve SL, Stupar RM, Vance CP (2011) Phenotypic and genomic analyses of a fast neutron mutant population resource in soybean. Plant Physiol 156:240–253

Bolon Y-T, Stec AO, Michno JM, Roessler J, Bhaskar PB, Ries L, Dobbels AA, Campbell BW, Young NP, Anderson JE, Grant DM, Orf JH, Naeve SL, Muehlbauer GJ, Vance CP, Stupar Robert M (2014) Genome resilience and prevalence of segmental duplications following fast neutron irradiation of soybean. Genetics 198:967–981

Du J, Grant D, Tian Z, Nelson RT, Zhu L, Shoemaker RC, Ma J (2010) SoyTEdb: a comprehensive database of transposable elements in the soybean genome. BMC Genom 11:113

Grant D, Imsande M, Shoemaker R (1996) SoyBase, a soybean genome database. Soybean Genet Newslett 23:51–53

Grant D, Nelson RT, Cannon SB, Shoemaker RC (2010) SoyBase, the USDA-ARS soybean genetics and genomics database. Nucleic Acids Res 38:D843–D846

Hyten DL, Choi I-Y, Song Q, Specht JE, Carter TE Jr, Shoemaker RC, Hwang E-Y, Matukumalli LK, Cregan PB (2010) Map of soybean and the development of a 1536 universal soy linkage panel for quantitative trait locus mapping. Crop Sci 50:960–968

Jaiswal P, Ware D, Ni J, Chang K, Zhao W, Schmidt S, Pan X, Clark K, Teytelman L, Cartinhour S, Stein L, McCouch S (2002) Gramene: development and integration of trait and gene ontologies for rice. Comp Funct Genom 3:132–136

Karp PD, Paley SM, Krummenacker M, Latendresse M, Dale JM, Lee TJ, Kaipa P, Gilham F, Spaulding A, Popescu L, Altman T, Paulsen I, Keseler IM, Caspi R (2010) Pathway Tools version 13.0: Integrated software for pathway/genome informatics and systems biology. Brief Bioinformat 11:40–70

Kent WJ (2002) BLAT—the BLAST-like alignment tool. Genome Res 12(4):656–664

Munger P, Bleiholder H, Hack H, Hess M, Stauss R, van den Boom T, Weber E (1997) Phenological growth stages of the soybean plant (*Glycine max* L. MERR.): codification and description according to the BBCH scale. J Agron Crop Sci 179(4):209–217

Pfeil BE, Schlueter JA, Shoemaker RC, Doyle JJ (2005) Placing paleopolyploidy in relation to taxon divergence: a phylogenetic analysis in legumes using 39 gene families. Syst Biol 54(3):441–454

Ribault (2015) IBP: Integrated Breeding Platform. http://www.integratedbreeding.net

Schlueter JA, Dixon P, Granger C, Grant D, Clark L, Doyle JJ, Shoemaker RC (2004) Mining EST databases to resolve evolutionary events in major crop species. Genome 47(5):868–876

Schmutz J, Cannon S, Schlueter J, Jianxin M, Mitros T, Nelson W, Hyten DL, Song Q, Thelen JJ, Cheng J, Xu D, Hellsten U, May GD, Yu Y, Sakurai T, Umezawa T, Bhattacharyya MK, Sandhu D, Valliyodan B, Lindquis E, Peto M, Grant D, Shu S,

Goodstein D, Barry K, Futrell-Griggs M, Abernathy B, Du J, Tian Z, Zhu L, Gill N, Joshi T, Libault M, Sethuraman A, Zhang X-C, Shinozaki K, Nguyen HT, Wing RA, Cregan P, Specht J, Grimwood J, Rokhsar D, Stacey G, Shoemaker RC, Jackson SA (2010) Genome sequence of the palaeopolyploid soybean. Nature 463:178–183

Severin AJ, Woody JL, Bolon YE, Joseph B, Diers BW, Farmer AD, Muehlbauer GJ, Nelson R, Grant DM, Specht JE, Graham MA, Cannon SB, May GD, Vance CP, Shoemaker RC (2010) RNA-Seq Atlas of *Glycine max*: a guide to the soybean transcriptome. BMC Plant Biol 10:610

Shoemaker RC, Grant D, Olson T, Warren WC, Wing R, Yu Y, Kim H, Cregan P, Joseph B, Futrell-Griggs M, Nelson W, Davito J, Walker J, Wallis J, Kremitski C, Scheer D, Clifton SW, Graves T, Nguyen H, Wu X, Luo M, Dvorak J, Nelson R, Cannon S, Tomkins J, Schmutz J, Stacey G, Jackson S (2008) Microsatellite discovery from BAC end sequences and genetic mapping to anchor the soybean physical and genetic maps. Genome 51:294–302

Stein LD, Mungall C, Shu S, Caudy M, Mangone M, Day A, Nickerson E, Stajich JE, Harris TW, Arva A, Lewis S (2002) The generic genome browser: a building block for a model organism system database. Genome Res 12(10):1599–1610

Zhang P, Dreher K, Karthikeyan A, Chi A, Pujar A, Caspi R, Karp P, Kirkup V, Latendresse M, Lee C, Mueller LA, Muller R, Rhee SY (2010) Creation of a genome-wide metabolic pathway database for *Populus trichocarpa* using a new approach for reconstruction and curation of metabolic pathways for plants. Plant Physiol 153(4):1479–1491

FSC
www.fsc.org
MIX
Papier | Fördert
gute Waldnutzung
FSC® C083411

Kolibri 360 GmbH
Ferdinand-Jühlke-Straße 7
99095 Erfurt, Deutschland
produktsicherheit@kolibri360.de